River Basin Development and Human Rights
in Eastern Africa — A Policy Crossroads

Claudia J. Carr

River Basin Development and Human Rights in Eastern Africa — A Policy Crossroads

Springer Open

Claudia J. Carr
Environmental Science, Policy
 and Management
University of California, Berkeley
Berkeley, CA
USA

ISBN 978-3-319-50468-1   ISBN 978-3-319-50469-8 (eBook)
DOI 10.1007/978-3-319-50469-8

Library of Congress Control Number: 2016959399

© The Editor(s) (if applicable) and The Author(s) 2017. This book is published open access.

Open Access This book is distributed under the terms of the Creative Commons Attribution-NonCommercial 2.5 International License (http://creativecommons.org/licenses/by-nc/2.5/), which permits any noncommercial use, duplication, adaptation, distribution and reproduction in any medium or format, as long as you give appropriate credit to the original author(s) and the source, provide a link to the Creative Commons license and indicate if changes were made.

The images or other third party material in this book are included in the work's Creative Commons license, unless indicated otherwise in the credit line; if such material is not included in the work's Creative Commons license and the respective action is not permitted by statutory regulation, users will need to obtain permission from the license holder to duplicate, adapt or reproduce the material.

This work is subject to copyright. All rights are reserved by the Publisher, whether the whole or part of the material is concerned, specifically the rights of translation, reprinting, reuse of illustrations, recitation, broadcasting, reproduction on microfilms or in any other physical way, and transmission or information storage and retrieval, electronic adaptation, computer software, or by similar or dissimilar methodology now known or hereafter developed.

The use of general descriptive names, registered names, trademarks, service marks, etc. in this publication does not imply, even in the absence of a specific statement, that such names are exempt from the relevant protective laws and regulations and therefore free for general use.

The publisher, the authors and the editors are safe to assume that the advice and information in this book are believed to be true and accurate at the date of publication. Neither the publisher nor the authors or the editors give a warranty, express or implied, with respect to the material contained herein or for any errors or omissions that may have been made.

Photographs by Claudia J. Carr

Printed on acid-free paper

This Springer imprint is published by Springer Nature
The registered company is Springer International Publishing AG
The registered company address is: Gewerbestrasse 11, 6330 Cham, Switzerland

This book is dedicated:
to 'Jolish'—the young leader who worked tirelessly to inform his own people and neighboring ones about the crisis at hand and who lost his life in this struggle—and to the young people of these Ethiopian, Kenyan and South Sudanese border lands who carry on the struggle to direct their own futures

to my son Calen and niece Leah—and those who follow them as agents of change for a better world

and

to Marion Hall, who taught me to see whole systems of nature and society and then sent me on my way.

Preface

For countless centuries, river basins have been fundamental to the survival of peoples residing in the vast drylands of Sub-Saharan Africa. The water, soils and living resources of these river systems—punctuating what are otherwise aridic and often harsh conditions—provide for human settlement, livestock grazing, seasonal flood (recession) agriculture, wild food harvesting, fishing and a host of other activities central to livelihoods. What happens to river basin systems determines the fate of millions of people. Predominantly pastoral in history economy and culture, the extensive drylands of eastern Africa are also where some of Africa's most ambitious economic development programs, including hydrodam, irrigated plantation, mineral, oil and gas projects, are being implemented.

One cannot experience the rhythm of daily life among pastoral villagers for any length of time without realizing that such developments have profound impacts on these longstanding survival systems and that there is an obvious disconnect between the life conditions of these pastoralists and the decision-making in the financial centers of Washington D.C., Brussels, and Beijing, as well as in the major cities of their own nations. Almost inevitably, one begins to question how these traditionally oriented peoples can possibly survive in the face of such development pressures and whether or not they can truly have a voice in determining their own futures.

This book is the outcome of a lengthy effort to answer these and other difficult questions as they pertain to sweeping changes in the semi-arid borderlands of Ethiopia, Kenya and South Sudan—changes already extending to the broader eastern Africa region. It is here that the Africa's largest hydrodam to date, the Gibe III dam, has recently been completed, on the Omo River in southwestern Ethiopia, and is moving into early operation. This megadam, together with its closely linked and extensive irrigated agricultural enterprises and a hydroelectricity transmission system for power export to the eastern Africa region, amounts to a multi-billion dollar development that is radically transforming the entire transboundary human and environmental systems. More than 500,000 indigenous pastoralists, agropastoralists and fishers reside in the lower Omo River basin, around Kenya's Lake Turkana in the easternmost segment of South Sudan's Ilemi Triangle. Most of them face partial or complete destruction of their means of survival, with no available livelihood alternatives. Already among the most marginalized peoples in the continent, the multiple ethnic groups in this region face impacts that are unimaginable to most outsiders.

From early on in the endeavor to understand changes underway in the region, the questions became more detailed. For example, what are the types of economies, or survival systems, in this vast, tri-nation transboundary region and how do they interact? How adaptive are these systems to changes in their environments and what are their main vulnerabilities in the face of major shifts in the resources available to them? What are the specific forces of government, international aid and private development now impacting the region and how have these come about? From what institutions and social priorities have they emerged? What account of the hundreds of thousands of indigenous residents has been taken—with what concerns and accuracy? What impacts have unfolded so far and how have the pastoralists attempted to cope with them? What human rights are pertinent to these changes and do the developments underway constitute violations of those rights? Finally, is there a positive way forward for such peoples to have a genuine voice in what economic development and other changes will be brought to their lands in the name of 'progress'?

I first came to know the peoples of this region many years ago, as a young ecologist with the international paleontological Omo Expedition, led by F. Clark Howell, Richard Leakey and Yves Coppens, in the lowermost Omo River basin of southwest Ethiopia. Basing a good deal of my work in pastoral villages in order to learn about the region's ecological change and its relationship to the indigenous land use patterns, I learned in the most concrete terms about the inseparability of 'environmental' and 'social' realities. Much of this effort is summarized in my book, *Pastoralism* in Crisis: the *Dassanetch of Southwestern Ethiopia*, and in several papers. Moving on to research and practical policy work elsewhere in eastern Africa (within Ethiopia, Somalia, northern and coastal Kenya and elsewhere), I experienced multiple contexts where river basin developments, including hydrodams, have fundamentally transformed local socioeconomic and environmental systems and influenced entire nations. Everywhere I engaged in policymaking circles—from African ministries, international aid organizations and the U.S. National Academy of Sciences to remote administrative offices and grassroots organizations struggling to effect change, the enormous impacts—and often, the conflicts generated—from major river basin developments were apparent.

When invited to return to southwestern Ethiopia in 2008, I eagerly accepted the opportunity—this time with private foundation support to investigate the social and environmental conditions there. Surprised to learn about the virtually unprecedented development planned for this region, my efforts evolved into a multi-year, intensive investigation of the changes underway and their likely impacts on the region. Early on, it was necessary to form a research team, identified as the South Omo/North Turkana Research Team (SONT), with local residents from two of the region's major ethnic groups—the Dasanech and the (northern) Turkana. SONT was able to work cooperatively with elders from many locales throughout the transboundary region. Meanwhile, I and several colleagues co-founded the Africa Resources Working Group (ARWG)—an informal network of scientific and policy focused professionals with experience in the region. Both of these efforts proved essential to the complex tasks at hand.

Conditions for field-based research in the border region were difficult, both logistically and politically. Within Ethiopia, it was necessary to carry out all investigations with extreme care due to the Ethiopian restriction of independent investigators from the region, as well as extensive government political surveillance and repression—measures generating pervasive fear among villagers throughout the area. In Kenya, the situation was far less difficult, with community members far more able to participate openly in our work. Information gathering included settlement area mapping, ecological reconnaissance, village survey, male and female household head interviews, recording of elder life histories and recording of livelihood shifts and available resources. Local government authorities, aid officials and technical personnel active in the area (e.g., in fisheries/Beach Management Unit, water development, health relief work) provided vital information and perspective. In 2009, the Africa Resources Working Group released a preliminary report concerning the Gibe III dam and its likely impacts; I subsequently released a lengthy report on the matter, based on my investigations with SONT and the ARWG (Carr 2012)—a report first posted at the ARWG website and later at www.academia.edu.

It is my hope that the information and perspectives presented in this book will promote further understanding of the sweeping changes underway in this eastern Africa region and their significance, as well as contribute to discourse and possible solutions. If it is useful to inquiring government and aid officials, villagers, nongovernmental organizations, students, concerned scholars, and other citizens within Africa and abroad, it will have been worth the effort.

It has been a profoundly moving experience and an honor to work toward accomplishing these tasks in the company of so many wise, persevering and courageous people.

Acknowledgments

This book was made possible by the contributions of many people—both in and out of eastern Africa. My deepest thanks are to the villagers in both the lowermost Omo River basin of Ethiopia and the northwestern lands of Kenya, west of Lake Turkana. The conditions of working in these two regions are vastly different. Villagers throughout the northern Turkana localities visited were not only welcoming to me and my African colleagues but also eager to discuss matters of their changing survival systems, livelihood, and political conditions in the region—broadly defined and their concerns about the major developments they knew are being undertaken in the region. This eagerness was prevalent, even with concern over the fact that such discussions are a sensitive matter for Kenyan authorities. While equally welcoming, villagers along the Omo River within Ethiopia live under conditions of pervasive political surveillance and repression by their government and cannot openly discuss their declining conditions for survival or their fears and opposition concerning the radical changes underway along the Omo River. Nevertheless, many retain the courage to speak about these matters and learning from them has been a most humbling experience. I must trust that their descriptions and concerns are accurately summarized in accounts presented in this book. The identities of all informants in these areas of Ethiopia and Kenya have been protected and will remain so.

It is my privilege to have helped form and direct the South Omo/North Turkana Research Team (SONT). The team was comprised primarily of Turkana individuals from local villages, participating in nearly all aspects of the field-based research for this project within Kenya, as well as Dasanech individuals from the west bank and delta region. The SONT 'experience' successfully gave form to what is often empty rhetoric of "locally based and participatory" research and offered new skill learning to those joining in the effort. I am grateful to all those who took part and have protected their identities as well. My greatest debt in the SONT effort is to 'Jolish' (whose traditional name must remain protected), the co-coordinator of field operations with me and a wonderful colleague in both investigative and interpretive phases of our work. His exceptional intelligence, cross-cultural outlook, and deep humanity, combined with his clear passion for retaining traditional cultural values, surely contributed to the remarkable respect he enjoyed throughout a vast area of northern Turkana. 'Jolish' tragically lost his life in the course of his unrelenting efforts. Working alongside him and sharing the ups and downs of this complex work and the people it concerns has made a major mark on my life. I am certain that countless others join me in being forever grateful to him for his uncompromising and inspirational efforts. The dedication of this book to him is but one small expression of this gratitude.

A number of local officials within northern and central Turkana were extremely helpful in this work, as were a very large number of council of elders members—especially along the northwestern shoreline of Lake Turkana. My thanks to them for their immensely valuable contributions. Many other individuals in Ethiopia and Kenya provided major logistics and other support. Most prominent among them was Elias H. Selassie during early phases of the field investigation and analysis. T. Solomon of Ethiopia provided much parallel support in the southwestern region there.

Jeffrey Gritzner, a well-known scholar of African drylands and expert in African resource policy, including as a former senior staff member at the U.S. National Research Council, has contributed greatly to this investigation. His synthetic view of environmental change, indigenous knowledge systems, and the intricacies of environmental recovery and restoration have impacted this work, as have his comments on draft sections of this book. Dr. Gritzner cofounded the Africa Resources Working Group (ARWG) with me in 2008. This informal network is comprised of 11 African and non-African individuals from the physical science, social science, public policy, and diplomatic professions—all with substantial experience in the eastern Africa region. Its members have contributed in fundamental ways to this work, both with information when requested and with comments on particular written sections. While the analysis and conclusions drawn in this book undoubtedly do not reflect the viewpoint of each member of the ARWG, I have endeavored to take all of their comments and criticisms into account and deeply appreciate their efforts. Most of these individuals are still actively engaged in the region and so need to remain anonymous.

Laura M. Daly, a young California-based professional in Geography and GIS technique, has produced the 'major' graphics for this book (indicated in the List of Figures). Ms. Daly spent countless hours patiently working with me to translate specific ground-based information and understanding into graphic representation in one of the most difficult regions in the world. Applying her skill and artistry as well as her own propensity for perfection, Ms. Daly somehow tolerated my unending requests for "just one more change" in our efforts to represent complex patterns and processes. Her close associate, Bertrand Johnson, generously assisted with a good deal of technical work on my large set of photographs for the book.

James Lindsay, a former Australian diplomat and director of a nonprofit organization focused on alternative energy in Africa, has been a major contributor to the progress and completion of this book. His contributions have included direct assistance with systematizing information from environmental and social impact assessments, assisting with matters of energy and other technical systems, and performing supplemental graphics and photo work. Mr. Lindsay has also provided substantial material support for both field-based and book preparation activities—a necessary supplement to private foundation funding provided. I extend my deep thanks for his willingness to assist in all ways.

The contributions of George Leddy to this book cannot be overstated. Dr. Leddy's longterm analytical background in international development matters and his generosity in critiquing the chapters of this book, including with his experience as an assistant executive editor of a major international journal, were of great help throughout this effort. I am I am deeply indebted to him, including for his unwavering recognition of the magnitude of the crisis unfolding and the importance of tracing its origins and essence.

The seismic discussion in Chap. 3 was grounded in a detailed siesmic report for the Gibe III dam region in Ethiopia by Steve Walter, a seismologist with the Earthquake Science Center at the U.S. Geological Survey. This report (included in Carr 2012) detailed the sobering reality of a 20 % probability that the dam will experience at least Intensity VII shaking within 50 years and possibly as high as Intensity VIII. For practical reasons, the report was not able to be incorporated into the present book and I am grateful for Dr. Walter's encouragement and continued advice for the present, somewhat extended study.

Joshua Dimon, an expert on extractive industry development in eastern and southeastern Africa and completing his doctoral degree at the University of California at Berkeley, played a major role in the extended seismic study for this book (Chap. 3). Mr. Dimon's careful investigative and synthetic work is greatly appreciated. He is also the primary author of Appendix A in this book which is an overview of the incursion of the oil industry in the Ethiopia-Kenya-South Sudan transbournday region and part of our jointly executre investigation of the industry's growth in eastern Africa. I extend my deep thanks to him for his excellent contributions.

James Carr, an environmental consultant and former senior officer of a nongovernmental organization promoting global fisheries reform—who is also my brother—offered major advice and assistance concerning the book's emphasis on indigenous fishers' knowledge of their own resource base, as well as other aspects of the investigation. He has long been a guiding force in my intellectual pursuits and perspectives, and continues to be so in his work as an environmental policy analyst.

My efforts to navigate the difficult terrain of international human rights standards and approaches to the matter of rights violations at both national and institutional levels benefitted greatly from the advise of a number of outstanding attorneys. Dan Siegel, Peter Weiner, and Michael Freund offered valuable advice at the outset of this portion of my investigation as it applies to the the transboundary region. Several attorneys highly experienced in international human rights issues in developing countries, including within Africa and Latin America, offered critical perspective and information, as well as comments on my drafts of the human rights section of the book. They are Laurel Fletcher and Alexa Koenig, both faculty at Boalt Law School at the University of California, Berkeley, and Naomi Roht-Arriaza, a faculty member at the University of California Hastings Law School in San Francisco. I am most grateful for their help. I also wish to offer special thanks to Michael Hannigan, President and Cofounder of the innovative company, Give Something Back, for his crucial insight toward making this book widely available and helping to implement that objective.

My deep appreciation extends to other professionals who have contributed to this volume. Jason Gritzner, a Watershed Program Manager for the National Forests Program and an individual experienced in East African river basin development, offered important information and perspective regarding hydrological and riverine ecology dynamics. Justin Fong, a public policy specialist with strong knowledge of the rising importance of China in African economic development offered many comments on book chapters as well as numerous thoughts on the political shifts underway.

Daniel Lavelle contributed valuable document research and comments concerning the physical character of Lake Turkana and patiently entered into valuable dicussions of the issues involved. Richard Brenneman offered insightful and penetrating comments on an early version of the book, contributing his expertise as an investigative journalist and his passion for in-depth analysis in social matters of consequence. Kathy Sheetz offered assistance in both investigative and presentational dimensions of the early phase of this book, with suggestions from her professional and political experience in health related issues Africa, Haiti, and elsewhere. Ann Kroeber offered valuable perspective concerning visual presentation of the materials at hand.

Among my faculty colleagues at UC Berkeley, I want to thank Andrew Gutierrez, who lent his support and expertise to this work from his knowledge of biological systems in the Ethiopian region. Miguel Altieri's many years of leadership in the field of agroecology and indigenous systems provided a critical source of information and perspective regarding the agropastoral component of livelihoods in the transboundary region. I greatly appreciate their efforts, including within the University context. My community-based colleague and friend in Berkeley, 'Samir' (Aka 'F.D.'), has also offered much encouragement.

In the world of education, the earliest major 'teacher' in my own life—Marion T. Hall—not only taught me about 'systems' ranging from plants to people, but also about the importance of explicitly considering those systems when trying to concretely address the real problems in this world. Many of my international development and natural resource policy students and young colleagues at UC Berkeley participated in various phases of this book's development. In its earliest phase, Adam Gray was particularly key in our investigation of the history of oil leases in eastern Africa, with strong focus on the Horn region. Crossley Pinkstaff made important contribtions to the economic and policy investigations of the eastern Africa region and has remained a valuable source of perspective and information. Andy Kreamer assisted with major bibliographic work and early project organizing toward field investigations. Michal Karmi, Juan Ramos, Dylan Kasch. Chase Livingston, Ian McGregor, Kaleigh Rhodes, Judi Li, Nicholas Calderon, Yuki Jiang, and Dawning Wu worked together to update operations of the oil and gas industry in eastern Africa. Alicia Krueger, Regina Clincy, Phoebe Song, Vickie Duong, Corey Wood, Lexi Spaulding, and Gyöngyi Gózon all donated their time to do fact checking and related work, making valuable suggestions for improvement. To all of these individuals, I offer my sincere thanks and hope that they learned from our experience together. I certainly did. If I have failed to mention others students' efforts, I hope that I shall be forgiven. The long and arduous times synthesizing information and its meaning throughout this work would not have been possible without the sustainment of music—especially by Dasanech and Turkana villagers and 'desert' dwellers throughout the region, but also the magical and energizing sounds of Ali Farka Toure', Dadi Kouyate' and Miriam Makeba.

Finally, I would never have been able to carry out this work without my family, to whom I give my deepest love and appreciation. In addition to my brother James, my son Calen provided unwavering encouragement throughout. His caring for people from many cultures and his uncanny ability to always look forward have been a constant source of inspiration and strength for me. My niece Leah and her two daughters, Marin and Jenesea, with their high spirits and caring for all forms of life have provided ongoing motivation. I thank them for their love and patience throughout this effort.

Much of this work derived financial support from a private foundation concerned with development, conservation, and resource rights in Africa and elsewhere. The foundation was flexible and helpful in all regards and my thanks to them for letting this entire effort move forward. In preparing for this book's publication by Springer-Verlag, I have had the incredible good fortune to work with Neil Olivier and Diana Nijenhuijzen. I deeply appreciate their assistance, patience and good cheer throughout the preparatory process and it is a great privilege to have worked with them. The final production team added a major boost to the book as well, with exemplary flexibility and professionalism.

Contents

1 **At Stake with River Basin Development in Eastern Africa** 1
   River Basin Development in Africa: Development Versus Disaster 1
   The Transboundary Character of Emerging Crisis in Eastern Africa 4
   The Gibe III Dam and Linked Agricultural and Power Export Developments 11
   Literature Cited 21

2 **The Persistent Paradigm for 'Modernizing' River Basins: Institutions and Policies in Ethiopia** 23
   Early River Basin Development in Ethiopia 23
   The Resurgence of Western Dominance in River Basin Development 30
   'Fast Track' to the Gibe III Megadam 32
   A Nexus of Public Policy Institutions for River Basin Development: Collaboration with Complicity 36
   Literature Cited 40

3 **The Seismic Threat to the Gibe III Dam: A Disaster in Waiting** 43
   High Seismicity in the Gibe III Dam Region 43
   Reservoir Seepage and Landslide Danger at the Gibe III Dam 47
   Failed Government and Development Bank Seismic Review 50
   Literature Cited 51

4 **Transboundary Survival Systems: A Profile of Vulnerability** 53
   Indigenous Livelihoods and Survival Strategy Systems 53
   Pastoral Dispossession and Rising Dependence on the Omo River and Lake Turkana 62
   Environments in the Transboundary Region: From Pristine to Degraded 68
      The Lowermost Omo River Basin and Environs 69
      Lake Turkana and Environs 70
   Cross-Border Conflict and Diminishing Resources: The Ilemi Triangle Ingredient 72
   Literature Cited 73

5 **Components of Catastrophe: Social and Environmental Consequences of Omo River Basin Development** 75
   Radical Reduction of River and Lake Waters by Omo Basin Development 75
   Consequences for the Lowermost Omo River Basin 80
   Consequences for the Lake Turkana Region 81
   Consequences for the Ilemi Triangle and the Broader Region 82
   Literature Cited 84

6 **The Rush to Rationalize: Public Policies and Impact Assessments** 85
   Launching the Gibe III Dam—And a System of Bias 85
   The Myth of Flood 'Disasters' as Rationale for Megadam Development 87
   Invalidity of the Ethiopian Government's Downstream Impact Assessment 93

	The False Promise of an Artificial Flood 'Solution'	99
	Multilateral Development Banks and the 'Complicity Treadmill'	101
	Literature Cited	110
7	**The Dasanech of the Lowermost Omo Basin: From Adaptation to Development Debacle**	111
	Dasanech Pastoral Decline: Roots and Responses	111
	Adapting from Upland Pastoral Life to Diversified Economy at the River	117
	Last Resort Survival: Desperate Dependence on Omo River Annual Flood	125
	Ethiopian Expropriation and Political Repression of Riverine Communities	135
	Literature Cited	144
8	**Nyangatom Livelihood and the Omo Riverine Forest**	145
	Nyangatom Omo Settlements and Dependence on Riverine Resources	145
	Fate of the Forest: Nyangatom Survival and Ethiopia's Heritage	151
	Literature Cited	156
9	**Turkana Survival Systems at Lake Turkana: Vulnerability to Collapse**	157
	Northern Turkana Pastoralists: The Long Decline and Migration to the Lake	157
	Adaptation from Pastoral to Fishing Livelihood	162
	Fishing Shoreline Communities: Household Practices and Resources	171
	Fish Species and Critical Habitats	179
	Counting the Discounted: Northern Turkana Population at the Lake	180
	Literature Cited	189
10	**Human Rights Violations and the Policy Crossroads**	191
	The Crisis Unfolding and the Human Right to Context	191
	The Ethiopian Government's Violations of Human Rights in the Transboundary Region	196
	Emerging Human Rights Violations in Kenya's Lake Turkana Region	199
	International Development Bank Collaboration with Human Rights Violations	201
	The Stark Policy Choice: Catastrophic Level Destruction *or* Sustainable Development Within a Human Rights Framework	211
	Literature Cited	214

Appendix A: Activation of Oil Exploration and Development in the Ethiopia–Kenya–South Sudan Transboundary Region 217

Appendix B: Species Collected in the Lower Omo River Basin and Transborder Region 223

Appendix C: Reference set of Selected Major Figures 231

Acronyms

AFD	French Development Agency/*Agence Française de Développement*
ACHPR	African Commission on Human and Peoples Rights'
AFDB	African Development Bank
ARWG	Africa Resources Working Group
CRBM	Campagna per la Riforma della Banca Mondiale (CRBM)
DFID	Department for International Development, United Kingdom
EAPP	Eastern Africa Power Pool
EEHP	Eastern Electricity Highway Project
EEPCO	Ethiopian Electric Power Company
EIA	Environmental Impact Assessment
EIB	European Investment Bank
ELPA	Ethiopian Electric Light and Power Authority
EPA	Environmental Protection Authority, Ethiopia
EPRDF	Ethiopian People's Revolutionary Democratic Front
ESIA	Environmental and Social Impact Assessment
ESMF	Environmental and Social Management Framework
EUR	Eurodollars
FAO	Food and Agricultural Organization of the United Nations
FoLT	Friends of Lake Turkana
GCI	Global Consulting Industry
GOE	Government of Ethiopia
GOK	Government of Kenya
IDA	International Development Association ('Soft Loan Window') of the World Bank
KMFRI	Kenya Marine Fisheries Research Institute
mg/L	Milligram(s) per liter
MW	Megawatts
kV	Kilovolts
MoFA	Ministry of Foreign Affairs (Ethiopia)
MOFED	Ministry of Finance and Economic Development (Ethiopia)
MOU	Memorandum of Understanding
MOWR	Ministry of Water Resources (Ethiopia)
NEMA	National Environment Management Authority (Kenya)
NGO	Nongovernmental organization
PBS	Promotion of Basic Services
ppm	Parts per million
SNNPR	Southern Nations Nationalities and People's Region (Ethiopia)
SONT	South Omo/North Turkana Research Project
TOR	Terms of Reference
UNEP	United Nations Environment Programme

UNESCO	United Nations Educational Scientific and Cultural Organisation
USD	U.S. Dollars
USGS	United States Geological Survey
USAID	United States Agency for International Development

List of Figures

Figure 1.1	Location of the Gibe III dam on the Omo River in the tri-nation transboundary region	5
Figure 1.2	Omo River delta expansion from 1973 to 2006 with the river's terminus in Kenya	6
Figure 1.3	Ethnic groups in the tri-nation transboundary region	7
Figure 1.4	Indigenous Dasanech and Nyangatom residents in the lowermost Omo River basin	8
Figure 1.5	Turkana villagers at Lake Turkana	9
Figure 1.6	Interethnic exchange network in the transboundary region	10
Figure 1.7	Major livelihoods in the transboundary region	12
Figure 1.8	Peoples of the tri-nation border region	13
Figure 1.9	Habitat variation in the transboundary region	14
Figure 1.10	The 'Gibe cascade' of hydrodam and electricity projects along the Omo River	15
Figure 1.11	Gibe III dam site and construction	17
Figure 1.12	Meandering section of the lower basin Omo River	19
Figure 2.1	Awash River basin in context of all Ethiopian river basins	25
Figure 3.1	The East Africa Rift System with Main Ethiopian Rift (MER)	44
Figure 3.2	Earthquake risk in Africa: Modified Mercalli Scale	45
Figure 3.3	Seismic activity within 1600 km of the Gibe III dam: 1963 to present	48
Figure 3.4	Seismic activity within 700 km of the Gibe III dam: 1906 to present. Indicated by magnitude of earthquake	49
Figure 4.1	Pastoral life along the Omo River west bank and Kibish River	55
Figure 4.2	Northern Turkana pastoralists in upland plains	56
Figure 4.3	Riverside settlement and secondary production along the lowermost Omo River	58
Figure 4.4	Northern pastoral Turkana dependency on Lake Turkana for watering and grazing	59
Figure 4.5	Northern Turkana fishing villagers at Lake Turkana	60
Figure 4.6	Indigenous village relocation (migration) to the Omo River and Lake Turkana: 1960 to present	61
Figure 4.7	Seasonal dependence on Omo River and Lake Turkana resources by pastoral, agropastoral, and fishing indigenous communities	65
Figure 5.1	Elevation of Lake Turkana relative to lake volume	76
Figure 5.2	Progressive retreat of Lake Turkana caused by the Gibe III dam and dam enabled irrigated agriculture (Base map from Hopson 1980)	77
Figure 5.3	Expanding armed conflict from effects of Gibe III dam and dam-linked development	79
Figure 5.4	Summary of the humanitarian catastrophe and conditions for armed conflict from the Gibe III dam and its linked large-scale irrigated agricultural development	80
Figure 5.5	Ilemi triangle border area with Ethiopia	83
Figure 6.1	Dimensions of invalidity in the Ethiopian government's downstream impact assessment of the Gibe III dam	88
Figure 6.2	Satellite images before and after the August 2006 Omo River flood	91
Figure 7.1	Dasanech herders and livestock	113
Figure 7.2	Pasture deterioration phases in upland plains. Dominant and comparable soil type in Dasanech region of the lower Omo basin (sandy-silt soils on relict beach/interridge areas)	114
Figure 7.3	Phases of ecological decline in lower Omo basin pastoral lands	115
Figure 7.4	Dasanech pastoral villagers and activities	116
Figure 7.5	Seasonal movement patterns of six Dasanech settlement areas west of the Omo River	117
Figure 7.6	Dasanech settlement migration from upland plains to Omo riverine zone: 1960–2011	118

Figure 7.7	Major Dasanech livelihood decline from upland pastoral economy (west side of the Omo River)	120
Figure 7.8	Dasanech woman making dugout canoe from a tree trunk in a small clearing in the Omo riverine forest	121
Figure 7.9	Zone of Ethiopian Government Expropriation of Dasanech villages and livelihood areas	122
Figure 7.10	Dasanech male elders in the Omo riverine zone	125
Figure 7.11	Flood recession agriculture and Dasanech planters	126
Figure 7.12	Dasanech Life along the Lower Omo River	128
Figure 7.13	Calves in starvation condition grazing in stubble of riverside farm plot	130
Figure 7.14	Desiccation of modern Omo delta and northern end of Lake Turkana predictable from Gibe III dam and dam-linked irrigation systems	131
Figure 7.15	Wetlands at the Omo delta terminus at Lake Turkana. Dasanech cattle grazing	132
Figure 7.16	Dasanech village complex at shoreline near northwestern extreme of Lake Turkana, close to Omo delta wetland	132
Figure 7.17	Dasanech crossing Omo River at high flood stage for transactions between west bank residents and Omorate traders	133
Figure 7.18	Planned and potential irrigated agriculture in the lower Omo River basin	137
Figure 7.19	Cracking silty clays in relict floodplains, near planned irrigated commercial farm	143
Figure 8.1	Nyangatom trek from Omo River villages to Kibish River and Ilemi Triangle settlements	147
Figure 8.2	Nyangatom man and woman at (5 m deep) watering hole dug in Kibish riverbed during the dry season	148
Figure 8.3	Nyangatom family in village alongside the Omo River's west bank	149
Figure 8.4	Nyangatom in agropastoral villages along west bank of the Omo River	150
Figure 8.5	Omo riverine forest in Nyangatom region	152
Figure 8.6	Transition zone between the Omo riverine forest and adjacent drylands	153
Figure 8.7	Riverine vegetation types along the lower Omo River	154
Figure 8.8	Riverine forest development along the lowermost Omo River	155
Figure 9.1	Pastoral life in northern Turkana	161
Figure 9.2	Northern Turkana livestock herds watering at the lake	163
Figure 9.3	Turkana fishing villagers along northwestern shores of Lake Turkana	165
Figure 9.4	Fishing and mixed fishing/pastoral Turkana village areas at Lake Turkana	166
Figure 9.5	Turkana fishing village activities at Kalokol and northward along the lake	167
Figure 9.6	Northern Turkana Fishing Villagers	168
Figure 9.7	Turkana girl and villagers with dried fish pallets set at roadside for transport to market	170
Figure 9.8	Turkana life at Ferguson's Gulf and northward along shoreline	178
Figure 9.9	Turkana woman drying fish for marketing with secondary geese/chicken—raising	179
Figure 9.10	Bathymetric representation of Lake Turkana retreat from Gibe III Dam and linked irrigation agriculture	182
Figure 9.11	Desiccation of Ferguson's Gulf and the modern Omo delta: projected from Gibe III dam and irrigated enterprises along the Omo River	183
Figure 9.12	Indigenous map of Turkana villages at Ferguson's Gulf	185
Figure 10.1	The planned eastern Africa electricity export and distribution system	204
Figure 10.2	Chinese-financed 51-km transmission connection between the Gibe III dam and the Ethiopia-Kenya Energy Highway Project	205
Figure A.1	Concessions for oil and gas exploration in the Ethiopia–Kenya–South Sudan transboundary region—2014	219
Figure A.2	Cumulative concessions for oil and gas exploration in Eastern Africa	220

List of Tables

Table 2.1	Three key components of the institutional policy nexus for major river basin development, with selective list of participants. Ethiopia, 1955–2013	39
Table 3.1	Earthquakes within 300 km of the Gibe III (3) dam since 1906	47
Table 4.1	Diversified food production among transboundary ethnic groups	63
Table 4.2	Major food related exchange: Turkana-Nyangatom and Dasanech-Turkana ethnic groups	64
Table 6.1	Recent Omo River flood history from a SONT 2009 survey of village elders	92
Table 6.2	Dimensions of invalidity of the Ethiopian government's environmental and social impact assessment (GOE 2009b)	96
Table 6.3	European Investment Bank and African Development Bank impact assessments	105
Table 7.1	Dasanech village complexes along the Omo River: 2009–2010	123
Table 7.2	Dasanech household wealth status and livelihood change. Households from west bank of the Omo River: 1972 versus 2009	124
Table 7.3	Selected irrigated agricultural enterprises in the lower Omo basin	138
Table 7.4	Partial list of Ethiopian government evictions and expropriations of Dasanech villagers	140
Table 8.1	Nyangatom livelihood activities dependent on riverine habitat	151
Table 9.1	Annual seasons described by northern Turkana fishers	169
Table 9.2	Household survey in Lake Turkana fishing communities: A Summary	172
Table 9.3	Lake Turkana fish species and habitats of importance to Turkana fishing communities	181
Table 9.4	Population estimates from Ferguson's Gulf region	186
Table 9.5	Population estimates for towns and village complexes along Lake Turkana's western shore from Kalokol to the Kenya-Ethiopia border	187

At Stake with River Basin Development in Eastern Africa

In the time of our fathers and grandfathers, our land was the land of good grass and it was big. The grass was tall for our cows and we moved our herds apart when danger came. The river gave us what we needed —water and grass for our animals. Our life was good. But look at our land now! It is bare and you can find our dying animals everywhere. Look at those carcasses! Our fathers and grandfathers did not know this hunger—they did not know this life. We have had to bring our villages to the river to find grass for our animals and to plant so we can feed our children. The poorest of us are even fishing. Now we are afraid that we will lose our river waters. What is happening to our river and our lands—do you know?

[Dasanech male elder in the western Omo River delta, within Ethiopia]

Abstract

River basin development in Africa has nearly unparalleled significance for the future of entire nations. Most major hydrodam projects undertaken in the continent have produced intense controversy, particularly over their socioeconomic and environmental impacts. In eastern Africa, river basin development is producing a major humanitarian and human rights crisis for a half million indigenous people in the border region of Ethiopia, Kenya and South Sudan. This crisis stems from developments in the Omo River basin of southwestern Ethiopia, with major international support. Construction of the Gibe III hydrodam—one of the world's tallest—is primarily geared to the production of hydroelectricity for the benefit of commercial and financial interests within Ethiopia and energy export throughout eastern Africa, as well as to major irrigated commercial agricultural development along the Omo River. The crisis at hand is an international one, especially since the Omo River is a transboundary watercourse—flowing from the Ethiopian highlands to its terminus within Kenya at that nation's Lake Turkana, where it provides most of the lake's water. The combined hydrodam and large-scale irrigation agriculture development would cause radical reduction of both river flow volume and lake level—thus destroying pastoral, agropastoral and fishing livelihoods of hundreds of thousands of indigenous people dependent on these waters. Catastrophic level collapse of survival systems in the region would usher in major new inter-ethnic, cross-border armed conflict as communities are forced to fight over the region's vanishing resources. Major human rights violations involving national governments and key international aid agencies are already underway.

River Basin Development in Africa: Development Versus Disaster

➢ Few development initiatives in Sub-Saharan Africa have the economic and political importance of river basin developments, whether in terms of the sheer size of undertaking, magnitude of social and environmental transformation, impact on the future of entire nations or controversy regarding beneficial *versus* destructive outcomes. The range of people and institutions frequently engaging in heated controversy over these developments—from World Bank executives and Prime Ministers through multiple levels of technical specialists, water resource/engineering consultants and non-governmental or civil society organizations to villagers directly impacted—reflects the significance of such developments within African nations.

For African heads of state, the nearly unparalleled capital intensity of major dam and physical infrastructure projects, along with linked agricultural and industrial development, requires major infusions of international capital—nearly always as loans, viewed as a most powerful river impetus for economic growth and 'transformation' and inevitably termed "*in the national interest.*" Assurances of major 'trickle down' of wealth and other benefits to the general population from such huge developments, moreover, are extended as promise of a 'better future'—while serving as a means of staving off increased or potential social unrest, as viewed by existing leaders.

International aid organizations, particularly multilateral development banks and major bilateral agencies, have for years held comparable enthusiasm for large dam and dam linked basin projects in Africa. The large size of loans extended for megadam development—well into the billions of dollars when dam-enabled projects are included, as well as the conditions attached to them—provide major donor countries and banks with opportunity for continued or expanded economic and political leverage in African nations along with major investment opportunities, access to natural resources and other benefits. Such developments are sometimes closely related to geostrategic objectives vital to donor country foreign policy.

Finally, rapid and definitive regional economic 'integration'—a major objective in foreign aid policies—can be rapidly advanced by dams generating hydroelectricity, along with major new infrastructure construction serving new industrialization, agro-industry and expansion of extractive industry interests.

Although most African river basins have been considered with reference to individual countries, a high proportion of them are actually transboundary in character—far more than the United Nations has officially recognized. In addition to major rivers that flow across national boundaries in eastern Africa, numerous others terminate at the Indian Ocean where their altered character has profound impacts on the biophysical, fisheries and other dimensions of coastal systems that extend across national boundaries.

➢ **A major boom in river basin development unfolded within Africa in the post World War II years—one primarily under the auspices of international aid programs as the region's nations became independent from their colonial administrations.** This boom in development was dominated by the United States for years, since its economy and domestic experience at the end of the war fostered strong incentive as well as potential to move its dam-building approach and expertise to Africa, Asia and Latin America. This dominance in African basin development lasted until a number of European nations also took on major roles in dam-building in the 1970s.

The U.S. modeled its approach after its own Hoover dam and Grand Coulee dam (at the Nevada-Arizona border and in Washington state, respectively) and its Tennessee Valley Authority—a U.S. government owned entity created by the U.S. Congress in 1933 to oversee river basin development.[1] A highly controversial and centralized planning effort introduced as part of President Franklin D. Roosevelt's New Deal, the TVA was broad spectrum in its roles which ranged among major dam construction, water and land policy coordination and management, and numerous other basin development initiatives, including agricultural ones.

These developments were part of a new centralized planning under President Roosevelt's New Deal that included elevating water resource policy to a key executive concern, with strong emphasis on 'harnessing' energy and promoting economic development. The rise of a new *'techno-managerial' approach* to water resources and river basin development emerged as a major fulcrum of economic expansion in a market economy. This new wave had clearly already swept up some of the most prominent water resource economists, geographers and other scientists and technicians.[2]

➢ **The most ardent critics of major dam and basin developments have long pointed to their major social and environmental problems and injustices, including the human rights violations that have accompanied them.** Denial of U.N. recognized human rights—including rights to water, health, livelihood, and freedom from forced relocation and political repression is widespread in river basin developments within Africa. In many ways, they parallel the many controversies occurring since the 1980s, in particular, between river basin developeers and their critics. However, these rights violations take an extreme form in the lowermost Omo River basin and northern Lake Turkana region, as detailed in the next chapters.

---

[1] Two comprehensive histories of international dam and river basin development are by Patrick McCully (2001) and Leslie (2005)—McCully taking a worldwide view and Leslie emphasizing cases in India, South Africa and Australia.

[2] Among the rising figures in this realm was Gilbert F. White, who was elevated to an executive water resource advisory position as a young man in his twenties by President Franklin D. Roosevelt. A biographical account of Gilbert White's life from the New Deal 1930s to the 2000s (Hinshaw 2006) provides an excellent view of the origins of these policies and their general application in international development. After playing a major role in megadam projects throughout much of Africa, Southeast Asia and the U.S., White became more equivocal (from the late 1980s onward) about the 'techno-managerial' approach in river basin/large dam development.

Key components of the 'techno-managerial approach' to major river basin developments—are 'scientific' reports (mandated by international and national development agency operational principles) includes feasibility and baseline studies as well as impact assessments—all typically commissioned by the very governments, aid organizations and transnational companies participating in the intended development. These are generally produced by firms and individuals within the global consulting industry—a multi-billion dollar industry whose members are 'trusted' repeat contractors. Documents produced by the global consulting industry generally endorse the overall thrust of the intended developments, with limited consideration of social and environmental potential impacts and with 'suggestions' for risk-minimization, mitigation and monitoring measures. These are typically based on the assumptions that the projects will continue to completion and that any projected social or environmental impact problems can be "solved" by technical and managerial action. Such legitimation efforts generally lead to project approval in eastern African nations and may other development country contexts, whereas in developed countries, criticism and opposition by experienced constituencies frequently delays or even blocks such projects.

This clash of perspectives between large dam developers and critics has nowhere been clearer than in the proceedings of the World Commission on Dams (WCD), with its comprehensive report of 2000. By now, the results of the detailed WCD studies and deliberations have been extensively reported on and evaluated. The optional guidelines produced by compromises among Commission members are just that: optional. The rush to megadam development and major river basin industrialization has continued, however—in fact, recently intensified.

The World Bank has in 2013 stated its intention to rejuvenate, or increase its emphasis on large dams in developing countries, and both environmental and social 'safeguards' of the multilateral development banks—notably the World Bank as the lead institution, and the African Development Bank are actually in the process of being substantially weakened, not strengthened.

One of the strongest indictments of large dams has been produced by economists Ansar, Flyvbjerg and associates at the University of Oxford's Said School of Business—in a sense, using the dam developers' own criteria of 'success.' The study was based on an investigation of the *real costs* of 245 large hydrodams—all built between 1934 and 2007 in 65 countries on five continents. This study represents the "largest and most reliable data set of its kind" with portfolios "worth USD 353 billion in 2010 prices" (Ansar et al. 2014). Ethiopia's Gibe III dam was included in this study, as noted below.

Using measures including the gap "between managers' forecasts and actual outcomes related to construction costs, or the cost overrun, and implementation schedule, or schedule slippage," Ansar, Flyvbjerg and associates concluded the following:

"We find that even before accounting for negative impacts on human society and environment, the actual construction costs of large dams are too high to yield a positive return" (ibid.).

Ansar and colleagues specifically cite Ethiopia's [Gilgel] Gibe III dam, along with other large dams in planning stages — stating that these are *"likely to face large cost and schedule overruns, seriously undermining their economic viability."* The matters of:

- Environmental and social impacts on local populations
- Risk of dam collapse, and
- Major debt incurred by African peoples

are noted as important additional considerations by the Oxford business group.

➢ **It is clear that something has gone terribly wrong in the realm of development policy formulation and implementation in terms of the gap between the rhetoric and the reality of macroeconomic and social/environmental outcomes.** Megadams and linked river basin developments have frequently been documented to cause extensive livelihood destruction with widespread malnutrition, disease contraction, irreversible natural resource losses and increasing economic inequality, yet they persist as cornerstones of aid funded national development programs in Africa, including in the eastern region.

In order to answer the question of how such a gap exists between the evidence for negative effects of large dam and basin developments, on the one hand, and the surge of megadam and linked river basin development, on the other, it is necessary to take full account of the political and economic *origins, objectives* and *'marketing'* of policies within Africa and internationally.

This book will consider these issues for the transboundary region of Ethiopia-Kenya-South Sudan/Ilemi Triangle (a contested area), where the Omo River—a major river originating in southwestern Ethiopia—flows into Kenya's Lake Turkana (Fig. 1.1). The Ethiopian government's interests in the Omo basin center on the construction of the Gibe III megadam, an extensive system of dam-dependent irrigated agricultural enterprises and an electricity export transmission line from Ethiopia to Kenya —the first step in a multi-nation, eastern Africa energy 'highway' promoting industrialization economic expansion. The pages to follow trace this development, its origins and unfolding through the matrix of national and international policy-making, and its predictable impacts on hundreds of thousands of indigenous people in the Omo basin within Ethiopia and a segment of Kenya, within the broad expanse of lands around Kenya's Lake Turkana and within the Ilemi Triangle/South Sudan.

The Transboundary Character of Emerging Crisis in Eastern Africa

The Omo River is an international river that flows from its source waters in the western highlands of Ethiopia to its terminus at Kenya's Lake Turkana. Lake Turkana is Kenya's largest freshwater body and derives about 90 % of its waters from the river. Prior to emptying into Lake Turkana, the Omo River meanders through a broad, semi-arid expanse of lowland plains. The Omo's annual flood delivers a major 'pulse' of freshwater, sediment and nutrients to the Lake Turkana —vital contributions to the physical and biological integrity of otherwise saline waters.

The transboundary character of the Omo River stems from recent expansion of the Omo's delta into Kenya's Lake Turkana —placing the river's terminus and much of the delta well within Kenya's national boundaries (Fig. 1.2). In less than a half century, the active Omo delta has grown from a 'birdfoot' profile to an area of over 500 km^2—a major biodiversity area of wetlands and a mosaic-like pattern of other vegetation types.

The lower Omo River basin and Lake Turkana region is one of Sub-Saharan Africa's most culturally diverse areas—home to at least thirteen distinct ethnic groups speaking languages of Cushitic, Eastern Nilotic, Omotic and Afroasiatic origin (Fig. 1.3).

Those indigenous groups nearest the tri-nation border area—the Dasanech, Nyangatom and northern Turkana—are primarily pastoral or agropastoral by tradition (Figs. 1.4 and 1.5) and have remained so for centuries, although with many adaptations. The harsh, semi-arid environment in which they reside has contributed to their complex and flexible survival strategies for coping with changing environmental and social conditions.

The Omo River and Lake Turkana are core components of the survival systems for much of the region's indigenous groups, most of whose livelihood systems are transboundary in nature. The survival of more than 500,000 of the region's pastoralists, agropastoralists and fishers requires access to adequate water and living resources from the Omo River or Lake Turkana.

> **The region's ethnic groups most heavily dependent on the Omo River or Lake Turkana are the Mursi, Bodi, Kwegu, Suri, Kara, Nyangatom, and Dasanech, in the lower Omo River basin, and the Turkana, El Molo, Rendille, Samburu, Gabbra (and some Dasanech) along the shores of Lake Turkana (Fig. 1.3).**[3]

Transboundary ethnic groups are linked together by a network of interethnic material and social exchange relations —a system essential to the survival of each individual group (Fig. 1.6). In recent years, however, nearly all transboundary groups have had to cope with a sharp decline in their capacity to sustain a pastoral life. This stress has reverberated to all groups—friends and enemies alike. The precipitous decline of the herding economy and pastoral life—virtual collapse in some areas—is largely the product of dispossession of resident groups at the hands of powerful external economic and political forces. Although they have sustained complex and flexible strategies for coping with periods of prolonged drought,

---

[3]Considerable literature has generated over the past half century regarding the environment and ethnic groups in the lower Omo basin, northern Turkana region and the borderlands of the Ilemi Triangle. Numerous selections are noted in later chapters.

Fig. 1.1 Location of the Gibe III dam on the Omo River in the tri-nation transboundary region

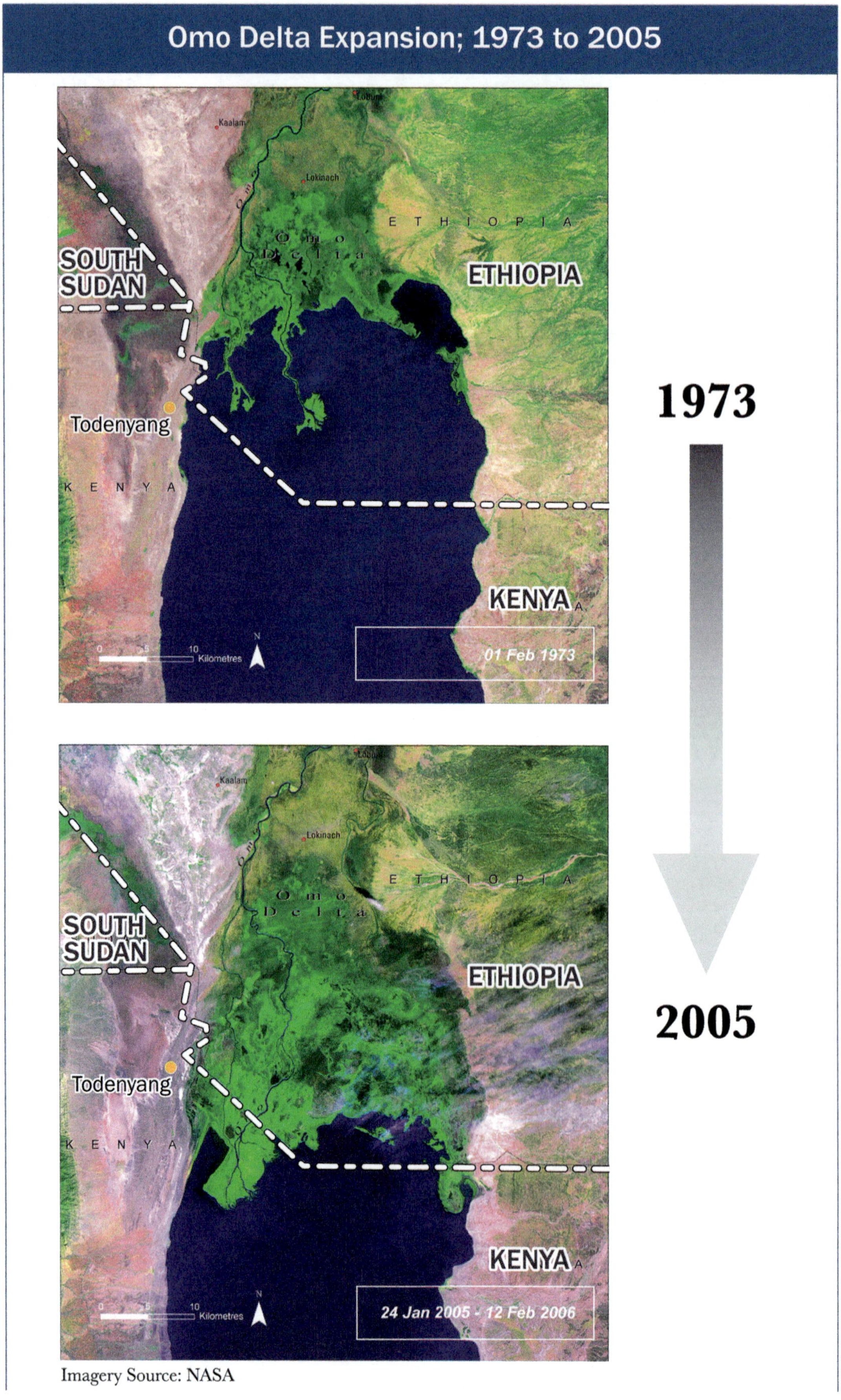

Fig. 1.2 Omo River delta expansion from 1973 to 2006 with the river's terminus in Kenya. *Source* original images from National Aeronautics and Space Administration (NASA). National borders and contested Ilemi Triangle section are added

Fig. 1.3 Ethnic groups in the tri-nation transboundary region

Fig. 1.4 Indigenous Dasanech and Nyangatom residents in the lowermost Omo River basin. *Top left* Dasanech elder woman (pastoral village). *Top right* Dasanech man at major ritual (*dimi*). *Bottom left* Young Dasanech girl. *Bottom right* Nyangatom man (agropastoral village at Omo River)

Fig. 1.5 Turkana villagers at Lake Turkana. *Top left* elder woman along northwestern shoreline. *Top right* fishing family near Kenya-Ethiopia border. *Bottom* Young boys on non-motorized sailboat

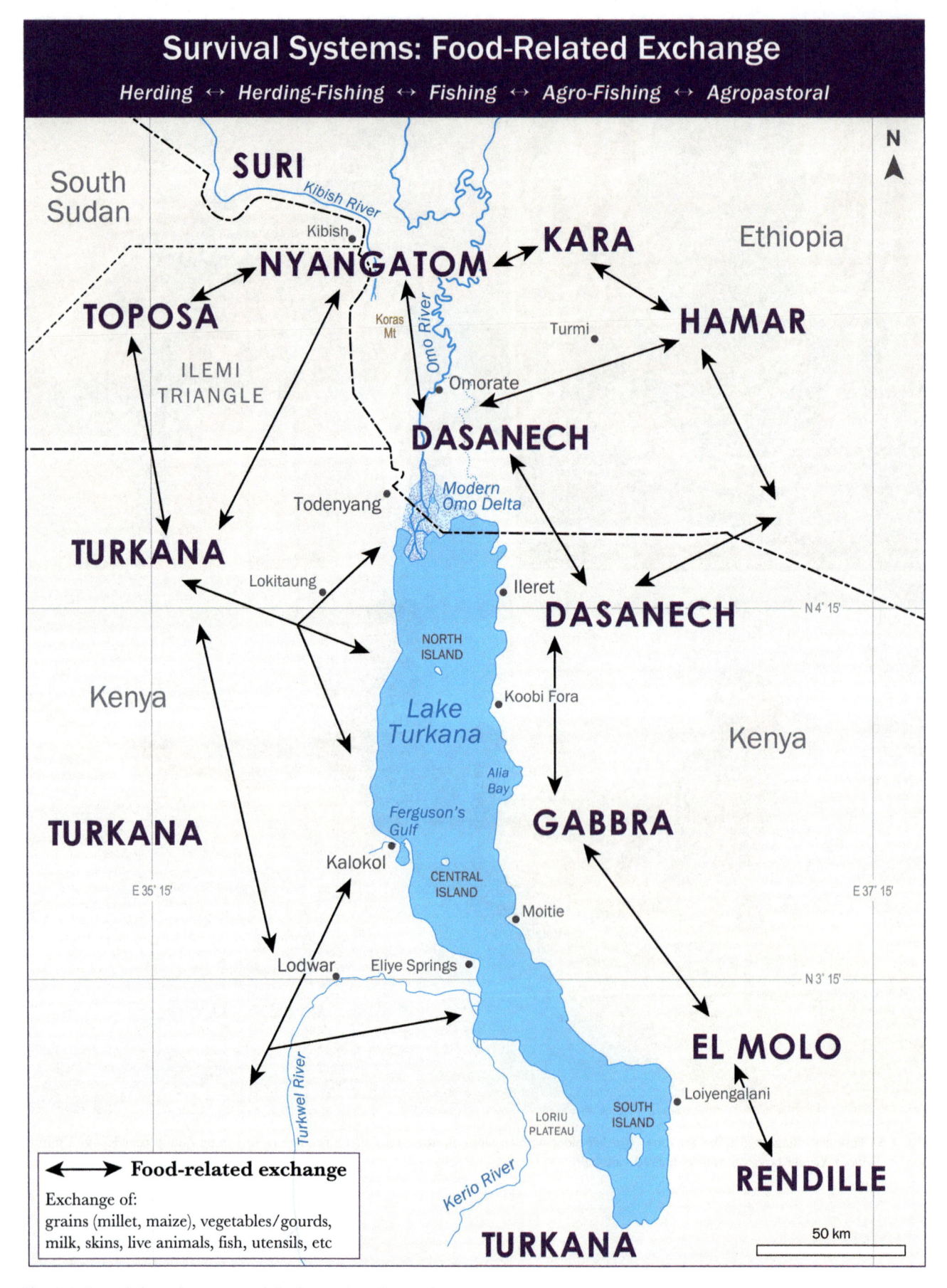

Fig. 1.6 Interethnic exchange network in the transboundary region

disease epidemics social conflict and other hardships for centuries, the region's ethnic groups no longer have the conditions necessary for recovery from these major stress periods. For individual groups, these conditions center on access to sufficient land, pasturage and water resources in order to implement long-term livestock herd diversification, overall production diversification, and maintenance of strong internal reciprocity relations, among others.

Whole segments of these indigenous groups have been forced to migrate to lands along the Omo River and around the shores of Lake Turkana in recent years. Tens of thousands of Dasanech and Nyangatom residents have settled along the Omo River, where they rely primarily on different combinations of flood recession agriculture, livestock raising, fishing, wild food gathering and hunting. Thousands of other Dasanech have migrated from their Ethiopian lands to join Dasanech settlements along the extreme the northeastern shores of Kenya's Lake Turkana (Fig. 1.3), so the group is clearly cross-border in extent. In Kenya, hundreds of thousands of central and northern Turkana have coped with herd losses by moving toward Lake Turkana where they take up fishing, fishing/herding and a variety of secondary activities.

For many of the poorest pastoral and agropastoral households, livelihood activities that once served as temporary measures —for example, as a means of recovery after major livestock or crop losses or as secondary production—have become their principal means of survival.[4] These activities include fishing, wild food gathering, chicken raising and household commodity production. Livelihood activities of the Dasanech Nyangatom and Turkana ethnic groups include those in Figs. 1.7 and 1.8. To a large extent, these production activities reflect major environmental variations created by the punctuation of the arid region by the Omo River, Lake Turkana, ephemeral streams, volcanic rock and tuffaceous outcrops, ancient beach ridges and floodplains, salt springs and other anomalous features (Fig. 1.9).[5]

➢ **The transboundary region was long an area of little concern to policy-makers in Ethiopia, Kenya and distant capitals, other than for determining national boundaries and securing borders. Recently, it has recently become the focus of major hydropower development, dam enabled irrigated commercial agriculture and oil/gas exploration.** These developments either necessitate regulation of the Omo River or at least benefit from it in terms of infrastructure building and securing the region in political and military terms.

Oil and gas concessions are active throughout much of the transboundary region. Ethiopian and Kenyan governments frequently announce 'new' oil and gas discoveries through their national media. In reality, exploration began decades ago as part of a broad cooperation program among government, corporate and aid sectors. Systematic exploration has been underway for a half century in the eastern portion of Africa, from the Red Sea to southern Africa. Exploration activity in the transboundary region has recently escalated, both in geographic extent and activity level. Oil company work parties in the region are highly visible to indigenous communities, with some local groups vigorously opposing their incursion—for example, in Turkana lands.[6] Appendix A of this book, written by J. Dimon with this writer's assistance, is a brief description of oil industry exploration in the transboundary region.

The Gibe III Dam and Linked Agricultural and Power Export Developments

➢ **The Gibe III dam represents a milestone project in Ethiopia's energy development.** It was planned as the nation's first megadam—both to double the nation's power generation for domestic development (mostly urban and industrial) and to launch the export of Ethiopian hydrodam generated electricity to surrounding countries for a planned East Africa energy network supported by international aid. The Gibe III dam is part of a planned "cascade" of hydro projects along the Omo (Fig. 1.10). The relatively small Gibe I dam was completed and is operating, while the associated Gibe II power generating project has been under extensive repair following two tunnel collapse disasters. The Ethiopian government and international planners are planning two additional dams—the Gibe IV and Gibe V.

---

[4]Hunting was one of these opportunistic or temporary means of food-getting. Recent depletion of wildlife in both the lower Omo basin and the Lake Turkana region, particularly with the pervasive use of firearms, has eliminated this option, for all practical purposes.
[5]See the geological, geochemical and geomorphological studies by Brown et al. (2006), Bruhn et al. (2011), Cerling (1986) and Butzer (1970, 1971).
[6]For the most part, only a small number of local residents—mostly young men who have traveled among South Sudanese military groups and later returned to their home areas—are familiar with oil development within South Sudan, including its effects on local communities and as a cause of conflict.

Fig. 1.7 Major livelihoods in the transboundary region. *Top left* Turkana fishing boat departing for expedition. *Top right* Dasanech pastoral village milking time. *Center left* Drying fish catch in a lakeside Turkana village. *Center right* Young Dasanech boy herding goats at Omo River with flood recession agriculture on nearby riverside flat. *Bottom left* Cattle grazing in annually flooded Omo delta wetlands. *Bottom right* Dasanech girl in flood recession farm plot

Fig. 1.8 Peoples of the tri-nation border region. *Top left* Dasanech agropastoral family at the Omo River. *Bottom left* Turkana children near fishing village at Lake Turkana. *Top right* Nyangatom woman at hand-dug well along seasonal Kibish River. *Bottom right* Turkana pastoral women in upland plains

Fig. 1.9 Habitat variation in the transboundary region. *Top left* Meandering Omo River section in the lowermost Omo basin with forest and recession agriculture on the river's inside bend. *Top right* Salt springs with doum palms along the river. *Center* Upland plains with complex vegetation/soil patterns and volcanic highlands. *Lower left* Modern delta at the Omo River's inflow to Lake Turkana. *Lower right* Lake Turkana with extremely shallow western shoreline

Fig. 1.10 The 'Gibe cascade' of hydrodam and electricity projects along the Omo River. *Source* GOE (2009a)

At 243 m height—when completed, the Gibe III would rank as one of the tallest dams in the world. It is located in a deep gorge of the Omo River in Ethiopia's highlands, 464 km southwest of Addis Ababa (Fig. 1.11), with the coordinates 6.8472°N × 37.3014°E. Until the 6000 MW Grand Ethiopian Renaissance Dam (GERD) on the Blue Nile is completed, the Gibe III would be Africa's largest dam. Initially designed as a rock-filled structure, this ill-advised Gibe III plan was later replaced by its present roller-compacted concrete (RCC) gravity design.[7] The dam's reservoir is 150 km long, with an area of 210 km^2 and a storage capacity of 14,700 cm^3—a volume equal to about two years of the Omo River's flow at the dam site. (According to GOE data, even the reservoir's live storage of 11,750 million cubic meters is greater than the second largest lake in Ethiopia—Lake Abaya).[8]

The dam is designed to generate 1870 Megawatts (MW) of electricity (6500 GWh per year) from ten turbines, each with 187 MW capacity. From the beginning, the Gibe III dam was planned for power production geared to both the domestic Ethiopian market and export markets in the eastern Africa region (see Chap. 2)—primarily by direct transmission to Kenya. By 2005, this export plan was explicitly viewed as key to the formation of the Eastern Africa Power Pool (EAPP) —a system designed to link the electricity grids of East African countries. A 2006 Inter-Governmental Memorandum of Understanding (MOU) formalized this export agreement when it was signed by the ministers of energy of both Kenya and Ethiopia (KETRACO 2006).[9] The international development banks have played the key role in these arrangements, including through the signing of the MOU.

The official cost of the Gibe III dam is variously stated as EUR 1.47 billion to EUR 1.7 billion, with subsequent unofficial estimates of well above EUR 2 billion. A Chinese loan of USD 500 million accounts for a substantial portion of this cost, with government and investor funding making up the difference, after international development banks backed away from funding due to violations of their required procedures for project support (see Chaps. 2 and 10, where 'fungible' funding to the Ethiopian government from the international development banks is described). The dam's stated cost excludes those of major infrastructural construction costs for the dam and dam enabled agricultural and other development, as well as the cost of loans to the GOE for domestic and international power transmission systems. This cost also excludes the inestimably huge costs of disastrous human and resource consequences of the project as well as the GOE's alleged mitigation and management plans.

Although the Gibe III dam is clearly regarded by the Ethiopian government as primarily for power production, it is also publically described as 'essential' for the elimination of 'excessive floods' that are "destructive of human life and property"—statements dealt with in Chap. 6.

The World Bank, African Development Bank (AFDB) and European Investment Bank (EIB) have been major contributors to the Omo River basin's commercial development planning, feasibility, and early dam construction. Their internal operational policies for funding projects have by all reports prevented them from funding the Gibe III dam directly, however, because of the GOE's no-bid contracting, its failure to establish impact assessments prior to project inception and other violations of procedural requirements. The Chinese are widely viewed to have rescued the project, however, with their major loan approval.

Other development bank support for the Gibe III and Omo basin development include:

- Funding, coordination and encouragement of support for Ethiopia's bureaucratic apparatus spearheading this development.
- Funding for the GOE's establishment and maintenance of the political and economic context for dam and irrigated agricultural development—commonly dubbed 'basic services' provision, and most prominently.
- Funding for the major power transmission line from Ethiopia to Kenya for an East African 'energy highway' program—a system widely known to be planned for the inclusion of Gibe III generated electricity despite an approximately 50-km 'gap' between the transmission line and the dam for which Chinese funding was secured.

---

[7]A consultant to the African Development Bank issued the first major criticism of the rock-filled structure (Mitchell 2009) and a subsequent assessment by a European Investment Bank consultant raised similar criticism (EIB 2010). Insurance coverage for the rock-filled version was not forthcoming—likely a factor in Gibe III planners' shift to RCC construction (Hathaway 2008).

[8]Lake Abaya is recorded as having a live storage of 8200 million cubic meters.

[9]The other signatories were of Burundi, DRC, Egypt, Rwanda, and Sudan.

Fig. 1.11 Gibe III dam site and construction. *Top* Contour map indicating steep slopes/topographic features at the Gibe III dam site and components of the project complex (GOE 2009a). *Bottom left* dam in progress, with access road. *Bottom right* Idealized version of Gibe III plan. *Source* www.gibe3.gov.et

Longstanding plans for export of Gibe III generated electricity contributed to signing of a power purchase agreement whereby Kenya will buy approximately 400 MW of electricity from Ethiopia for a period of twenty years (beginning in 2017).[10] The plan for export of Ethiopian power, the details of which are outlined below, is part of the multi-billion dollar

---

[10] Among many reports of this agreement is that by Kenya Power and Lighting Co., Ltd. in its Kenya (2011/2012) Annual Report and Financial Statements.

East Africa Power Pool (EAPP), spearheaded and largely funded by the World Bank and African Development Bank (AFDB). GOE officials and numerous aid officials have described the Gibe III dam as "the first step" in this program.[11]

> The World Bank and the African Development Bank (AFDB) have both disclaimed any connection between the Gibe III dam and the major transmission line for power export from Ethiopia to Kenya that they are funding. These assertions are in line with the banks' internal regulations precluding their funding of the dam. This claim is not supported by the reality of the development, however, since **the effects of these developments are *cumulative* and *synergistic***—a matter detailed in Chaps. 6 and 10. The same situation exists with regard to the AFDB's claim of "no relation" to the irrigated agricultural development along the Omo River, since the Gibe III and major dam enabled agricultural enterprises also have cumulative and synergistic effects.

➢ **The Gibe III dam and the irrigated agricultural enterprises it enables would usher in massive scale indigenous livelihood and natural resource destruction in the lowermost Omo basin, Kenya's Lake Turkana region and overall in the tri-nation border region.** The dam would radically reduce the Omo River's downstream flow volume by at least 60–70 % and cause a precipitous decrease in inflow to Lake Turkana during the reservoir-filling period and beyond.[12] Such loss of river flow volume would destroy critical fish reproductive and life cycle conditions and desiccate riverine and Omo delta habitats for flood recession agriculture, livestock watering and 'last resort' grazing. It would also eliminate the pristine Omo riverine forest, which is the last of its type in Sub-Saharan Africa and one of the richest remaining wildlife areas of Ethiopia. Figure 1.12 indicates the rich forest development bordering the Omo River in the semi-arid lowland portion of the river—as well as lowlying locales for recession agriculture along annually flooded riverside sand spits and waterside flats. These environments are vital to the survival of indigenous residents—pastoralists, agropastoralists and fishers. This destruction would be greatly worsened by dam enabled large-scale irrigated agricultural enterprises planned along the Omo River, many of which are already under construction.

The radical decrease in Omo River inflow to Lake Turkana would cause major lake likely retreat—most prominently in the shallow northern and central portions of the lake. The northern shoreline of the lake would recede by around 8–10 km, for example, with multi-kilometer retreat of lake waters throughout much of the lake (see figures in Chaps. 4 and 5). Such retreat would cause the desiccation of near shore fish reproductive habitat leading to destruction of fish stocks that are essential to the survival of hundreds of thousands of fishing and pastoral/fishing indigenous residents, as well as destroy access to adequate and potable water by that population and the region's watering and lakeside grazing resources for untold the region's livestock.

Catastrophic level human destruction would have already occurred by even the end of the dam's closure for reservoir filling—itself a multi-year process. The GOE's 'artificial flood' program—espoused as a 'solution' *after* the filling period —would be entirely inadequate to prevent massive human destruction among Ethiopian and Kenyan communities residing below the dam. As detailed in Chap. 6, such programs are frequently 'promised' but never implemented in large dam development within Africa. The Ethiopian government itself has noted the real 'possibility' of no such artificial flood program or one of limited duration. Even the increased flood level suggested by a development bank consultant (EIB 2010) would be entirely inadequate for the bare survival of the region's resident population and the obstruction of remaining river waters by irrigated agricultural plantations would only worsen the crisis.

> The destruction of pastoral, agropastoral and fishing livelihoods would swiftly produce a major humanitarian disaster, with widespread conditions of starvation, disease and spiraling interethnic armed conflict in the tri-nation border region as groups desperately compete for vanishing resources. Policies to effect the Gibe III dam and its attendant irrigation agriculture violate the U.N.-recognized human right to adequate water and its associated right to livelihood for hundreds of thousands of indigenous residents.

---

[11] Impact assessments and related studies, produced by global consulting firms and individuals through their contracts with governments and aid agencies (GOE, GOK, World Bank, AFDB, EIB, etc.) are referenced in this book to the contracting government or agency, not the global industry consultants. Exceptions to this are documents with multiple contracting agencies or non-agency documents produced by GCI firms or individuals.
[12] This period is dubiously defined by the GOE (2009b) as requiring one to three years—a length of time highly unlikely to be sufficient, according to Africa Resources Working Group (ARWG) physical scientists working in the area. This is due to the fractured nature of the reservoir's rock walls providing for major leakage from the reservoir. This matter is outlined in Chap. 3. No data has been recorded by the GOE for downstream Omo River flow volume or lake inflow prior to initiation of Gibe III dam construction, or in ensuing years.

Fig. 1.12 Meandering section of the lower basin Omo River. *Top* Sand spit with indigenous flood recession agriculture along (annually flooded) waterside flat and forest on higher (*not* flooded) natural levee. *Bottom* Well-developed forest (trees to 27-m height) on inside bend of meander *versus* semi-arid vegetation along the 'outside' bend

A major earthquake occurrence of 7 or 8 magnitude is predictable for the Gibe III dam portion of the Ethiopian Rift Valley within the next fifty years (see Chap. 3). According to available documents, dam collapse—whether caused directly by an earthquake or by a seismic event combined with landslides or sediment buildup behind the dam—is certainly plausible and would produce cataclysmic destruction of human life both in the lowermost Omo basin and around Kenya's Lake Turkana. In addition to unprecedented human destruction from such an event, the financial costs of a major dam disaster would add incalculable burden to the citizens of both nations. Chapter 3 details this threat to the peoples downstream in Ethiopia and Kenya.

➢ **The Ethiopian government is already implementing its plans for an extensive system of dam enabled, large-scale irrigated commercial farms along the lower Omo River accompanied by GOE evictions of tens of thousands of indigenous residents.** As detailed in later chapters, hundreds of thousands of hectares are allocated by the GOE for these state owned and private enterprises which are being constructed on the traditional lands of multiple ethnic groups. Massive land areas and resources essential to the livelihoods of these groups are being expropriated, with accompanying eviction of tens of thousands of villagers.

- Major abstraction of river waters for these irrigated large-scale commercial farms along the Omo would both multiply and indefinitely extend the radical reduction of Omo flow volume and lake inflow during the Gibe III reservoir-filling period. In combination, these two major sources of water denial to the downstream system and the elimination of the river's annual flood which is vital for sustainment of downstream survival systems would create catastrophic losses for hundreds of thousands of pastoral, agropastoral and fishing residents.

> Large-scale irrigated farm enterprises along the Omo are dependent on the Gibe III dam's regulation of the river—a fact ignored or directly misrepresented in government and development bank assessments of 2009 and 2010 (see Chap. 6). The fixed irrigation systems and high-value crops of these large-scale farms necessitate a predictable and calibrated abstraction of river water in order to maximize agricultural productivity.[13]

- Widespread Ethiopian government expropriation of indigenous villagers for the establishment of irrigated commercial private and government plantations has been underway for at least five years in both the higher altitude Mursi portion and the Dasanech/Nyangatom portion of the lower Omo basin. This process is documented by international human rights organizations for the traditional lands of the Mursi and their neighbors (Fig. 1.3). To the extent possible, SONT researchers have documented the expropriation process in the less accessible Dasanech (and Nyangatom) lands in the lowermost Omo basin (see Chaps. 7 and 8).
Most of the Dasanech communities that have been expropriated from their traditional lands along the lowermost Omo have no option but to take refuge within (or nearby) the modern Omo delta. This involuntary movement into the delta region greatly worsens the already crowded conditions there for agropastoralists and fishing communities already settled there.[14]
- The Ethiopian government forces respond with repressive measures—often, with brutality—when Dasanech or Nyangatom communities resist expropriation from their riverine lands. Fear among villagers throughout the region is intensifying. Such political repression and the GOE's denial of residents' access to river waters essential to their survival constitute human rights violations, as defined by the United Nations and other international bodies. A culture of fear pervades villages throughout the region where communities are already desperately searching for a means of survival.

The response of the Kenyan government (GOK) to the survival plight of indigenous pastoralists and fishers around Lake Turkana indicates a predominant attitude of indifference. This view is reflected in World Bank statements in internal documents, for example, citing the high likelihood of the GOK "overlooking" impacts of lake retreat on its indigenous population in favor of Gibe III dam electricity generated that is contracted for export to that 'power deficit' nation.

---

[13] A large sugar plantation already under development (the Kuraz plantation) in traditional Mursi lands (Fig. 1.3) has a regulated water source—through a weir already constructed in that locality.
[14] Some households with remaining livestock relocate to degraded grazing lands around the northeastern shoreline of Lake Turkana—risking major hostilities with Turkana pastoralists.

Resistance among Turkana communities to the Gibe III dam—and other developments in the region—met with increasing militarization by the GOK. Community leaders report government warnings to activists and threats of reprisal for protest of GOK policies.

The matter of violation of U.N. defined human rights—especially the *human right to adequate water* with its accompanying rights to livelihood, health and freedom from political repression—is outlined in Chap. 10.

Literature Cited

Brown, F.H., B. Haileab, and I. McDougall. 2006. Sequence of Tuffs between the KBS Tuff and the Chari Tuff in the Turkana Basin, Kenya and Ethiopia. *Geological Society of London* 163(1): 185–204.
Bruhn, R.L., F.H. Brown, P.N. Gathogo, and B. Haileab. 2011. Pliocene volcano-tectonics and paleogeography of the Turkana Basin, Kenya and Ethiopia. *Journal of African Earth Sciences* 59(2–3): 295–312.
Butzer, K.W. 1970. Contemporary depositional environments of the Omo delta. *Nature* 226: 15.
Butzer, K.W. 1971. Recent history of an Ethiopian Delta. University of Chicago Department of Geography Papers No. 136.
Cerling, T.E. 1986. A mass-balance approach to basin sedimentation: constraints on the recent history of the Turkana basin. *Palaeogeography, Palaeoclimatology, Palaeoecology* 54: 63–86.
Ethiopia, Government of, Ethiopian Electric Power Corporation (EEPCO). 2009a. CESI, Mid-Day International Consulting Engineers (MDI), Gibe III Hydroelectric Project, Environmental and Social Impact Assessment, Report No. 300 ENV RC 002C Plan.
Ethiopia, Government of, Ethiopian Electric Power Corporation (EEPCO). 2009b. Agriconsulting S.P.A., Mid-Day International Consulting, Level 1 Design, Environmental and Social Impact Assessment, Additional Study of Downstream Impacts. Report No. 300 ENV RAG 003B.
Ethiopia, Government of, Ethiopian Electric Power Corporation (EEPCO) n.d., A. Samuel, The first hydro power station in Ethiopia. www.eepco.gov.et/corporationhistory.php.
European Investment Bank (EIB). (2010, Mar). Sogreah Consultants, Independent Review and Studies Regarding the Environmental & Social Impact Assessments for the Gibe III Hydropower Project, Final Report, 183 pp.
Hathaway, T. 2008. What cost Ethiopia's dam boom? A look inside the expansion of Ethiopia's energy sector. *International Rivers*. http://www.internationalrivers.org.
Hinshaw, R.E. 2006. *Living with Nature's Extremes: The Life of Gilbert Fowler White*. Johnson Books, 338 pages.
Kenya, Government of 2011/2012, Annual Report and Financial Statements.
Ketraco (Kenya Electricity Transmission Co. Ltd.). 2006. Eastern Africa Interconnector (Ethiopia—Kenya). http://www.ketraco.co.ke/projects/ongoing/ethiopia-suswa-k.html.
Leslie, J. 2005. *Deep Water. The Epic Struggle Over Dams, Displaced People and the Environment*. Farrar, Straus and Gireau, New York.
McCully, P. 2001. *Silenced Rivers. The Ecology and Politics of Large Dams*. New York: Zed Press.
Mitchell, A. 2009. Gilgel Gibe III dam Ethiopia: technical, engineering and economic feasibility study report. Presentation transcript, submitted to African Development Bank. Personal communication, manuscript.
St. Petersburg Times. (1945, Sept. 7), Ethiopian Emperor Grants Sinclair Oil Concessions.
United States, Government of NASA Johnson Space Center Collection. 2009. Earth observation of Omo River Delta, Lake Turkana. http://www.nasaimages.org.
World Commission on Dams (WCD). 2000. *Dams and Development*. London: Earthscan. 198 pages.

Open Access This chapter is distributed under the terms of the Creative Commons Attribution-NonCommercial 2.5 International License (http://creativecommons.org/licenses/by-nc/2.5/), which permits any noncommercial use, duplication, adaptation, distribution and reproduction in any medium or format, as long as you give appropriate credit to the original author(s) and the source, provide a link to the Creative Commons license and indicate if changes were made.

The images or other third party material in this chapter are included in the work's Creative Commons license, unless indicated otherwise in the credit line; if such material is not included in the work's Creative Commons license and the respective action is not permitted by statutory regulation, users will need to obtain permission from the license holder to duplicate, adapt or reproduce the material.

The Persistent Paradigm for 'Modernizing' River Basins: Institutions and Policies in Ethiopia

Abstract In view of the nearly unprecedented level of destruction of human life and natural resources from dam development, the question arises as to how such a policy could come into play without major questioning and challenge. The following is an effort to answer this question, by tracing the trajectory of the river basin development policy involving the Ethiopian state and international development forces—from its beginnings to the Gibe III dam. Citing alleged priorities of the 'national interest,' and programs for economic 'growth and transformation,' the Ethiopian government and international development banks, together with Kenyan government cooperation, have for decades actively pursued Omo basin hydrodam and dam enabled agricultural development. The nexus of institutions most central to the design, implementation and legitimation of major hydrodams and their associated river basin development is structured in such a way as to prioritize macro economic and political objectives with the externalization of the well-being—in the Ethiopia case, the very survival—of local residents who are among the most marginalized and vulnerable to destruction from this development.

Early River Basin Development in Ethiopia

➢ **Plans for the Gibe III dam and Omo River basin development emerged from decades of Ethiopian government efforts to develop its water resources along with major involvement on the part of international aid and investment institutions.** Focus on large dam and dam-enabled enterprises—especially hydroelectricity and large-scale irrigation agricultural projects—persisted throughout the post WW II period, with continuity of purpose and approach, from the years of Haile Selassie rule with major western nation support, through self-proclaimed 'socialist' authoritarian rule by the regime commonly referred to as the Derg, to the current centralized EPRDP-dominated government with its return to market-driven 'modernization.'

As early as 1912, Emperor Menelik II constructed the first hydropower plant on the Akaki River—a tributary of the Awash River, in order to power the palace and a group of small factories in Addis Ababa (U.S. Government 1916). Following his accession to the throne of his uncle, Menelik, Haile Selassie continued the interest in developing Ethiopia's water resources, including hydropower—development likely facilitated by his consolidation of power in the monarchy with the Constitution of 1933 (Habte Selassie 2013). He reportedly authorized a redevelopment of the Akaki River, in 1932, for example, where the Ethiopian air force, radio communications and other developments were established. His plans were interrupted, however, by Mussolini's brutal invasion from Italian controlled Eritrea and Somaliland in 1936, with bombing of the Akaki and other areas (Nicolle 1997)—followed by the years of Italian occupation.

Having made the mistake of putting his trust in the 1928 Treaty of Friendship between his monarchy and Mussolini of Italy, despite both nations being members of the League of Nations, which mandated non-aggression among its members, Haile Selassie was exiled. After a failed plea for help from the League of Nations, the emperor began a five year residence in

England. With the beginning of World War II in 1930, the exiled monarch was catapulted onto the world stage and for many, almost instantly became a "prophet." Habte Selassie wrote, in 2014, "For people in the African Diaspora, particularly in the United States and the Caribbean region, his name became a talisman for Africa's liberation". The Allied powers' resounding counter to the Italians in Ethiopia and the rest of the Horn ultimately restored Haile Selassie to power via his passage to Sudan and the actions of liberation forces, bolstered by the British (Vestal 2011). Meanwhile, the monarch had established strong ties with the Americans—connections that were to lead to a major U.S. role in Ethiopia in the years to follow.

During their years of occupation, the Italians built some generators and extended the power supply to some major towns as one dimension of infrastructure and commercial development. Almost immediately after their occupation of Ethiopia, the Italians began building the Aba Samuel dam/power plant on the Akaki River, with a 40,000 m^3 reservoir. Although initially relatively small (6 MW), the power plant was expanded (in the 1950s. The Italian dam thus initiated the harnessing of a tributary of the 1200 km Awash River, which was to become the centerpiece of hydropower development in Ethiopia during the early period of post-WW II modernization under Haile Selassie.[1]

Reinstated to the throne in Ethiopia, Haile Selassie began negotiating with the allied powers for Ethiopia's full sovereignty and became a signatory of the Peace Treaty concluding the war. The monarch then brought Ethiopia into membership at the United Nations in 1945—the first African head of state to do so. Despite his highly popular and revered image abroad, the emperor faced considerable unrest at home and needed major assistance to stabilize his position. This suited the allies and particularly the U.S., which Haile Selassie viewed as less threatening in imperialistic terms (Habte Selassie, op. cit.). For their part, U.S. foreign policy analysts regarded Ethiopia as an excellent *entre'* point to the still largely colonial African continent. Moreover, Ethiopia was perfectly situated for U.S. geostrategic interests in the early years of the Cold War. Indeed, in subsequent years Ethiopia became a major recipient of U.S. military and economic assistance in Africa and fully embroiled in the regional strategies of Cold War superpowers. These factors combined to bring about major 'modernization' efforts in select regions of the empire and the beginning of a virtually permanent presence of western aid and investment institutions.

The World Bank—active for only a few years since its creation as a Bretton Woods institution—was already engaged in economic planning in many African countries. Ethiopia became its first African member (and client) state with an official Bank mission established and loans issued. From the beginning, the Bank's key planning and coordination of the 'modernizing' effort within Ethiopia were done in concert with U.S. involvement. A U.S. Technical Mission, for example, carried out detailed study of Ethiopia's potential economic development prior to the World Bank's first mission to the country in 1950, contributing to the Bank's issuance of its first loans to Ethiopia that very year.

The first World Bank loans to the monarchy were in fact the first loans by the Bank to Africa. Two loans issued in 1950, following the Bank's first mission to the country, were for the establishment of an Ethiopian Development Bank. These were mostly to promote agricultural and industrial enterprises, and road construction, including 'feeder roads' to the Imperial Railroad from Addis Ababa to neighboring Djibouti, Ethiopia's only shipping outlet at the time (World Bank 2013).[2] Built during Menelik's rule by the French, the railroad traversed the broad, semi-arid Awash River basin in northeastern Ethiopia (Fig. 2.1) that was soon to be the focus of major water, power and irrigated agricultural investment and development. Additional funding in the 1950s included finance for telecommunications and road construction.

Following the 1960 establishment of the Bank's soft loan window, the International Development Association (IDA), the Ethiopian monarchy received loans on a consistent basis and the World Bank continued its dominance in Ethiopian development policy.

> Generally speaking, the World Bank's early prominent role in economic planning and financing for the monarchy during the early post-war years—closely coordinated with U.S. assistance and Haile Selassie's own enthusiasm for establishing Ethiopia as a 'modern state' in economic and political terms—effectively set the conditions for major river basin development in the empire.

---

[1]The dam became non-functional in the early 1970s. Recently, the GOE (EEPCO) began plans to refurbish the Aba Samuel.
[2]Haile Selassie's disregard for the U.N. mandate for a popular referendum in Eritrea to determine its possible autonomy or affiliation with Ethiopia, following its liberation from Italian occupation, led to his brutal actions to annex Eritrea—regarded as a critical shipping outlet at the Red Sea, a source of resource appropriation and of major interest to the Cold War focused U.S. for new military installations and surveillance of the Soviet Union.

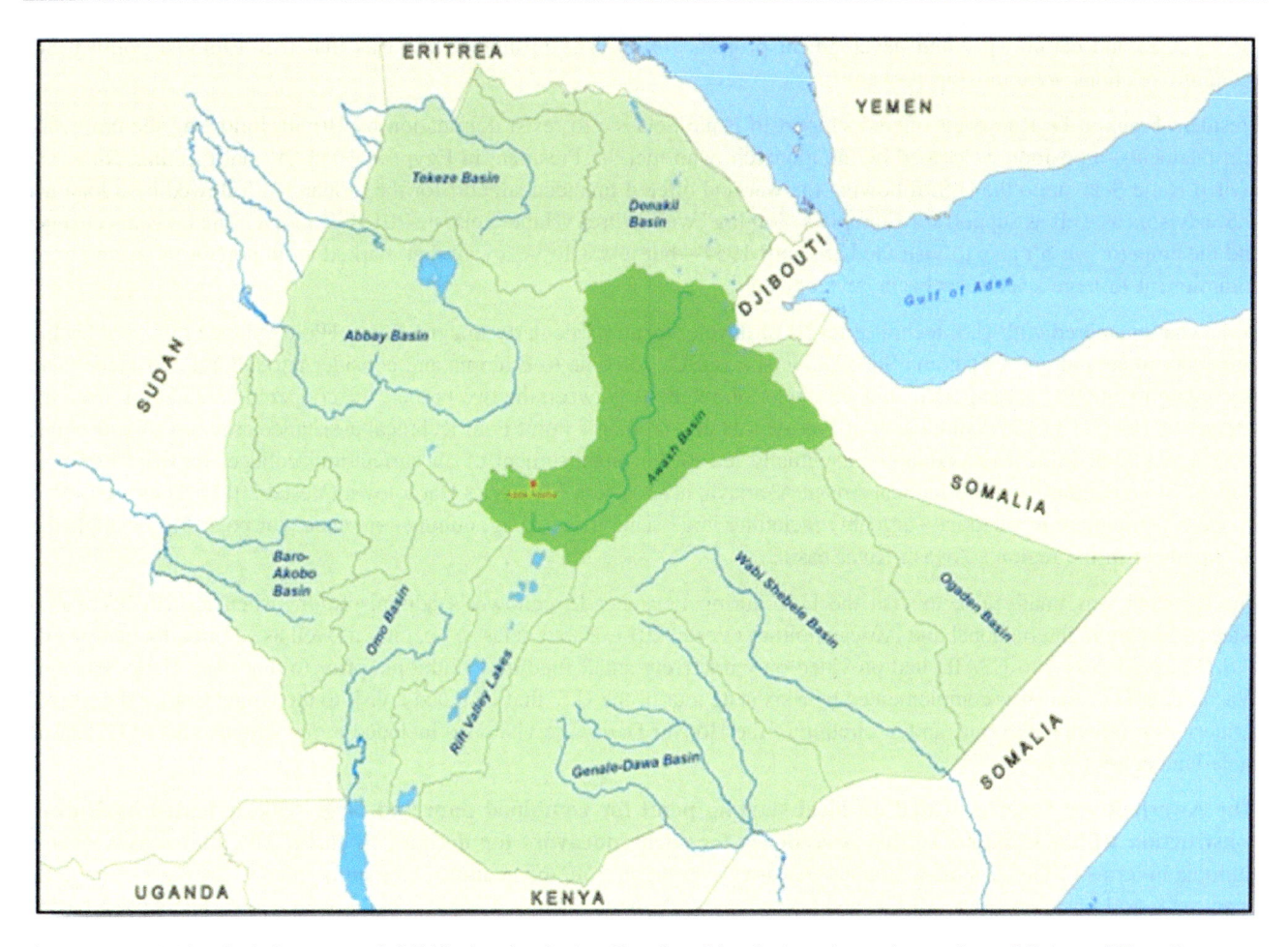

Fig. 2.1 Awash River basin in context of all Ethiopian river basins. *Note* Omo River basin to the southwest. *Source* Ministry of Water Resources, Government of Ethiopia

Establishment of an administrative apparatus for river basin development began early in Ethiopia, with the 1956 formation of the nation's first agency to promote hydrodam development—the Ethiopian Electric Light and Power Authority (EELPA). EELPA was charged with responsibility for planning and implementing hydrodam and dam enabled, large-scale agricultural and other industrial development. This included overseeing the generation, distribution, and sale of electricity, with emphasis on commercial and industrial sales.

➢ Official relations between Ethiopia and the United States actually dated as far back as 1903. A U.S. mission headed by Robert P. Skinner was sent to Ethiopia in order to negotiate a commercial treaty with Emperor Menelik (Skinner 2003). The mission ended on a positive note with establishment of a personal relationship between Menelik and U.S. President Teddy Roosevelt.[3] Ethiopia offered a unique opportunity for U.S. presence in Africa since it was not under colonial domination from Europe (Vestal op.cit.)

Relations between the two countries remained positive and in active during the early to middle 1930s and again in the 1940s. In 1935, more than half of Ethiopia was deeded to the Anglo-American oil corporation. According to news reports, the emperor's calculation was that this action would block a feared Italian invasion failed (St. Petersburg Times 1945). Following the defeat of the Italians and Haile Selassie's reinstatement in Ethiopia, the emperor firmed up a nothing short of colossal new oil arrangement with the American oil company, Sinclair, whereby Sinclair was granted a long term concession over the entire nation of 350,000 square miles, with the right to explore, produce and remove oil from the empire in return

---

[3]The 'good will' aspect of the U.S.-Ethiopia connection established was symbolized by Menelik's gift of some lions and ivory to Roosevelt. In a sense, this good will was sustained and once again emerged under U.S. President Franklin Roosevelt in the 1940s.

for royalties and certain specified development projects (*ibid.*). The upshot of this was that U.S.-Ethiopia political and economic relations were moving forward.

President Franklin D. Roosevelt offered support to Haile Selassie to resist domination by Britain following the emperor's reinstatement—a position welcomed by the monarch, who met the President in Egypt in 1945. At that meeting, Roosevelt invited Haile Selassie to the U.S. following the war and offered financial support for the monarch—followed by a loan and U.S. advisors as well as support for Ethiopia's "joining" with Eritrea (Habte Selassie 2014). Roosevelt died, events changed and the emperor wasn't able to visit the U.S. until 1954—but it was to be a visit that marked an important milestone in his commitment to large scale river basin development.

Somewhat enamored with U.S. technical levels of development, particularly in agriculture, Haile Selassie invited a leading university in agriculture, Oklahoma State University (OSU), to come to Ethiopia and consider agricultural assistance there. According to an OSU participant in and chronicler of, relations between the two parties, the cooperation emerging from this interaction became an early component of President Harry Truman's Point Four technical assistance program. A subsequent 1952 agricultural team from Oklahoma eventually led to the establishment of an agricultural college, Imperial Ethiopian College of Agriculture and Mechanical Arts at Alemaya, near eastern Ethiopia's Harar town (Vestal 2011). Thus the birth of 'modern' agriculture in Ethiopia—arguably including large-scale irrigated agriculture—an endeavor poised to be applied to development in the region's Awash River basin.

The monarch was finally able to visit the U.S. during President Eisenhower's administration, when he had become an esteemed figure in international and African politics (Vestel 2011, Habte Selassie op. cit.), as well as admired by many in the U.S. The predecessor to USAID had provided only relatively small funding to Ethiopia prior to that visit. Haile Selassie's Washington D.C. tour was complemented by a six state tour of the U.S. that included a visit to the Grand Coulee dam, an oil refinery in southern California, and a 'circling of the Hoover Dam. The visit also included a momentous visit to Oklahoma State University.

➤ **The Awash River basin provided an ideal starting point for combined dam and large-scale irrigated agriculture construction within Ethiopia, setting precedence for such endeavors for decades to come.** The third largest river in Ethiopia in terms of the catchment area, the Awash River flows from the central Ethiopian highlands eastward through the open, semi-arid lowlands of the Middle and Lower Awash Valley, where it dissipates in the Afar desert. The Awash basin's outstanding potential for large-scale irrigation agriculture and its proximity to the developing commercial and industrial enterprises—including the railroad to the seaport in Djibouti—were all attractive qualities to both the monarchy and western agencies.

Haile Selassie himself took a personal interest in developing the Awash Valley. By then, he had proclaimed that "Ethiopia is the water tower of Africa"—a phrase welcomed by development officials bent on developing the region and viewing Ethiopia's potential for aiding that process with its abundant water resources. The extensive prior U.S. agricultural assistance in the eastern region as well as the World Bank's continual promotion of development, were important forces promoting the Awash development, as were technical contributions by United Nations.

Events moved quickly and the government began construction of the Koka dam in 1957—in Oromo lands, about 75 km from Addis Ababa. Completed in 1960, the dam was viewed as a "large" dam at the time (43 MW), with a height of 48 meters (compared with over 240 meters planned for the Gibe III dam), a power generating capacity of 43 MW and a reservoir of 180 km^2. Beyond the rationale that the construction of the dam was being paid for by Italy as part of war reparations, the Koka was rationalized as a major means of fostering economic growth, thus reducing Ethiopia's glaring poverty through electrification and provision of water for "all who needed it" in Addis Ababa and several other major towns. In reality, electricity generated by the dam was overwhelmingly allocated to industry around the capital, along with agribusiness and enterprises in the Awash Valley. The cost of the dam was covered by Italy as part of its war reparation agreement, and construction was awarded to Italian contractors.[4]

In 1962, the monarchy, with aid agency advice and funding, established the Awash Valley Authority (AVA) as the office responsible for the development and management of the river basin. The AVA was fashioned after the Tennessee Valley Authority in the U.S. and was the forerunner of successive river basin development authorities in Ethiopia. It had the

---

[4]The Italian war reparations payment for the dam was pointed out to me by Elias Habte Selassie, former senior legal officer at the Awash Valley Authority.

contradictory responsibilities of promoting irrigation and commercial agricultural development in the Valley, on the one hand, and administrating the indigenous groups in the Valley for their 'welfare', on the other.[5]

The structure and function of the AVA largely paralleled river basin authorities being established in other developing countries during these years. Like them, it attempted to preside over political protest generated by the radical reduction of river flow volume, dispossession of major indigenous populations, and inevitable interethnic conflict increase as groups were forced to compete for shrinking resources.

The AVA regarded the Awash Valley to be largely "unoccupied" and "underutilized", and developers tagged the peoples of the Middle and Lower Valley—including Afar, Somali and Oromo pastoralists and agropastoralists—"backward" and "primitive." The vast majority of government and development specialists shared this perspective. Hence, the livelihoods and survival strategies of the pastoralists were of no consequence in the rush to develop hydropower and major commercial agriculture.[6] There were implementing partners in the formation of the AVA, including the Victoria Water Resources Commission of Australia, contracted by the Food and Agriculture Organization (FAO) of the U.N. to help implement water resource development and agribusiness development. As per their Terms of Reference (or 'scoping' specifications), consultants for feasibility studies took no account of the Valley's indigenous population—assessing only irrigation agriculture and water resource development potential. Despite multiple volumes of feasibility studies for agricultural development that this writer has inspected, none contain any substantive treatment (even identification) of the indigenous populations in the region—let alone their settlements, livelihood needs, or resource dependency on riverine or other resources in the Awash basin.

➣ **The Koka Dam and its accompanying large-scale irrigated commercial agriculture caused major economic decline among the Awash Valley's indigenous Oromo, Somali and Afar pastoralists and agropastoralists residing there for centuries.** Numerous reports of these effects have been produced over the years, including by Bondestam (1974), Emmanuel (1975), Carr (1978) and Kloos (1982). These groups depended on Awash River resources for grazing and watering their livestock, for food gathering and hunting during times of severe drought, and in some instances, for flood recession agriculture. Sharply increased vulnerability to malnutrition and disease among the indigenous population brought both heightened and new diseases (Kloos 1982), especially those accompanying the commercial farms (for example, tuberculosis, malaria, and schistosomiasis).

The monarchy continued with its planning and construction of hydrodams and agribusiness development, with strong international support. The monarchy built two additional dams along the Awash: Awash II (32-MW) in 1966 and Awash III (32-MW) in 1974 (along with transmission facilities)—both well downstream along the 1200 km river, in traditional Afar lands, and both with financing by the World Bank. These too have produced intense conflict with local residents, since they were directed primarily to corporate-style irrigated agricultural enterprises (World Bank 1969).

In the Middle Awash Valley, Ittu and Kerrayu Oromo pastoralists lived side by side. Both experienced major economic decline from the Koka dam and new irrigation agricultural enterprises along the Awash River. This writer conducted a U.S. National Science Foundation funded socioeconomic study among the Ittu Oromo in 1975–76, documenting the impacts of both the dam and the major sugar cane plantation (H.V.A., or Hangler Vondr Amsterdam) established on traditional Ittu lands and with a major plantation along the river since 1966 (Carr 1978). Middle Awash Valley pastoralists were expropriated from their riverside Awash Valley lands by large foreign-owned, irrigated commercial farms (Carr 1978; Bondestam 1974); Afar sultanates, however, controlled significant areas of irrigated agriculture in the Lower Awash Valley following the construction of the Koka Dam. The biggest losers from these agricultural developments were the pastoralists and agropastoralists, since their livelihood depends on access to the riverine zone for livestock watering and last option grazing—especially during prolonged drought periods—as well as for flood recession agriculture, wild food gathering and on occasion, hunting. The pastoralists were essentially cut off from their traditionally available resources along the Awash River and had to risk hostility and violence from the plantation workers in order to gain access to the highly polluted (with

---

[5] I am grateful to Elias Habte Selassie, former senior legal officer at the AVA, for his insights regarding the roles and impacts of the AVA on the Awash Valley's population.
[6] There were exceptions. One FAO consultant's report detailing the land tenure crisis for the pastoralists, for example, was suppressed by the U.N. agency and the AVA. A senior staff member of the Land Reform Ministry in the middle to late 1970s—during the Derg period—produced a second report on the plight of the pastoralists, but this too was suppressed. The director of the legal division of the AVA also raised the issue of Afar and neighboring ethnic group expropriation by the government's concessions to irrigated farming interests, although marginalization of these peoples continued. This writer acquired these reports with AVA and Ministry personnel and discussed the problems of dissent during research in 1975 in the Middle Awash Valley (see below).

pesticides and fertilizers) and stagnant plantation waters in order to obtain water for their own household use.[7] Government repressive measures were commonplace, along with increasing militarization of the region and periodic flare-ups of violence between the disenfranchised pastoralists and the farm laborers—most of whom had been transported from politically volatile regions in the highlands.[8]

By the early 1970s, 37 % of Ethiopia's labor force was in agriculture, and of this, 70 % were employed at H.V.A. facilities (Araia 1995)—many of whom were at the Metahara farm in the middle Awash Valley. The expulsion of the Kerreyu and Ittu pastoralists' and agropastoralists' from their Awash lands in favor of large-scale irrigated commercial farms generated major indigenous resistance.

Only one more relatively large dam was constructed during that period—the 134-MW dam in the Abay River ("Blue Nile") basin of western Ethiopia. Like the Koka and later dams along the Awash River, the Finchaa dam was planned with large-scale agricultural enterprises in mind—especially sugar. The Finchaa dam was financed by the World Bank as well, after arrangements between the monarchy and USAID deteriorated over the matter of tied aid for U.S. contractors and suppliers.

By the end of the 1960s, the World Bank had functioned as the key force in planning, coordinating or funding hydrodam, road, telecommunications and power production in Ethiopia. The combined (IBRD and IDA) Bank loans extended for these developments in Ethiopia between 1950 and 1969 totaled USD 84 million (World Bank 1969; 1969 currency values).

➢ **Following the overthrow of the monarchy in 1974, the military regime—or Derg[9]—was consolidated under Mengistu Haile-Mariam and derived support primarily from the Soviet Union, until the late 1980s.** What is less well-known, however, is that the vast majority of aid to Ethiopia during this period came from the West, including virtually all of the relief aid to the country—especially during but not limited to, the crisis level hunger years of the mid-1980s. The World Bank, for example, produced numerous economic reports and missions during these years of the Soviet-backed Mengistu regime.

From 1981 on, the Bank extended several hundred million dollars of loans to the Soviet-backed regime and turned a blind eye to Mengistu's repressive policies. It also allocated USD 39 million to Ethiopia's Ministry of Agriculture, for example, while the Ministry was directing a massive program of evictions and forced relocations throughout the country. The Bank had adopted a strategy of continued presence in the USSR-backed, so-called 'socialist' state of Somalia since 1969—a strategy that it considered successful toward building the western powers' agenda. It again utilized this strategy in Ethiopia during the Mengistu 'socialist' period.

One large dam was completed under the Derg—the 153 MW Melka Wakena dam on the Shebelle River—along with a variety of water management schemes and small dams, mostly in the Gambella area of western Ethiopia. The Shebelle is a transboundary river, originating in eastern Ethiopia and flowing through semi-arid Somalia to its terminus at the Indian Ocean. The Somali government (then headed by U.S.-backed President Siad Barre) viewed the Melka Wakena and its accompanying irrigation agricultural development with hostility since it radically reduced the river's downstream flow to Somalia's primary agricultural area—the Lower Shebelle Valley.

Mengistu initiated one giant agro-industrial project, the Tana-Beles development—an endeavor that failed entirely and produced widespread hunger, disease and death among project settlers. The Tana-Beles plan was for more than 200,000 ha of farming land and a 'resettled' (by force) population of 75,000–80,000. The construction company for this EUR 150 million project was Salini Costruttori—a family owned and globally active company engaged in Ethiopia since Haile Selassie's time.[10]

---

[7]This writer held a number of discussions with the manager of the HVA sugar plantation. In these discussions, the manager frankly stated that the protection of local residents' access to the river or to water, as well as the prevention of pollution in the plantation's open irrigation channels was "nowhere in the company's agreement" with the GOE and that these matters were "none of their concern." Heavy government police presence in the region prevented effective protest, although on occasion, the Ittu and Kerrayu resisted. One dramatic instance of this involved a 200 km march by Ittu Oromo elders and their supporters to the Parliament building in Addis Ababa.

[8]In recent years, some Kerreyu and Ittu pastoralists in the Middle Awash Valley have initiated small irrigation agriculture on available lands along the river—partly in defense of further expansion of the commercial farms. The human ecological and survival strategy relations of these Ittu communities and the impacts on them by these developments are described in this writer's 1978 report.

[9]The Derg (or 'Dergue'), a short name for the Coordinating Committee of the Armed Forces, Police and Territorial Army—later retitled the Provisional Military Administrative Council, ruled Ethiopia rom the overthrow of Haile Selassie until 1987. In 1987, it was then 'dissolved' by the its head, Mengistu Haile Mariam, who held on to power until his government was defeated in 1991.

[10]Salini became the primary construction firm contracted by the GOE for the series of Gibe dams (described below).

Tensions between Ethiopia and Egypt were increasing during the Derg period over Ethiopia's stated intention to dam the Abbay River (Blue Nile), which Egypt viewed as a dire threat.[11] Ethiopia had, in fact, been insisting on its "sovereign right" to develop its rivers at a number of international meetings, (including a major United Nations international water conference in 1977)—an insistence continuing until the present.

There was in fact international development bank support for this dam undertaken by the Derg. While the Soviets and Ethiopians progressed with plans to implement the project, the development banks were conducting feasibility studies for the damming of the Abbay River. The African Development Bank (AFDB) was reportedly intending to fund Mengistu's project until Egypt successfully maneuvered its blockage in 1989.[12]

Ethiopia remained at least formally identified with the Non-Aligned Movement (NAM)—established in 1955 in Bandung and attended by Haile Selassie's government at the first Summit in 1961. The Derg continued to actively engage in NAM, but with the approaching fall of the Soviet Union and major need of economic and other support, the Mengistu regime became more disengaged.

➢ **By the late 1980s, Soviet aid under Gorbachev's "perestroika" was waning and the Ethiopian regime faced continued economic decline. Mengistu began shifting policies to adopt a series of Western economic 'reforms.'** The new policies included lifting price controls, privatizing farms, and allowing free marketing of some goods.

An important institutional change at this time was the formation of a new agency to oversee river basin and power development. The government, under international agency 'advice', restructured the Awash Valley Authority—widely considered to be a failure—to form the Ethiopian Valleys Development Studies Authority (EVDSA) in 1987. The EVDSA's mission to "identify development projects," including their irrigation potential, and to "ensure environmental protection" for all Ethiopian river basins was a component of Mengistu's interest in continuing hydrodam development with Western support. Like the AVA, the new agency did not take into account the livelihood needs of indigenous peoples who resided throughout the lowlands downstream from the dams already built or under consideration. The regime's massive scale relocation of highlanders onto pastoral lands in eastern and southwestern Ethiopia, coupled with its repression of populations resisting these actions, were evidence enough of such disregard—actions that remained unchallenged by the Soviets and the western banks.

➢ **The African Development Bank (AFDB), meanwhile, was continuing the plan for 'modernizing' Ethiopia's river basins by funding the country's first comprehensive Master Plan for river basin development in 1990.**

- The AFDB sponsored plan for Ethiopian river basin development, entitled "Preliminary Water Resources Development Master Plan for Ethiopia," was released in 1990. This was a multi-volume desk study produced by WAPCOS—a public sector Indian water, power and infrastructure global consulting firm that was formed in 1969.
- The Master Plan identified and detailed the hydrodam and irrigation agricultural development potential for all major Ethiopian river basins, including the Omo River basin.
- As with feasibility studies for the Awash River basin development, no account was taken of the indigenous populations within these river basins, including the predictable dam and irrigated agriculture impacts on their livelihoods.

> Given the fact that the GOE had already begun construction of the dam almost three years prior to any downstream impact assessment, the development banks and their contracted consulting industry analysts complete the three-way system of collaboration and complicity that illustrates the institutional nexus for river basin development policy.

---

[11]The conflict building between the two nations was more complex, including Ethiopia's objection to Egypt's plan to allocate water to the Sinai 'New Lands' project.
[12]This writer's personal communication with an African Development Bank official in Dhaka, Bangladesh, in 2001—an individual previously stationed in Ethiopia.

The Resurgence of Western Dominance in River Basin Development

➢ **The Ethiopian People's Revolutionary Democratic Front (EPRDF)—a coalition of opposition forces led by the Tigrayan Peoples Liberation Front (TPLF) chairman Meles Zenawi—overthrew the Mengistu regime in 1991.** Reversing Ethiopia's prior 'socialist' orientation, the EPRDF relied on Western powers for political, military and economic support.

The EPRDF government's emphasis on economic growth and export-oriented development suited western interests. The U.S., for example, has viewed Ethiopia as a key factor in its geostrategic interests within eastern Africa, especially after the collapse of Somalia. Together with its European allies, the U.S. has considered Ethiopia to be not only a partner in 'anti-terrorism' and security activities, but also a main source of power for economic development and western investment in eastern Africa (including oil and gas industry expansion—see Appendix A). These interests have reinforced their "turning a blind eye" approach to the EPRDF's policies of political repression, land expropriation, and other types of human rights violations.

Under the EPRDF, development bank hydrodam planning continued, with large-scale infrastructure projects, including railways and major roads. The GOE consistently proclaimed (as it continues to) that large dams are a necessity for Ethiopia's 'national security' and bring 'economic progress for all.' GOE and aid agencies alike quote low levels of electrification—including in the poorest rural areas—as supporting this development element, even though large dam electricity primarily benefits urban industry, large-scale irrigation agriculture (when it accompanies dam construction) and export marketing. As in the Haile Selassie and Derg periods, the EPRDF omitted consideration of the impacts that the planned dams would have on indigenous peoples in the lowlands downstream from the projects. Funding and procedural requirements of international finance organizations, however, necessitated more sophisticated efforts to legitimize these developments.

➢ **Soon after the EPRDF came to power, a major turning point occurred regarding the development potential of the Omo River's hydropower.** Hydrodam constructions in the lower Omo basin quickly became a central concern in GOE and development bank planning. The hydroelectricity generating potential of the Omo River was deemed outstanding, based on exceptionally high flow volume measurements in a deeply gorged section of the river—the present site of the Gibe III dam. This potential had, in fact, been assessed years earlier, under the Derg, by both World Bank consultants and North Korean specialists.

In 1993, the African Development Bank (AFDB) contracted the U.K.-based global engineering consulting firm, Richard Woodroofe & Associates to prepare a Master Plan Study for the Omo-Ghibe River basin. This lucrative contract of USD 6.4 million specified a broad range of issues for consideration, including baseline description of the basin's physical and social character, water and irrigation agriculture potential. The nine- volume Master Plan recommended "*future development to the year 2030, considering the basin in two [administrative] parts*" (Woodroofe & Associates 1996).

The AFDB's Terms of Reference limited the study and Master Plan to Ethiopia—despite the transboundary nature of the Omo River and the obvious implications of dam building for the Kenyan economy and environment—and excluded consideration of downstream indigenous peoples' livelihood systems. The economies, resource dependency patterns, and vulnerability of populations to the impacts of dam and irrigated agricultural development were treated as 'externalities'—an approach that has largely persevered to the present. The Woodroofe & Associates Master Plan noted the "importance" of "further study" of these issues, but it did not question basic GOE plans for major dam and irrigated agricultural development.

Acres International Ltd., a Canadian consulting firm long associated with the international banks and Ethiopia's government, produced a Power System Planning Study for EELPA in 1996. Acres prepared an Ethiopia Power System Expansion Master Plan for EEPCO in 2000, and yet another update of the plan in 2005—the latter with the objective of tripling Ethiopia's power supply in five years, to 2842 MW. The planned increase in power generation far exceeded projections of domestic needs, with the surplus of at least 50 % likely to be exported to neighboring countries. In 2004, Acres International engaged in consulting on the Gibe dam developments but was debarred by the World Bank, based on corruption charges.

➢ **In order to rationalize and promote the aggressive new hydro development, in 1997 the GOE established the Ethiopian Electric Power Corporation (EEPCO), to promote and coordinate hydrodam and electricity projects.** Ethiopia's top executive office issued regulation No. 18, corporatizing EEPCO for "indefinite duration and conferring it with the duties and powers of the previous Ethiopian Electric Light and Power Authority" (EELPA). With its huge budget of ETB 6.1 billion (about USD 890 million in 1997), EEPCO took the lead in developing and managing electricity, including its transmission and distribution within Ethiopia, as well as overseeing the development of electricity export to neighboring countries.

> Energy export, with Ethiopia as the **"tower of energy"** in Africa, was substantially advanced by this institutional change—one that had been planned for years but with ominous implications for hundreds of thousands of indigenous residents in the lower Omo.

As one of Ethiopia's most prominent parastatals, EEPCO has enjoyed a central role in the highly politicized arena of hydropower development. A stalwart of Ethiopia's major privatization and economic growth and development "at all costs" approach, EEPCO has been far stronger than its predecessor, with an large budget and more than 12,000 employees (only exceeded by the Ministry of Education). It was defined to be directly accountable to the Prime Minister, who appoints its top executive. As part of their responsibilities to implement the policies of the Prime Minister's office. EEPCO officials have handled many of the government's responses to international critics of the Gibe III dam.

> Immediately after its creation, EEPCO engaged in planning the first dam in the Omo-Gibe River basin—an effort on the agenda of the government and the international banks since the comprehensive Master Plan of 1990/91.

Other small GOE agencies were established to have responsibility for regulation and oversight of energy and environmental issues: the Ethiopian Electricity Agency (EEA), Environmental Monitoring Unit (EMU), and Environmental Protection Authority (EPA). These were created largely in response to international aid organization procedural requirements for consideration of funding. All are subordinated to the agencies they are allegedly overseeing and are clearly controlled by the Prime Minister's office, where all major energy policy is formulated or approved.

The Ethiopian Electricity Agency (EEA) was created at about the same time as EEPCO, with responsibility for "review" of EEPCO policies—notably, tariffs. However, the EEA had no enforcement powers and was reduced to a purely advisory role, with some political legitimation value to the GOE and the international banks.

An Environmental Monitoring Unit (EMU) was established in 2002 by the GOE's 'Proclamation for the Establishment of Environmental Protection Organizations.' This occurred while the Gibe I dam was already under construction—yet it was to 'oversee' any problematic impacts of the project. However, the EMU has shown no divergence from the policies or practices of EEPCO, and it is structured to act in an advisory capacity and to prepare technical reports for EEPCO. These included the 2009 '*Environmental and Social Impact Assessment*' for the planned **Gibe III-Sodo 400 kV Power Transmission Lines Project**. Significantly, Sodo—a town about 50 km from the Gibe III dam site—is a hub for the planned 'Power Pool' or 'Energy Highway' between Ethiopia and neighboring countries.

> Since the internal procedural requirements of the AFDB and World Bank precluded them from supporting the Gibe III dam, the 50 km power line from the Gibe III dam to Sodo was arranged to be funded largely by the.[13] The falsity of the banks' claim of non-involvement in the Gibe III dam is a matter revisited in Chaps. 6 and 10.

➢ **The Environmental Protection Authority (EPA) was established as a monitoring and regulatory agency in 2002, by Proclamation 295—largely at the behest of international funding agencies. It replaced a predecessor body embedded in a Ministry.** Although it is designated as an "autonomous" or "independent" agency, the Authority is directly responsible to the Prime Minister's office (the Prime Minister designates the EPA head or Environmental Council member and those from the regional governments, as well as reviews its policies). Among the 26 Powers and Duties of the EPA are its establishment of "a system for environmental impact assessment of public and private projects, as well as social and economic development policies…," including hydroelectric and other major capital projects. The Authority's responsibilities largely mirror those in major donor country governments and include shaping environmental legislation, preparing the GOE's State of the Environment reports, and 'ensuring' that environmental policies are implemented by:

---

[13] Contrary to statements by later 'independent' development bank consultants that they were 'unaware' of the plans for major energy export, rather than water releases for downstream small-scale irrigation projects for 'social development' in the pastoral and agropastoral lowlands of the Omo basin, the GOE and development banks' intentions were fully evident throughout the 2009 report and other accessible technical reports.

- Overseeing environmental impact assessments.
- Issuing or withholding approval on EIAs—with timely receipt of them for review.
- Guaranteeing "participatory environmental management".

> The gap between EPA rhetoric and reality has remained starkly evident in the case of large dam development. Although all dam projects are legally required to undertake a 'full review' and license approval by the EPA prior to construction, in practice, this directive has been ignored and in any case, the Authority remains under esssential control by the Executive.

'Fast Track' to the Gibe III Megadam

> The combination of the GOE's disregard for pre-project impact assessment or project outcomes beyond its narrow objectives, the turning of a blind eye to the GOE's procedural violations on the part of international finance agencies, and the complicity of the global consulting industry has produced a 'fast track' approach to the Omo River basin development which ignored the impending dam impacts.

This approach to Omo River basin development has set the downstream Omo region with its major indigenous population along with Kenya's Lake Turkana area and its communities on the road to a social catastrophe and major environmental destruction. This destruction would be 'paid for' by the citizens and taxpayers of Ethiopia and Kenya.

➢ **The GOE heralded the Gibe I and Gibe II projects in the Omo River basin as the beginning of a Gibe cascade of dams** —all widely represented by the GOE and development agencies as in the 'national interest.' This cascade includes the Gibe I (184 MW), the Gibe II (420 MW), the Gibe III (1870 MW) being completed, and the planned Gibe IV (1472 MW) and Gibe V (560 MW) dams.

The Gibe I dam initiated the 'cascade' of dam development (Fig. 1.10) in the Omo, or Gibe-Omo River basin. It was constructed on the Gilgel Gibe River ("little Gibe" River) which originates in the highlands around Jimma, Ethiopia. The river flows northward from the Gibe I dam for about 25 km until it joins the larger Gibe River (from the highlands of Wollega) flowing southward. Downstream from this confluence, the Gibe River is referred to as the Omo River. The Omo River then continues southward to its terminus at Lake Turkana.

Construction of the Gibe I dam had begun much earlier, with a GOE contract to a North Korean engineering firm. The project was abruptly halted, for political reasons, until Meles' government reactivated it. The World Bank provided the major funding (USD 331 million) for the Gibe I dam with additional support from the European Investment Bank (EIB) and the Austrian government. Construction of the Gibe I was contracted to Salini Costruttori—a company long familiar to GOE officials and well regarded by them for its "timely completion of projects." Salini is a multi-billion dollar, privately owned, global corporation with close ties to Ethiopia, dating back to its contracted building of the Legedadi dam and treatment plant on the Akaki River, east of Addis Ababa, during the Haile Selassie era.

The EEPCO issued an Engineering Procurement Construction (EPC) contract to the Italian-based Salini Costruttori—a 'turnkey' contract for a completed project, eliminating any oversight during the construction period. By a special clause in the contract with the GOE, Salini also had no financial liability for the project. EEPCO repeated its turnkey contracting— *disregard for bidding process and project oversight*—in successive Gibe dam contracts with Salini.

While Salini was the main construction contractor, the firm's frequent partners—Studio Pietrangeli (also long involved in Ethiopian water projects) and ELC-Electroconsult (with *Coyne et Bellier*), designed the structures. As with Salini, none of the Spanish and German firms contracted for joint ventures or services were subject to external oversight.

Several international non-governmental organizations investigated the process of Gibe I dam guidelines compliance with international requirements and presented detailed accounts that squarely contradicted the World Bank's assertion that the project had "satisfied all required safeguarding of the environment and local human communities." Most detailed among these reports is that by the Rome-based non-governmental organization, *Campagna per la Riforma della Banca Mondiale* (CRBM), as part of its investigation of the European Investment Bank's funding procedures (CRBM 2008). The CRBM pointed out, for example, that the Ethiopian EPA did not produce an environmental or socioeconomic impact report (EIA) *prior to* the development, despite GOE and international bank requirements to do so.

➢ **The GOE also signed a no-bid EPC ('turnkey') contract with Salini for the Gibe II project—again without open bidding and oversight.** The Gibe II power station was a continuation of the Gibe I project and did not involve dam construction. Instead, water was to be transferred from Gibe I through a 26-km tunnel passing under Fofa Mountain to the Gibe II power plant. Studio Pietrangeli planned the overall scheme on behalf of Salini, according to EEPCO, as part of the ELC-Electroconsult contract. Salini also contracted its long-term partner, S.E.L.I. S.p.A, for EUR 37 million to perform the tunnel excavation.

EEPCO contracted the Gibe II project in 2004, the same year that Gibe I began operation. Funding for this 420 MW project was EUR 375 million: EUR 50 million from the European Investment Bank with additional funding approved by the Italian Directorate General for Development Cooperation (DGCS). Major controversy arose in the Italian government over serious irregularities in the contracting process. The financial and political complexities of this situation—exemplified by the Prosecutor's Office in Rome filing criminal charges against the Gibe II project contractors—complicated the involvement of Italian agencies in funding the Gibe enterprises (CRBM 2008).

➢ **Three different tunnel disasters occurred at the 26 km Gibe II tunnel project—in 2006, 2007 and 2010.** The first project breakdown occurred in October 2006 when an intake portal hit a fault that spewed mud under high pressure, causing an entire section of the tunnel to collapse.

> A second failure occurred in June 2007, when a section of a secondary tunnel collapsed—filling a substantial section of the main tunnel.
> Yet a third breakdown came about in 2010, when high pressure from a geologic fault caused a major break with mud and rock flowing into the tunnel.[14]

Evidence points to these disasters being linked to seismic occurrences. ARWG physical scientists working in the region for decades described the seismically active character of the region (ARWG 2009), and this has been substantiated by the research of Kinde and associates (see Chap. 3). This interpretation is bolstered by the geological profile of the Gibe II project, as well as the seismic record in the region. Salini reportedly had overseen poor seismic studies of the region, and without external oversight. Hence, conditions were ripe for such collapse. Of equal concern is the fact that Salini itself was in charge of repairing the tunnels and 'remedying' the failures. According to its EPC contract terms with EEPCO, *Salini bore no responsibility for these breakdowns in the tunnel*.

These failures by Salini and the design/construction engineering firms underscore the ominous situation regarding seismic and landslide danger at the Gibe III dam.

➢ **The formation of the Gibe III dam project reflected the conflicts of interest at play and the lack of oversight, as well as disregard for the environmental and social impacts of the Gibe I and Gibe II projects.** The sequence and timetable of GOE (EEPCO) dealings with Salini about the Gibe II illustrate the procedural violations of Ethiopian as well as international aid providers.

- Shortly *after* Salini began construction on the Gibe II project, EEPCO was negotiating with the company to conduct a technical feasibility study of the 1870 MW Gibe III project, even though Salini was obviously a contender for the construction contract. The GOE signed a feasibility agreement in October of 2005. Salini presented a preliminary design document in mid-January of 2006. Two weeks later, Salini and EEPCO signed a Memorandum of Understanding (MOU) for a 'full' feasibility study.[15]

---

[14] See Chap. 3 for more detail. These breakdowns are detailed in the Tunneltalk Forum (www.tunneltalk.com).
[15] The timing of these phases is outlined at the EEPCO website (http://www.eepco.gov.et).

- After only five months (in July 2006), the EEPCO signed a no-bid Engineering Procurement Construction Contract (EPC) with Salini for construction of the Gibe III dam, with no provision for external oversight. On both counts, the arrangement contravened both Ethiopian regulations and international development bank procedural requirements for project funding. EEPCO justified the no-bid award by citing Salini's past performance, particularly "its record of completing projects on time."
- The Gibe III dam contract was for EUR 1.47 billion. This figure excludes expenses for constructing the energy transmission and distribution system, as well as the "assured" mitigation and other measures to cope with negative socioeconomic and environmental impacts. Funding for the Gibe III dam continued with difficulty, as the GOE applied for support from a variety of multilateral and bilateral aid and financial agencies.

Construction of the dam began in December 2006, despite no environmental or socio-economic impact assessment. Ethiopian law requires that the assessment be approved prior to major capital projects being undertaken. Clearly, the GOE's insistence that required procedures were adhered to in the case of the Gibe III dam is incorrect—a fact well known to the World Bank (WB), the AFDB, the EIB, and major bilateral agencies.

➢ **Three years after Gibe III construction began, EEPCO released two environmental and socioeconomic impact assessments (ESIAs), both of which were fundamentally flawed.** The first impact assessment, entitled *Gibe III Hydroelectric Project, Environmental & Social Impact Assessment*, 300 ENV RC 002C, was released by EEPCO in January 2009 (GOE 2009a). The contract for this assessment was awarded to the Italian engineering firm, *Centro Electrotecnico Sperimentale Italiano* S.p.A (CESI) by Salini and EEPCO. As per its Terms of Reference, CESI addressed only the vicinity of the Gibe III dam under construction—omitting the entire downstream impact zone in both Ethiopia and Kenya. Even with this limited scope, the report contained major false assumptions and misinformation, as pointed out by numerous international critics (mostly non-governmental organizations and researchers, including those from the ARWG and Oxford University).

EEPCO released its second environmental and social impact assessment (ESIA) later in 2009—this one produced by another Salini partnering firm, Agriconsulting of Italy, in association with MDI Consulting Engineers. This report, entitled *Gibe III Hydroelectric Project: Environmental Impact Assessment—Additional Study on Downstream Impact* (GOE 2009b), is riddled with omissions, misrepresentations, and fabrications. Details of its failings are located in Chap. 6 of this book, and in Carr (2012).

The GOE's exclusion of transboundary impacts was clearly done with full awareness by the international development banks, which have taken the same approach in their Terms of Reference with consultants. There was clear precedence for this dismissal of transboundary effects of the planned Gibe III dam. The World Bank's (2004) statement on the matter is clear and must be regarded as both irresponsible and in violation of its own prescribed principles.

*The Omo basin in the southwest should be an early candidate for large-scale development. It produces an annual flow of some 17 BCM of water with considerable irrigation potential, estimated at 348,100 ha, mainly in Kolla (lowland zones). The Omo River flows in southern Ethiopia into Lake Turkana, most of which lies in a **sparsely inhabited region of northwestern Kenya**. The Omo River is particularly important, both for its large annual flow and its irrigation and hydroelectric potential, and its being one of the principal basins where **there is unlikely to be any objection by downstream countries**. [Emphases added]*

➢ **The GOE's no-bid contracts and lack of oversight for the Gibe III dam presented a challenge for the development banks, since these actions violated the banks' regulations for project funding.** This problem necessitated only indirect means of support for the project. The World Bank and African Development Bank have engaged in multiple types of 'backdoor' support, including:

(i) Coordinating funding and related support among multiple development agencies.
(ii) Funding 'independent' environmental and social impact assessments (EIA, SIA) of the dam project. As noted earlier, these are lucrative contracts issued to consultants whose past performance provides assurance of basic acceptance, with suggested 'improvements' such as further studies and possible (optional) mitigation components.
(iii) Funding for government 'services', including technical capacity-building, local governance efforts, 'basic services' support by the state and unspecified expenditures that can bolster the Gibe III project—through dam and electrical

power related infrastructure development or strengthening of police presence for quelling dissent or evicting and expropriating the resources of residents in the project area.

(iv) Funding for 'downstream' phases of the hydrodam development—especially power transmission and distribution systems—especially those serving bank plans for regional economic integration in eastern Africa.

> While these indirect funding pathways of backing the Gibe III may satisfy internal requirements of the banks and the scrutiny of policy-makers in major governments, they do not exempt them from responsibility for the impending human rights catastrophe produced by the dam's operation. The same can be said for global consulting industry firms and individuals.

➢ **In the face of growing opposition to mega-dam projects throughout developing countries, the continued legitimation of mega-dam projects—especially controversial ones like the Gibe III—relies heavily on the work of allegedly "independent" impact assessment consultants.** Like the Ethiopian government, the World Bank, EIB and AFDB (along with major bilateral agencies) require the services of 'trusted' consultants to produce these reviews. The reality of the bidding process is that only those consulting firms and individuals with a history of tacit support for large water development projects and a demonstrated 'understanding' that any serious reservations about a project in question will be raised only as 'concerns' or 'suggestions' for mitigation, monitoring or further study, are generally awarded the lucrative contracts.[16]

Following the Ethiopian government's request for funding from the European Investment Bank, the Bank contracted Sogreah Consulting for an 'independent' review of the GOE's environmental and social assessment. Sogreah had produced the feasibility study for water resource development, and irrigation agriculture in Ethiopia's Awash Valley. Founded in 1923, Sogreah is a global engineering, development and management firm supporting worldwide projects, including billions of dollars of hydrodam, energy and electricity developments. Other Sogreah contracts in the region had included feasibility and design of dams on the Atbara and Setit rivers in Sudan.

The AFDB contracted a consultant who had held a senior position for twenty-six years at Gibb Africa for assessment of Gibe III dam impacts on Lake Turkana. Gibb Africa was originally Sir Alexander Gibb & Partners, first active in Ethiopia in the 1960s. Sir Alexander Gibb & Partners became a limited partnership in 1986 and joined the U.S.-based Jacobs Engineering Group in 2001—contracting for water resource projects, including coordination of multi-investor projects for the World Bank that totaled hundreds of millions of dollars.

Both the EIB and AFDB assessments failed to challenge the most serious omissions, misrepresentations and fabrications of the GOE's downstream impact assessment (GOE 2009b). The GOE's failures included its dismissal of the major seismic threat to the Gibe III dam region, denial of the transboundary nature of the Omo River and its effects, misrepresentation of basic hydrological and ecological features of the region, and misrepresentation of the livelihoods and vulnerability of the indigenous population, and fabrication of "disaster" for local populations from floods (particularly the 2006 Omo River flood).

While the AFDB and EIB 2010 assessments of Gibe III impacts on Lake Turkana and the lower Omo River basin, respectively, did identify some major problems with the GOE's preparation and specific plans for the dam, their comments and conclusions primarily took the form of 'suggestions'—generally, for 'further study' or for new 'mitigation' components. Moreover, most of their suggestions and reservations about the dam and its effects were embedded in lengthy technical text that few would read, so these qualifications were sidelined by the development banks and the GOE—and they remain so. Much of this failure rests with the original 'bounding' or 'scoping' agreement between bank and consultant, as well as the absence of any *accountability* other than to the banks themselves—certainly not to the populations most affected by the development. Chapter 6 details these failures for both GOE and development bank assessments. The AFDB and EIB impact assessments fail to explicitly address the transboundary nature of the impact system—instead, only noting the limitations of their Terms of Reference, or 'scoping' agreement, for example. They also failed to explicitly treat the *cumulative and*

---

[16]Since the late 1980s, global consulting industry firms from India have remained active, WAPCOS' key consulting role in 1988–1990 The head of EEPCO encouraged Indian companies to invest in his country at a business meeting in March 2008, when the Indian NHPC and others 'expressed interest' in power project related roles (Thakkar n.d.). Indian investors have also invested in the irrigated plantations on expropriated lands in the lower Omo basin.

synergistic effects of the Gibe III dam and the dam enabled large-scale irrigated agricultural enterprises, despite full knowledge of the importance of irrigation system planning by the GOE.[17] Since the 2010 release of the two reports, GOE Executives, EEPCO and development bank officials continue to cite them as basically endorsing the Gibe III project—particularly in the case of the AFDB report regarding Lake Turkana.

> Given the fact that the GOE had already begun construction of the dam almost three years prior to any downstream impact assessment, the development banks and their contracted consulting industry analysts complete the three-way system of collaboration and complicity that illustrates the institutional nexus for river basin development policy.

A Nexus of Public Policy Institutions for River Basin Development: Collaboration with Complicity

➤ **A nexus of institutions engaged in public policy for river basin development emerged early in the post-World War II period—in Ethiopia, as well as elsewhere in Africa and other developing countries.** These institutions are central to the decisions regarding the *formulation* and *implementation* of policy—including its rationalization, promotion and 'evaluation.'

The three central components of this policy nexus are:

i. **The client recipient nation state(s)— primarily executive offices.** These offices of the state generally oversee all phases of the development.
ii. **International aid and finance—primarily international development banks.** Other multilateral organizations (regional development banks, U.N. offices), as well as and major donor country bilateral aid agencies, export credit associations, and related organizations.
iii. **The global consulting industry (GCI).** Within this multi-billion dollar industry, dominant firms and individuals operate transnationally, though sometimes focused in particular regions. In river basin programs, engineering and water resource firms predominate. 'Independent' oversight agencies for large basin development programs vary widely—including those rooted in scientific, multilateral development and academic organizational contexts.

Numerous ancillary institutions and individuals are engaged in formulation, implementation and evaluation phases of river basin development projects. These include a variety of technical offices, local administrative offices, police and military agencies, sub-contracted company consultants and others. Companies meet frequently and often informally with development bank and other key aid agency and government officials, where much of the decision making actually occurs regarding the basic character of the new or expanded projects planned—generally long before any information is released to the public or even to related professional circles. Such gatherings and informal relationships established are often key determinants of the projects or entire programs being shaped, including the scope of specific contract arrangements. This is particularly evident in the African context, where contract awards are frequently informally arranged and details are only nominally available.

Much of the policy discussion in the pages of this book concerns the roles of eastern African and donor states as well as the international finance institutions in river basin development. Some attention to the third component of the policy nexus as outlined above—the global consulting industry (GCI)—is essential to an understanding of the processes unfolding in the tri-nation transboundary region from the Omo River basin development.

Within Ethiopia, the global consulting industry (CGI) was evident by the 1950s, when some members were essential to Ethiopia's first large river basin development—in the Awash River basin. Two of the emerging consulting industry firms that were most active in the Awash basin projects were Sogreah Consultants SAS and Sir Alexander Gibb & Partners. Both were well known in the contract bidding process for major engineering and water projects, and both grew to handle hundreds of millions of dollars worth of development projects.

---

[17]As detailed below, cumulative and synergistic effects between the Gibe III dam and the electricity transmission system—*planned well before the GOE, EIB and AFDB impact analyses*—are also necessary, yet were excluded.

Sogreah Consultants, the France-based, international engineering consulting firm, signed a contract to survey available resources for irrigation agriculture in the Awash Valley, under the auspices of the FAO/UN Special Fund. They produced multiple volumes from this work in 1965—but these were released five years after the Koka dam construction. The Terms of Reference in Sogreah's contract limited the survey to soil, climatic, geomorphic and other physical factors deemed 'sufficient' for estimating the potential of irrigated agriculture. In subsequent years, Sogreah would remain active in Eastern Africa, and it reemerged as the European Investment Bank's consultant for an 'independent' review of the GOE's downstream assessment of the Gibe III's environmental and social impacts (GOE 2009b).[18] No substantive attention was given to characterizing the indigenous peoples in the Valley with respect to their resource needs.

Another major contract for Awash Valley development work was awarded to Sir Alexander Gibb & Partners, a U.K.-based multinational engineering firm that formed a key associate firm—Gibb Africa. As with Sogreah, Gibb Africa won numerous contracts for dam-related development within Eastern Africa. As indicated at the company's website, Gibb Africa has secured more than 1000 contracts for the region. Among these was the controversial Victoria dam in Uganda. A former associate partner of Gibb Africa later contracted with the African Development Bank to produce its assessment of the Gibe III dam's impacts on Lake Turkana (AFDB 2010).[19] Gibb Africa has its headquarters in Nairobi, Kenya.

As the hydro industry has constricted in many developed nations, many consulting (and construction) companies began increasing their roles in developing countries—particularly through relationships with international aid organizations spearheading massive scale dam, river basin and related water resource developments. The strongest 'take-off' period for global consulting industry (GCI) expansion was during the 1970s and 1980s—hastened by environmentalist actions in the developed countries which curtailed further expansion of the hydropower industry. In a sense, the major upswing in water resource and river basin development in developing countries—certainly including Ethiopia—'saved' the industry from possible demise. In a market economy and industry where a growth imperative prevails, consulting firms had little recourse but to operate with complicity relative to the content of work performed for the expanding hydrodam and agribusiness industries in developing countries.

➢ **The nexus of institutions spearheading and rationalizing much of the major river basin development within Africa and elsewhere in the developing world constantly adapt to changing political and economic circumstances, including opportunities to initiate, implement or evaluate major capital projects in river basins.**[20]

International development banks, major bilateral aid and other finance agencies simultaneously fashion their relations in this major transformative sphere of river basin projects to suit their geostrategic or investment interests within the region. Similarly, global consulting firms and individuals pursue avenues of technical specialization and geographic focus that maximize their revenue from contracts with South states and international finance organizations.

Global consulting industry firms are a critical component of the rationalization—or legitimation—of development projects and programs, including through 'scientific' and allegedly 'independent' feasibility studies, baseline studies, environmental and socioeconomic impact assessments (EIAs or ESIAs), evaluations of impact assessments, formulation of management or monitoring plans and other contracted efforts. Contract amounts for these different consulting efforts in river basin developments vary with a large number of factors. Hundreds of thousands of dollars are common contract amounts extended to GCI members for such functions where large capital projects such as megadams and major linked developments are involved. Even fractions of river basin developments can amount to such large sums—or larger. For example, the tender extended by the European Investment Bank (EIB) for the 'independent' assessment of the Gibe III dam—a contract awarded to Sogreah—was announced with a contract amount of up to *EURO 300,000*. Other contracts, including for eastern Africa hydrodam and river basin related development studies (for example, of major irrigation works) produce similar—sometimes, even multi-million dollar contracts. Considering that many of these are primarily desk studies, the profitability levels are obviously attractive to competing GCI members.

---

[18]The EIB review was conducted after the GOE requested funds for the Gibe III dam construction.
[19]The AFDB impact assessment for the Gibe III dam was conducted just after the Gibb Africa senior officer left the firm to form a new consulting group, the Nairobi-based Water Resource Associates—also with numerous experienced global industry consultants.
[20]Some of these firms, notably Salini that was to become so prominent in the Gibe sequence of dam-building, became both construction and consulting corporations—sometimes combining planning, impact assessment, construction and project 'oversight' through self-monitoring.

> The reliability and consistency of international consultants fulfilling these roles is a key factor in the establishment of a 'contract treadmill,' with a ***circle of complicity*** whereby the planning and implementation of major river basin projects —with 'preferred' consulting firms and individuals playing a crucial role.

The consequences of this circle of complicity are profound in developing countries, including in eastern Africa. In the service of major river basin developments, the institutional nexus skews the assessment (or evaluation) process, toward the following:

- An 'efficient' transformation of entire regions to suit the objectives of developing country Executive offices, international aid organizations, and private investment firms—objectives frequently diametrically opposed to the most basic survival needs of local residents, particularly those of highly marginalized populations.
- A net 'positive' or at least unchallenging assessment of the megadam or major river basin development plan in question —with 'suggestions' such as (later) additional studies, slight modifications of a technical nature, or mitigation 'concerns'.

A closed system of information, 'assessment' and 'safeguards' and, 'consultations', with local populations ("stake-holders") as a means of satisfying donor-required procedures—without truly independent evaluation or real accountability to the populations most directly affected by the development.[21]

The composition of the institutional nexus for a large dam and associated river basin development within Ethiopia is outlined in the table below. The specific institutions of such a nexus vary from one country to the next within eastern Africa and the continent more broadly.

What is effectively a 'revolving door' of contract signing between firms and individuals of the global consulting industry (GCI), on the one hand, and governments, international aid[22] organizations and transnational corporations, on the other, extends well beyond hydrodam and river basin development. GCI members are commonly able to successfully bid on lucrative contracts for major infrastructural projects, such as roads, ports and communications, as well as extractive industries—notably the petroleum industry——in Africa, the Middle East and elsewhere. Of the global consultants identified in Table 2.1, for example, Lahmeyer and Sogreah are actively engaged in oil and gas industry consulting in Africa and elsewhere; Gibb Africa is also engaged in oil work and major infrastructure consulting—the latter with an AFDB contract for the international highway project between Nairobi, Kenya and Addis Ababa, and the AFDB (2010) impact assessment author is reportedly a prominent consultant with one of the two main oil companies active in the Lake Turkana region— Tullow Oil.

➢ **The Peoples Republic of China basically came to the rescue of the Gibe III's financing gap in 2010 when the state-owned Industrial and Commercial Bank of China (ICBC) signed an agreement for a USD 459 million contract Dongfang Electric Corporation, backed up with a USD 420 million loan.** The World Bank and the AFDB were reluctant to *directly* finance the Gibe III dam electricity transmission systems, owing to Ethiopia's violation of their internal procedural requirements for funding. Instead, they have approved more than a billion dollars in funding for the 'Energy Highway' between the two nations but 'only' from Sodo, Ethiopia—a distance of slightly over 40 km from the Gibe III distributing station. As later chapters of this book detail, there is no way to separate definitively the Gibe III dam project from the transmission line extending from Sodo to Kenya, since Gibe III electricity would be entering the system at Sodo. While the international development banks were hampered by their internal funding requirements, the Chinese government again came to the assistance of the GOE. China agreed to a USD 96.7 million loan from its Export-Import Bank (EXIM) for the construction of the 'missing' connection—the 50 km, 400 kV electric transmission line from the Gibe III dam to Sodo. This arrangement left the GOE with only USD 13 million of additional finance needed for the project—an amount easily manageable from other credits (see Chap. 10).

---

[21]The socioeconomic consultants for a preliminary 2009 AFDB report on the Lake Turkana region stated that they had been instructed by the Bank to 'inform' local communities about the 'strong benefits' of the Gibe III dam (see Chap. 6).

[22]Contract procurement data is often inaccessible or obfuscated within government and development bank public documents. For all practical purposes, such information remains out of view for the general public, or for that matter—for parliamentary or congressional oversight committees, not to mention the communities being impacted.

Table 2.1 Three key components of the institutional policy nexus for major river basin development, with selective list of participants. Ethiopia, 1955–2013

ETHIOPIAN STATE* ⇒	INTERNATIONAL AID/FINANCE ⇔	GLOBAL CONSULTING INDUSTRY** ⇐
Including governments of Monarchy → Derg → Present: AVA EDVSA EELPA EEPCO EPA*** MME MOFED MOWR	**Multilateral Development Banks:** World Bank AFDB EIB **Bilateral Agencies:** USAID, DFID, AFD … **European Commission** **United Nations**	Sogreah Studio Pietrangeli Gibb Africa/Sir Alexander Gibb/Jacobs Richard Woodroofe & Associates Acres Lahmeyer Coyne et Bellier /Tractebel Centro Electrotecnico Sperimentale-Italiano (CESI) Agriconsulting of Italy, associated with MDI Consulting Engineers Water Resource Associates SELI WAPCOS MWH Global BRL (Ingéniere) Power Grid Corporation of India

*Government agencies identified in above text sections
**Individual consultants and Chinese firms not included

Although competitive with Western investment in hydrodam, major infrastructure, and extractive industry development within eastern Africa, Chinese finance is sometimes adaptive to development bank policies within the region. Chinese funding for the Sodo-Gibe III transmissions lines (and the Gibe III dam itself) is a case in point. The AFDB and World Bank claim that they are "not involved" in the Gibe III dam project, while Chinese corporations benefit from contracts for construction of the energy transmission system.

China's funding for the Gibe III project—both the dam and transmission lines—adds an extra dimension to its complex relationships of cooperation and competition with Western nations regarding power development in eastern Africa. Chinese financing of highly controversial dams within Ethiopia, including the Grand Millennium dam, continues this complex relationship. Its loans are typically extended without conditionality such as requirement of environmental and socioeconomic assessment or inclusion of safeguards—a laxity no doubt welcomed by the GOE. China is already a major force in extractive industry development, especially oil and gas, throughout much of eastern Africa where it has made major inroads into the system of leases and construction contracts (see Appendix A).

➢ **As part of its continuing economic liberalization, the GOE has 'modernized' some important government structures related to river basin development.** EEPCO was divided into two divisions in late 2013: a construction division (for building and managing power generating stations) and a utility division (for managing transmission and distribution—with marketing outcomes). EEPCO was then acclaimed as more "efficient" and "reliable"—assertions that were likely true from the sole standpoint of the management of dam construction, power generation, transmission, distribution, and international as well as domestic electricity marketing. The General Manager of the Gibe III project has recently been appointed as Director of the new Construction Division, to meet what the GOE, World Bank, and African Development Bank anticipate will be rapidly expanding energy import demand in Kenya and other eastern African countries (see Chap. 10). EEPCO's long-term director was escalated to the position of Energy Advisor to the Prime Minister.

The EEA (newly termed the Ethiopian Energy Authority) was assigned to 'oversee' private power companies newly permitted to invest in Ethiopia. However, the EEA remained largely in a 'review' position (e.g., around tariff proposals), subject to the Council of Ministers. The EPA was 'upgraded' in 2013, from a proposal by the Prime Minister' office, to become the Ministry of Environmental Protection and Forestry. The former head of the EPA became an advisor to the new environment minister. The former head's statement to Reuters News in August of 2013 illustrated the EPA's misinformation and lack of concern for human life and environmental conditions downstream from the dam when he asserted that major dam projects on the Omo River "*…will not cut off water supplies downstream nor worsen living conditions for local people.*"

> In the present advanced phase of the Gibe III controversy, the situation regarding accountability of the GOE and development banks policies with respect to the lower Omo River basin remains unchanged.

➢ **Reaction by civil society to the Gibe III project was swift and has expanded in recent years.** Critics first emerged following the GOE's initiation of the project without any account of the downstream human population and environment, and stepped up their criticism after the GOE's release of two allegedly 'objective,' or scientific impact assessments released two years after construction began (GOE 2009a, b). Criticisms and analysis emerged from a variety of non-governmental organizations—primarily international ones, given the politically repressive policies of the GOE. Among these organizations were Campagna per la Riforma della Banca Mondiale (CRBM), Survival International, International Rivers, a Kenya-based civil society organization closely affiliated with International Rivers—Friends of Lake Turkana, Human Rights Watch and the Oakland Institute.

Other critics emerging were researchers and policy focused individuals with established familiarity with the region. These included the world-renowned paleontologist and conservationist, Dr. Richard Leakey, Professor David Turton at the University of Oxford (and founder of www.mursi.org), this writer at the University of California, Berkeley, and other members of the Africa Resources Working Group (ARWG) in Africa, Europe and North America. Numerous organizations and individuals within Kenya, as well as in the Ethiopia diaspora have also questioned the project. Overt opposition to the Gibe III dam within Ethiopia is not politically feasible, given the high level of repression and retribution.

Numerous reports by these organizations have detailed the major procedural violations and the dire consequences of the Gibe III dam for half a million indigenous people residing downstream from the dam, within Ethiopia and Kenya. Gradually, these concerns have percolated through international policy circles and have been raised periodically by government representatives in Kenya, Europe and the United States. U.S. and European nations' diplomatic support and international aid to the Ethiopian and Kenyan governments, however, have effectively precluded effective examination of the humanitarian and human rights disaster looming on the horizon.

Literature Cited

African Development Bank (AFDB)., Kaijage, A.S., Nyagah, N.M. 2009. Report, socio-economic analysis and public consultation of Lake Turkana Communities in Northern Kenya. Tunis.

African Development Bank (AFDB). (2010, Nov) S. Avery, Assessment of hydrological impacts of Ethiopia's Omo Basin on Kenya's Lake Turkana Water Levels, Final Report, 146 pages.

Africa Resources Working Group (ARWG). (2009, January). A Commentary on the Environmental, Socioeconomic and Human Rights Impacts of the Proposed Gibe III Dam in the Lower Omo River Basin of Southwest Ethiopia. http://www.arwg-gibe.

Bondestam, L. 1974. Peoples and capitalism in the Northeast Lowlands of Ethiopia. *Journal of Modern African Studies* 12: 428–439.

Carr, C.J. 1976. Plant ecological variation and pattern, In: the lower Omo Basin, In: Earliest man and environments in the Lake Rudolf Basin, eds. Y. Coppens, F.C. Howell, L.L. Isaac and R.E.F. Leakey. University of Chicago Press.

Carr, C.J. 1978, The Koka Dam, agribusiness and marginalization of the Ittu Oromo Pastoralists in the Awash Valley of Ethiopia. Report to National Science Foundation, 170 pages.

Carr, C.J. 2012. Humanitarian catastrophe and regional armed conflict brewing in the border region of Ethiopia, Kenya and South Sudan: The Proposed Gibe III Dam in Ethiopia, Africa Resources Working Group (ARWG), 250 pages. https://www.academia.edu/8385749/Carr_ARWG_Gibe_III_Dam_Report.

Campagna per la Riforma della Banca Mondiale. 2008. The Gilgel Gibe Affair. http://www.crbm.org/modules.php?name=browse&grpid=60.

Campagna per la Riforma della Banca Mondiale, Ministry of Water Resources (MOWR). 1996. Woodroofe, R. & Associates with Mascott Ltd., Omo-Gibe River Basin Integrated Development Master Plan Study. Vols. I–XV.

Campagna per la Riforma della Banca Mondiale, Federal Negarit Gazeta. 2002. Proclamation No. 295/2002. http://www.ilo.org/dyn/natlex/docs/ELECTRONIC.pdf.

Literature Cited

Campagna per la Riforma della Banca Mondiale, Ethiopian Electric Power Corporation (EEPCO). 2009a. CESI, Mid-Day International Consulting Engineers (MDI), Gibe III Hydroelectric Project, Environmental and Social Impact Assessment, Report No. 300 ENV RC 002C Plan.

Campagna per la Riforma della Banca Mondiale, Ethiopian Electric Power Corporation (EEPCO). 2009b. Agriconsulting S.P.A., Mid-Day International Consulting, Level 1 Design, Environmental and Social Impact Assessment, Additional Study of Downstream Impacts. Report No. 300 ENV RAG 003B.

Campagna per la Riforma della Banca Mondiale, Ethiopian Electric Power Corporation (EEPCO) n.d., EEPCO origins, A. Samuel, The first hydro power station in Ethiopia. www.eepco.gov.et/corporationhistory.php.

Kloos, H. 1982. Development, drought, and famine in the Awash Valley of Ethiopia. *African Studies Review* 25(4): 21–48.

Mitchell, A. 2009. Gilgel Gibe III Dam Ethiopia: Technical, engineering and economic feasibility study report, Presentation Transcript, submitted to African Development Bank. Manuscript, personal communication.

Nathaniel, R. 2004. 50th Anniversary of his imperial majesty Haile Selassie I: First visit to the United States (1954–2004). Trafford Publishing.

Nicolle, D. 1997. *The Italian Invasion of Abyssinia 1935–36*. Osprey Press.

Norconsult. 1957. The Koka project, Hydroelectric development in the Awash River, Conditions and specifications. Oslo, Norway.

Thakkar, H. n.d., India's Dam Building Abroad: Lessons from the experience at home? http://www.indiaenvironmentportal.org.in/files/India%20dam_0.pdf.

United Nations, Food and Agriculture Organization (FAO). 1965. Report on Survey of Awash River Basin, Vol. I–III.

United Nations, Food and Agriculture Organization (FAO). 1994. Studies for integrated irrigation systems-Ethiopia-project findings and recommendations. Terminal Report of UNDP/FAO project ETH/88/001. www.fao.org/waicent/faoinfo/agricult/aquastat/Ethiopia.htm.

United Nations, Food and Agriculture Organization (FAO). 2014. Committee Decisions 37 com 7B.4 Lake Turkana National Parks (Kenya). N 801 press release. http://whc.unesco.org/en/decisions/5023/.

United States, Government of, Bureau of Foreign and Domestic Commerce. 1916. *Supplement to Commerce Reports: Review of Industrial and rade Conditions in Foreign Countries in 1914 by American Consular Officers*, Vol. II. Accessed through University of Chicago Library.

Vestal, T.M. 2011. *The Lion of Judah in the New World*. Greenwood Publishing Group.

WAPCOS. 1990. *Preliminary Water Resource Development Master Plan for Ethiopia*. Vol. III, Annex A, Hydrology and hydrogeology; Vol. V, Annex J, Hydropower.

Woodroofe, R. & Associates, with Mascott Ltd. 1996. Omo-Gibe River Basin Integrated Development Master Plan Study, Final Report. Vol. I–XV.

World Bank. 1970. *Economic Growth and Prospects in Ethiopia*, Vols 1–5, Eastern Africa series: no.AE9. Washington, DC: World Bank http://documents.worldbank.org/curated/en/.

World Bank. 2004. Ethiopia's Path to Survival and Development: Investing in water infrastructure, concept paper, Background Note for FY04 CEM. http://siteresouces.worldbank.org/INTETHIOPIA/Resources/PREM/CP-WaterMila01.pdf.

World Bank. (2013, Feb). Report No. 75145, Republic of Djibouti, Final report, Transport and logistics in Djibouti: Contribution to job creation and economic diversification, Policy note.

Open Access This chapter is distributed under the terms of the Creative Commons Attribution-NonCommercial 2.5 International License (http://creativecommons.org/licenses/by-nc/2.5/), which permits any noncommercial use, duplication, adaptation, distribution and reproduction in any medium or format, as long as you give appropriate credit to the original author(s) and the source, provide a link to the Creative Commons license and indicate if changes were made.

The images or other third party material in this chapter are included in the work's Creative Commons license, unless indicated otherwise in the credit line; if such material is not included in the work's Creative Commons license and the respective action is not permitted by statutory regulation, users will need to obtain permission from the license holder to duplicate, adapt or reproduce the material.

The Seismic Threat to the Gibe III Dam: A Disaster in Waiting

Abstract

The Gibe III dam is located near the Main Ethiopian Rift (MER), the seismically active northern arm of the East African Rift (EAR), which is capable of producing large magnitude, destructive earthquakes. The United Nations Office for the Coordination of Humanitarian Affairs Regional Office for Central and East Africa estimates there is a 20 % risk of 7 or 8 magnitude earthquakes occurring within the next 50 years in the MER. Earthquakes of these magnitudes pose significant threat to dams, through direct collapse or landslides triggering collapse. Collapse of the Gibe III dam would result in catastrophic loss of human life, livestock, wildlife and environments in the downstream riverine and Lake Turkana regions, exceeding the worst known dam failure in history—the Vaiont disaster in Italy. Even more moderate seismic events, combined with highly probable major landslides, sediment buildup and pressure from impounded water behind the dam threaten dam stability. The GOE discounts the seismic danger to the planned Gibe III dam, ignoring key geological information. The international development banks and bilateral agencies engaged with feasibility and impact studies as well as funding of the project—directly and indirectly—also ignore available data pointing to major seismic risk.

High Seismicity in the Gibe III Dam Region

The East African Rift (EAR) is a fault-bounded, long series of depressions extending about 5000 km from the Afar Depression in the Horn of Africa, southward through eastern Africa. For at least 40 million years, the African tectonic place has been splitting into two different plates (the Nubian and Somali). As a consequence, the rift has widened and down-dropped to form wide valleys, where rivers flow down from the bordering highlands to form lakes, including in the western and northern branches of the EAR (Fig. 3.1)—including Lakes Malawi, Victoria, Tanganyika, Albert and Turkana—the largest of eight Kenya Rift Valley lakes. This process is ongoing—contributing to the separation of the Main Ethiopian Rift (MER) portion of the EAR by 2–5 mm per year and driving the high levels of seismic activity in the tectonic province of the MER, the Afar Depression and the northern portions of the EAR eastern and western branches. An updated U.S. Geological Survey version was issued in 2014.

This tectonic province is the appropriate geographic scale for assessing earthquake probability at the Gibe III dam site. Almost all seismicity in eastern Africa is associated with these major tectonic boundaries. The abundance of recent earthquakes along the complete length of this tectonic margin demonstrates that these boundaries continue to be active zones of spreading and it is likely that large earthquakes will continue to persist along these zones (Fig. 3.2).

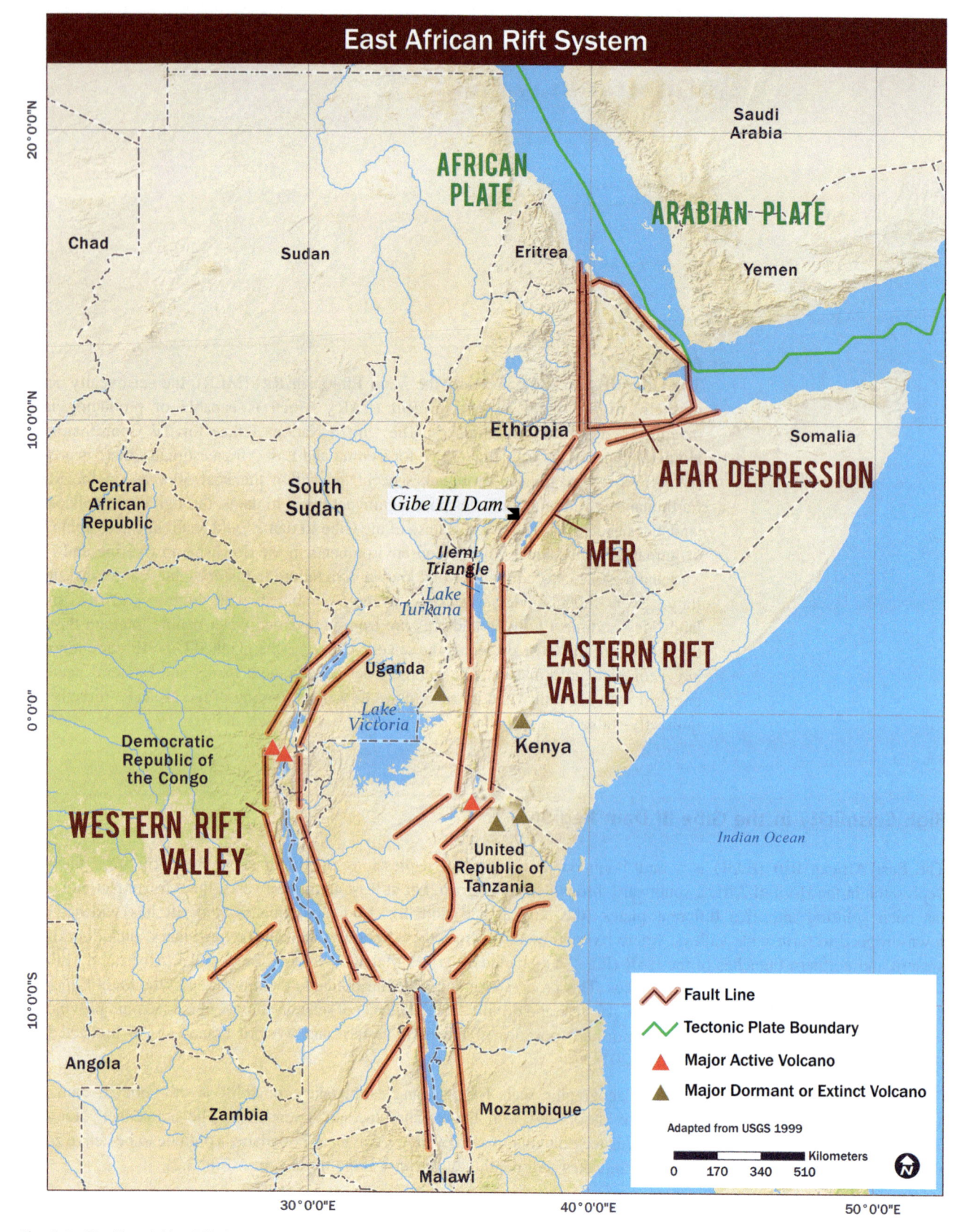

Fig. 3.1 The East Africa Rift System with Main Ethiopian Rift (MER). *Source* Map adapted by ARWG from U.S. Geological Survey 1999 (2012)

Fig. 3.2 Earthquake risk in Africa: Modified Mercalli Scale. *Source* U.N. OCHA-ROCEA (2007)

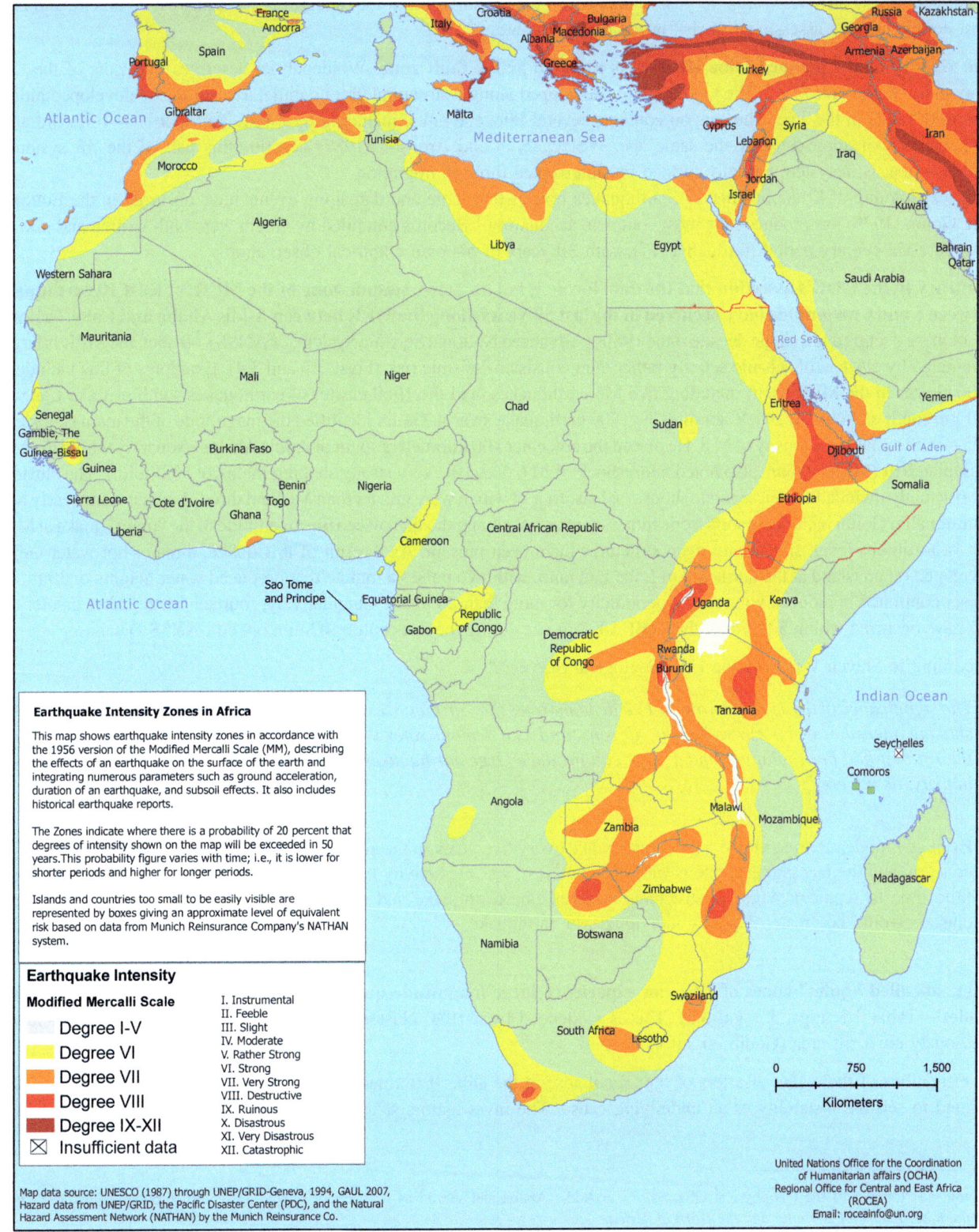

➢ **There is a 20 % chance of at least a magnitude 7 (M7) or 8 (M8) earthquake happening within the next 50 years in the MER region, in which the Gibe III dam is located** (Fig. 3.2). This estimate is based on data from the United Nations Office for the Coordination of Humanitarian Affairs Regional Office for Central and East Africa (OCHA-ROCEA) map of earthquake intensity zones (based on the 1956 Modified Mercalli Scale/MMS). No part of the MER has a predicted magnitude of less than M6 within 50 years. The OCHA report is a result of the first global earthquake assessment, and is the most comprehensive assessment of risk in the region to date.

The vicinity of the planned Gibe III dam is itself an active fault zone. Woldegabriel describes the walls of the Omo Canyon as "fault controlled". The topography is developed along a structural grain parallel to a rift that is developed along a system of faults in the area. The canyon contains several Pliocene units, including the Moiti Tuff that is ~4 Ma old and a volcanic rock of approximately the same age. Woldegabriel and Aronson (1987) describe this part of the rift system as "failed"—that is, a system partially formed but interrupted through migration.

Early cataloguing of Ethiopia's historic earthquakes reveals that large and damaging events have occurred in the Ethiopian Rift. Gouin (1979) has produced the most complete accounting.[1] Records compiled by Gouin were both written and oral and covered a six-century period (through 1977), with 30 years of his own empirical observations.

Contrary to the GOE's assertion that the dam locale is not an active seismic zone of the MER, at least 10 earthquakes between 5 and 6 magnitude have occurred in the last 50 years along the MER between Addis Ababa and Lake Turkana. Six of these occurred in a single decade—the 1980s (Advanced National Seismic System, or ANSS).[2] In fact, the MER is largely dominated by clusters of seismic activity rather than consistent seismic risk (Figs. 3.3 and 3.4). Examples of this include the 1906 cluster in the MER itself, including two M6 earthquakes, and the 1960 cluster of earthquakes south of Lake Langano, 150 km south of Addis Ababa. Regarding the 1906 earthquake swarm. Gouin identified the magnitude of the main shock as M 6.75—with a shaking intensity of M 8. He placed the epicenter 100 kilometers south of Addis Ababa—squarely within the Main Ethiopian Rift (MER). Gouin also noted a foreshock of M 6.6 and the very strong shaking intensity of 7 (ibid.). The cluster in 1906 actually included eleven separate shocks, at least four of which were greater than M6, and three of which were nearly M7s. Additionally, Gouin includes reports from witnesses near Lake Langano that placed the magnitude of the "main" quake at M9—that is, a ruinous level, should any main structures have been present. As a result of this quake, a major hot water Geyser developed on an island at the north end of Lake Langano, with two pulses a minute that produced water heights of 30 m. The geyser continued in reduced volume and periodicity for more than 20 years. Cumulatively, fourteen earthquakes greater than M5 have occurred within 300 km of the GIBE III dam site during the last century (Gouin op. cit.; ANSS Database).

According to Steven Walter of the U.S. Geological Survey:

> *The significance of the 1906 is twofold: 1) It demonstrates that earthquakes capable of producing destructive to ruinous shaking intensities can occur within the rift zone, and 2) it demonstrates that like the Juba, Sudan swarm, areas along the rift that had been relatively quiet prior can produce large earthquakes—present activity is no guarantee of future activity, as it were.*[3]

Even one moderate (magnitude 5 to 6) earthquake every five years in the zone of the Gibe III dam would be reason for concern, and the fact that the zone is subject to bursts of seismic activity poses an even greater risk, as there would be little time to repair damages to structures between one earthquake and the next. The cumulative impacts of these clusters would be greater than any individual M5 earthquake.

➢ **Even so-called "quiet" zones of rifts can experience large-magnitude quakes and thus cannot be trusted to remain "quiet"** (Table 3.1; Figs. 3.3 and 3.4). This is evidenced by a 1990 cluster of earthquakes near Juba in South Sudan, a previously quiet rift area (Gouin op. cit.).

Investigations of the recent disasters of tunnel collapse at the Gibe II (termed *GG II*) project by Kinde and Engeda (2010) pointed to seismic instability as an underlying cause. The investigators stated that "the failures reported in GG II always

---

[1] Gouin was founder of the Geophysical Observatory in Addis Ababa and served as its primary seismologist until just before his 1979 book, *Earthquake History of Ethiopia and the Horn of Africa*.
[2] Advanced National Seismic System (ANSS) of the U.S. Geological Survey.
[3] See report in Carr (2012). Several of the public documents referenced here were introduced in Walter's report.

Table 3.1 Earthquakes within 300 km of the Gibe III (3) dam since 1906

Date	Magnitude (NEIC)	Magnitude (Gouin)	Distance from Gibe III (km)	Distance from Plate boundary
25 Aug 1906		6.6	185	35
28 Oct 1906		6.?	?	?
14 Jul 1960		6.3	134	23
23 Jan 1968	5.1	5.1	207	152
2 Dec 1983	5.1		145	40
20 Aug 1985	5.4		196	42
10 Jul 1987	5.3		89	11
25 Oct 1987	5.6		171	19
28 Oct 1987	5.4		137	3
8 Jun 1989	5		63	15
13 Feb 1993	5		274	11
20 Jan 1995	5		130	17
19 Dec 2011	5.1		95	59

Sources Gouin (1979), Advanced National Seismic System (ANSS) database, Walter (in Carr 2012)

occurred near faults—even when no seismic event occurred, further highlighting the dangers these projects face in actual seismic events" (Kinde and Engeda 2010). Their report continued with the conclusion:

> [T]the region around the project area has a recorded history of significant seismic events. The outcome of our study shows that projects such as GGII, GGIII, and GG IV should consider seismicity into the design and construction processes. **Neglecting the risks posed by seismic hazards in such a region with** [sic], **we believe, has tremendous negative consequences in the future usefulness of these projects**. [Emphasis added.]

Reservoir Seepage and Landslide Danger at the Gibe III Dam

➢ **According to geologists working in the dam region for several decades, reservoir filling is likely to extend for years, due to heavily fractured volcanic rock throughout the Gibe III dam site, including in the reservoir's natural walls**. Woldegabriel and Aronson (1987) also document these fractures for the region. Basalts, particularly those from volcanic flows that cooled unevenly, can form both vertical fractures (vesticules) and large-scale lava tubes capable of transporting large volumes of water. Basalt is especially vesiculated and highly fractured at the top and bottom of individual flows, where cooling occurs more quickly. These factors can make basalts highly permeable—some of the most productive aquifers there are—"even akin to karst". (ARWG geologist, personal communication).

Basalts dominate the geology of the MER. This is due to three dominant eras of volcanic activity: the first during the Eocene, 35–45 Ma ago; the second a series of flood basalts erupting 29–31 Ma ago; and the third, more recent shield volcanic eruptions of more alkaline basalt. The basalts that dominate the MER are composed of transitional tholeiites, are more alkalic and contain more sodium than potassium (Rogers 2005).

➢ **Even if the roller compacted concrete (RCC) construction of the proposed Gibe III performs well—a challengeable assertion—the fractured and jointed volcanic rock at the dam location would produce seepage conduits**. Ongoing seismic activity will likely open up more fractures over time, so seepage from the fractured rocks in the reservoir's walls would likely slow down filling and early retention of waters in the reservoir by multiple years, according to ARWG geologists. This is especially true because of the high hydrostatic pressure from impounded water behind the dam that will promote reservoir seepage.

The resultant extended period of radically decreased river flow below the dam—already a disastrous situation from even the "up to 3 year" reservoir filling period predicted by the GOE and accepted in the AFDB's 2010 assessment—would be disastrous for indigenous communities downstream along the Omo River and around Lake Turkana. This major reduction in

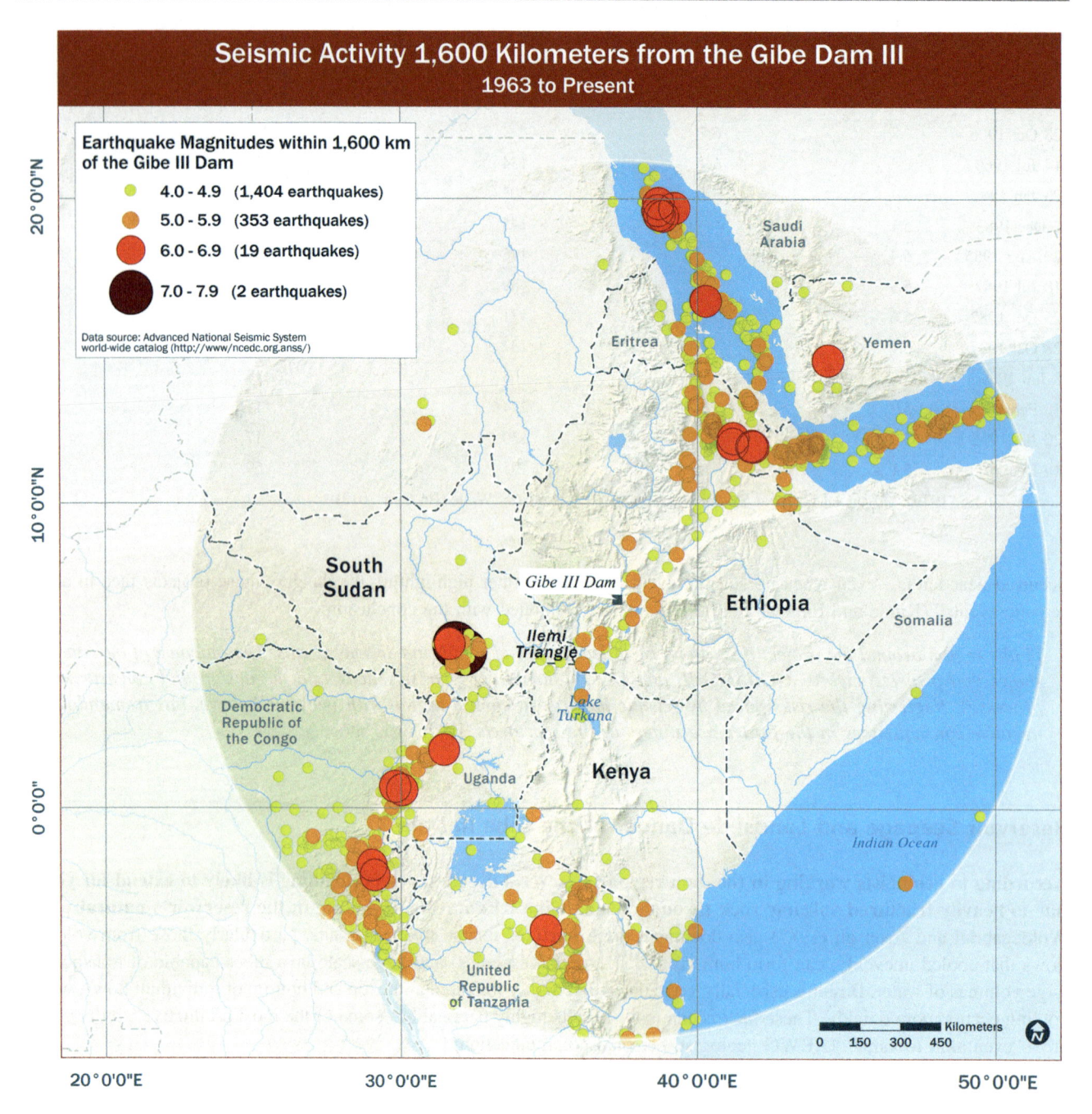

Fig. 3.3 Seismic activity within 1600 km of the Gibe III dam: 1963 to present. Indicated by magnitude of earthquake. *Source* data from the Advanced National Seismic System (ANSS) World Wide Catalogue (U.S. Geological Survey). Graphic by ARWG

Omo River downstream flow due to the Gibe III dam is greatly magnified by the radical abstraction of Omo River waters by major irrigated agribusiness enterprises the GOE is promoting throughout much of the lower Omo basin (Human Rights Watch 2012).

> **Even if the dam itself remains intact after a seismic event, it is highly plausible that springs, leaks and seepage developing in the adjacent rock abutments could cause a dam failure.** This danger stands despite the design of the Gibe III dam calling for 'grouting fractures' within the native rock around the dam.

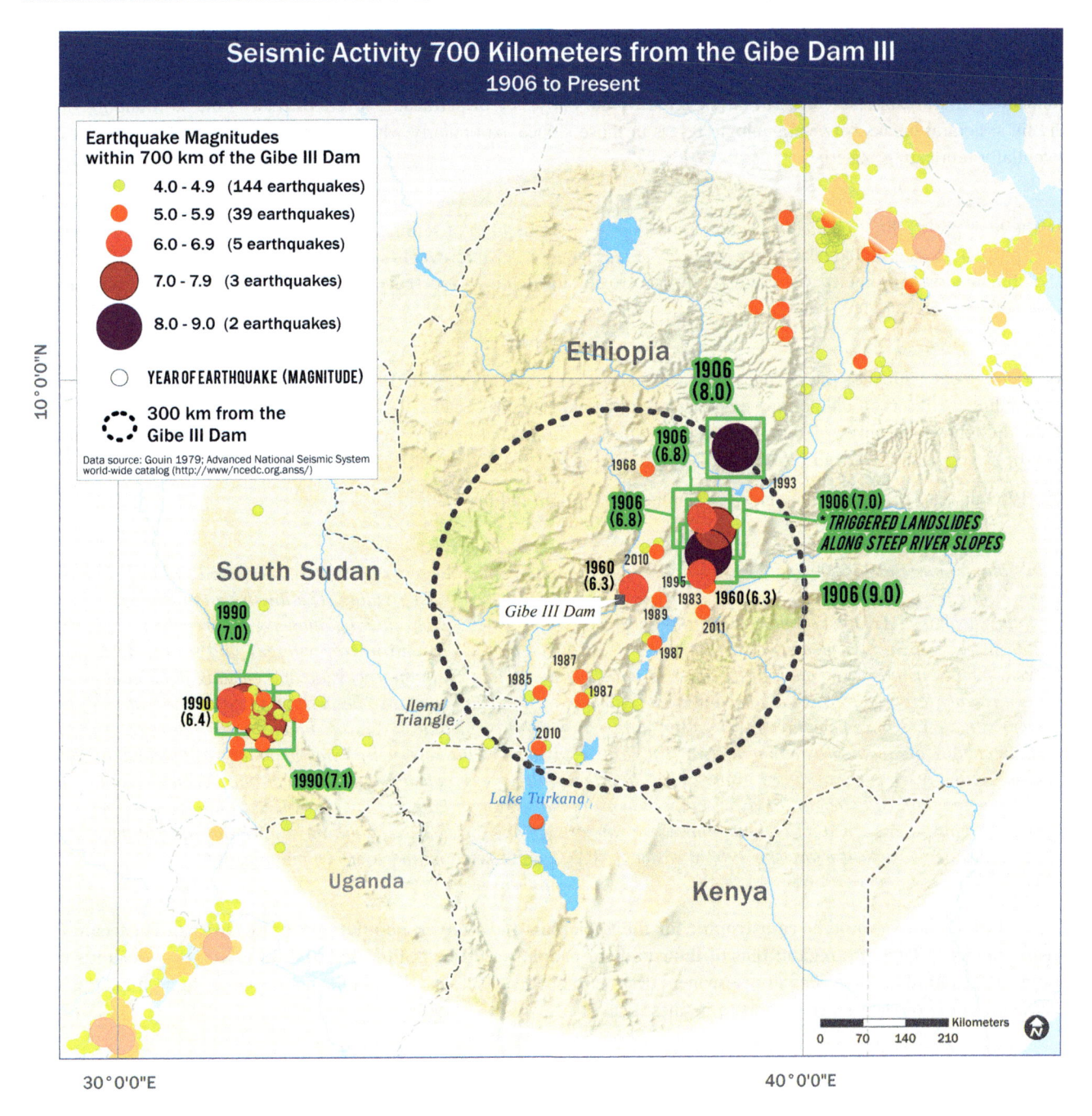

Fig. 3.4 Seismic activity within 700 km of the Gibe III dam: 1906 to present. Indicated by magnitude of earthquake. *Source* data from Gouin (1979), Advanced National Seismic System (ANSS), World Wide Catalogue (U.S. Geological Survey). Graphic by ARWG

➢ **Hot springs occur within the Lower Omo basin region and these are most likely generated by deep circulation along faults**. The well known 'forty springs' of Arba Minch result from fractures in volcanic rocks contiguous with those at the proposed dam site. These hot springs are most likely generated by deep circulation along faults. Similar springs have been documented in the Ugandan portion of the rift valley as caused by extremely high filtration of rainfall in the mountains through basaltic fractures to the base of the mountains.

➢ **Filling of the reservoir and early dam operation would promote the probability of a major landslide, particularly as within the context of the inherent instability of the steep, highly fractured 'natural walls' of the reservoir**. Impoundment at Gibe III will change the base level of the river in the immediate area: local aquifers are fracture controlled, and it is possible that some slip surfaces (landslide soles) may become lubricated so that rock masses are more likely to slide. Saturation of clay-rich soils along the canyons exacerbates the risk of swelling and landslides, as well.

Landslides in artificial reservoirs can be triggered by the buoyancy of permeable layers of rock in the surrounding slopes, and by structural breaks between geologic layers in those slopes, particularly when divided by thin layers of clay or other potentially lubricating material.

> The high probability of landslides, along with highly likely buildup of sediment behind the Gibe III dam - even with mild to moderate earthquake occurrence, let alone a major seismic event—presents a risk to the dam's integrity. Sedimentation is promoted, for example, by soil erosion upstream—exacerbated by deforestation and overexploitation of soils for agriculture.

As was evident with the Vaiont landslide dam disaster in northern Italy in 1963, initial geologic testing of the slopes surrounding the reservoir did not find evidence of potential problems because the clay layer between the top geologic layers of the dolomites was only ~1 mm thick in many places. The Vaiont event is discussed in detail by Semenza and Ghirotti (2000), Genevois and Ghirotti (2005), and Massironi et al. (2013).

The 1963 disaster of Vaiont illustrates the significance of dam failure.

> *In Vaiont, more than 2600 people were killed when 260 million cubic meters of a slide block moved suddenly into the newly filled Vaiont Reservoir – behind a very tall dam that blocked a deep valley. The intensive landslide occurred within seconds and displaced more than half of the reservoir's water, generating a giant wave that reached a height of 250 meters which then created an enormous wall of water that swept into nearby villages and towns, destroying everything in its path. The dam itself remained intact. In the Vaiont case, the rockslide and ensuing flood could have been readily foreseen by logical consulting. The cause of the landslide may also be pertinent to a consideration of the Omo situation. The sedimentary rocks of the Vaiont River Valley include layers of shale, a clay-rich rock. And the rocks comprising the nearby mountain (Mt. Toc) tilt steeply toward the reservoir. When the dam was finished in 1960, filling of the reservoir introduced groundwater into the shale layers, causing them to swell and become unstable. At first, the mountainside began slowly creeping down slope at a rate of half an inch per week. As filling continued and more groundwater seeped into the mountain, the rate of slippage increased to eight inches per day, and ultimately to 30 inches per day, just before the 1963 disaster. [ARWG earth scientist, personal communication]*

➢ **Gibe III dam failure would be catastrophic for the Ethiopian and Kenyan populations in the entire downstream Omo basin and Lake Turkana region: tens of thousands of people would be obliterated and hundreds of thousands more could face calamity**. The essentially permanent destruction of livestock and environmental resources in both nations from such an event is of such magnitude as to be inestimable.

Failed Government and Development Bank Seismic Review

➢ **The seismic threat, along with the related dangers of landslides, sediment buildup and seepage are entirely discounted by the GOE in both of its major impact assessments** (GOE 2009a, b). By adopting this approach, including with omission of available information and Literature, the GOE ignores the implications of such threat for its own indigenous citizens living downstream from the Gibe III dam, along with the vast indigenous Kenyan population residing along Lake Turkana.

The government's approach to the seismicity threat is 'bolstered' by a belated and dismissive seismicity report prepared by the contractor for the dam's construction—Salini (see last chapter). Salini's report remains unchallenged by the development banks although they are fundamental contributors to the long-term planning, financing and rationalizing of the Gibe III and/or its linked developments.

The GOE's impact assessments for the Gibe III refer to a GOE Seismic Hazard Assessment (2007) that considers seismic events only during the past century (GOE 2009b). They conclude that there is "no real evidence" of present seismic activity in the project area.

- *The project area (East-North East quaternary trending) does not seem to be active at this time, as illustrated in detail in the above-mentioned report.*
- *Despite the evidences that a certain seismic activity affected the region in historical times, according to the L1D Geological Report no evidences seem to exist of present seismic activity in the project area* (GOE 2009a).

As evidenced above, both the GOE timeframe and geographic scale are far too narrow to effectively determine seismic risk. Section 5.13 of the GOE's assessment, by noting only earthquake occurrences within the recorded time, clearly implies that this narrow window of time is indicative of the (lack) of potential for future earthquake activity. Typical seismic design, however, looks much further back in time to faults showing activity.

The appropriate geographic zone for assessing seismic risk in the Gibe III region is misrepresented in the GOE reports. Citing the International Commission on Large Dam's (ICOLD) guidelines for determining seismic risk, the GOE nevertheless ignores ICOLD's guideline that the tectonic province as a whole be considered when evaluating probabilities of tectonically linked seismic risk. As noted above, the tectonic province for this region should include the northern portion of the East African Rift Valley, including the northern ends of the Western and Eastern branches of the rift valley, the Main Ethiopian Rift, and the Afar Depression. These systems are linked, tectonically, and thus should be considered when evaluating seismic probabilities.

The suggestion in the downstream GOE impact assessment (GOE 2009b) for a 'warning system' with *scattered sirens* throughout the riverine zone downstream from the Gibe III dam, in order to warn of 'necessary releases' of flood pulses from the dam—or imminent dam failure—is nonsensical in the extreme. The area of impact of a major water 'release', dam overtopping or dam failure—one encompassing both the lower Omo basin and the Lake Turkana region—is so massive, with hundreds of thousands of agropastoralists, pastoralists and fishers confined to Omo riverside/delta lands, or the level plains around Lake Turkana, that such a warning system would be useless. Additionally, historically, these systems have rarely been implemented for other large dams, and sudden releases of water due to impending reservoir overtopping have often resulted in downstream drownings (for example, Cahorra Bassa in Mozambique).

Such an event would send enough water, sediment, and (in the case of dam collapse) material from the dam itself, into a massive wave, or series of waves through the lowermost Omo basin and into Lake Turkana—destroying human communities and biological systems in its path. These matters are essentially discounted in the European Investment Bank and both African Development Bank impact assessments (2009, 2010). The EIB (2010) report is dismissive of the seismic threat and responded to Africa Resources Working Group (2009) evidence summarized (in 2009) only with the statement, "Some risk is unavoidable." The AFDB (2010) impact assessment for Gibe III impacts on Lake Turkana offers only inconspicuous notation that a dam collapse would produce a two meter rise in lake level in the body of the text—omitting any reference to the cataclysmic force and impact of such an event and excluding the subject altogether in the assessment's Summary and Conclusions.

Literature Cited

Advanced National Seismic System (ANSS) World Wide Catalogue (U.S. Geological Survey).

Africa Resources Working Group (ARWG). 2009. In *A commentary on the environmental, socioeconomic and human rights impacts of the proposed Gibe III Dam in the lower Omo River Basin of Southwest Ethiopia*. http://www.arwg-gibe.

African Development Bank (AFDB). 2009. In Eds. A.S. Kaijage, N.M. Nyagah, Final Draft Report, Socio-economic analysis and public consultation of Lake Turkana Communities in Northern Kenya. Tunis, 189 pp.

African Development Bank (AFDB). 2010. In Ed. S. Avery, Assessment of hydrological impacts of Ethiopia's Omo Basin on Kenya's Lake Turkana water levels, Final Report, 146 pp.

Carr, C.J. 2012. Humanitarian catastrophe and regional armed conflict brewing in the border region of Ethiopia, Kenya and South Sudan: The proposed Gibe III Dam in Ethiopia, Africa Resources Working Group (ARWG), 250 pp. https://www.academia.edu/8385749/Carr_ARWG_Gibe_III_Dam_Report.

Ethiopia, Government of (GOE), Ethiopian Electric Power Corporation (EEPCO). 2009a. CESI, Mid-Day International Consulting Engineers (MDI), Gibe III Hydroelectric Project, Environmental and social impact assessment, Report No. 300 ENV RC 002C Plan.

Ethiopia, Government of (GOE), Ethiopian Electric Power Corporation (EEPCO), Ethiopian Electric Power Corporation (EEPCO). 2009b. Agriconsulting S.P.A., Mid-Day International Consulting, Level 1 Design, Environmental and social impact assessment, additional study of downstream impacts. Report No. 300 ENV RAG 003B.

European Investment Bank (EIB). 2010. Sogreah Consultants, Independent review and studies regarding the environmental and social impact assessments for the Gibe III hydropower project, Final Report, 183 pp.

Genevois, R., and M. Ghirotti. 2005. The 1963 Vaiont landslide. *Giornale di Geologia Applicata* 1: 41–52.

Gouin, P. 1979. *Earthquake history of Ethiopia and the Horn of Africa*. Ottawa: International Development Research Center.

Human Rights Watch. 2012. *What will happen if hunger comes? Abuses against the Indigenous Peoples of Ethiopia's Lower Omo Valley*. http://www.hrw.org/sites/default/files/reports/ethiopia0612webwcover.pdf.

Kinde, S, and S. Engeda. 2010. *Fixing Gibe II—Engineer's perspective*. http://www.digitaladdis.com/sk/Fixing_Gilgel_Gibe_II.pdf.

Massironi, M. et al. 2013. Geological Structures of the Vajont Landslide. *Italian Journal of Engineering Geology and Environment*. Presented at International Conference, Vaiont 1963–2013, October 8–10, Padua, Italy.

Rogers, N.W. 2005. Basaltic magmatism and the geodynamics of the East African Rift System. *Geological Society London, Special Publications* 259(1):77–93.

Semenza, E., and M. Ghirotti. 2000. History of the 1963 Vaiont slide: the importance of geological factors. *Bulletin of Engineering Geology and the Environment* 59: 87–97.

United Nations, Office for the Coordination of Humanitarian Affairs Regional Office for Central and East Africa (OCHA-ROCEA). 2007. Earthquake risk in Africa.

U.S. Geological Survey. Walter, S. 2012. Gibe 3 Seismicity review. Report prepared for C. J. Carr

U.S. Geological Survey. Hayes, G.P. et al. 2014. Seismicity of the Earth 1900–2013 East African Rift: U.S. Geological Survey Open-File Report 2010–1083-P, 1 sheet, scale 1:8,500,000, http://dx.doi.org/10.3133/of20101083p.

Woldegabriel, G., and J.L. Aronson. 1987. Chow Bahir Rift, a failed rift in Southern Ethiopia. *Geology* 15: 430–433.

Open Access This chapter is distributed under the terms of the Creative Commons Attribution-NonCommercial 2.5 International License (http://creativecommons.org/licenses/by-nc/2.5/), which permits any noncommercial use, duplication, adaptation, distribution and reproduction in any medium or format, as long as you give appropriate credit to the original author(s) and the source, provide a link to the Creative Commons license and indicate if changes were made.

The images or other third party material in this chapter are included in the work's Creative Commons license, unless indicated otherwise in the credit line; if such material is not included in the work's Creative Commons license and the respective action is not permitted by statutory regulation, users will need to obtain permission from the license holder to duplicate, adapt or reproduce the material.

Transboundary Survival Systems: A Profile of Vulnerability

> **Abstract**
> The transboundary region is culturally diverse, with indigenous languages of Cushitic, Eastern Nilotic, and Omotic and Afroasiatic origin. Several ethnic groups—the Nyangatom, Turkana and Toposa—are members of the Karamojong Cluster of cultures and speak mutually intelligible languages. The Dasanech, on the other hand, are Cushitic in linguistic affiliation. At the core of the region's indigenous economies are longstanding survival systems that are highly adapted to shifting environmental and social conditions, with ethnic groups linked through complex exchange networks. In recent decades, increasing dispossession and marginalization imposed by powerful external political and economic powers since colonial times have recently forced much of the region's population—particularly the Dasanech and northern Turkana—to settle at the Omo River or Lake Turkana as a last option means of survival. Despite centuries of resilience from even the most difficult times, these groups have now been pushed into extreme dependency on these two major water bodies and they have greatly increased vulnerability, even to stresses once familiar to them. They are now vulnerable in the extreme to massive scale destruction of their survival systems, with region-wide hunger and new mortality caused by the Gibe III dam and dam enabled irrigated agriculture along the Omo.

Indigenous Livelihoods and Survival Strategy Systems

> **The survival strategy systems of transboundary ethnic groups have emerged from centuries of indigenous knowledge and highly adaptive survival strategy systems.**

> The region's ethnic groups most heavily dependent on the Omo River or Lake Turkana are the Mursi, Bodi, Kwegu, Suri, Kara, Nyangatom, and Dasanech, in the lower Omo River basin, and the Turkana, El Molo, Rendille, Samburu, Gabbra and Dasanech along the shores of Lake Turkana (Fig. 1.3).[1] Numerous neighboring groups also rely on the

---

[1] Some of the most detailed ethnographic information in the region is that for the indigenous Mursi ethnic group—residing well upstream from lowermost Omo basin and Lake Turkana region that forms the core of this book. The majority of literature for the Mursi region has emerged from research by the anthropologist, Turton (1977, 1991, 1995, 2013) and his associates, who have written numerous pieces concerning the changes in that region—most recently the major scale land grab underway along the Omo River, for Gibe III enabled commercial scale irrigated agriculture plantations. This expanding crisis, particularly as it is affecting for the Mursi and Bodi peoples (Fig. 1.3), is relatively well reported and can be accessed both at the Oxford based website, www.mursi.org, and at a number of non-governmental and other websites web sites, including Survival International, International Rivers and the Oakland Institute. Other key literature for the Omo River region, near the related Mursi group, is that for the Suri agropastoralists (Fig. 1.3)—with particularly detailed accounts and interpretations of the region produced by Abbink (1997, 2000, 2003, 2009). For the lowermost Omo River basin where the Kara, Dasanech and Nyangatom reside, the literature is less extensive—in no small part owing to the physical and political difficulties accessing the region. Critical historical accounts of the Ethiopian monarchy's expansion into the southwestern region are presented in Donham and James' edited volume, *The Southern Marches of Imperial Ethiopia: Essays in History and Social*

river or lake—either for seasonal resource use or through exchange networks. These include the Me'en, Hamar, Dizzi, Chai, Toposa, Arbore and Pokot.

Ethnic groups in the transboundary region have historically been pastoral or agropastoral in emphasis, at least until recently, when very large numbers of them have had to resort to either strong reliance on agriculture, fishing, or both. Like pastoral peoples throughout semi-arid Africa, their survival systems comprise strategies for both risk minimization and recovery from livelihood setbacks. A striking exception to this is the El Molo ethnic group, a predominantly fishing people, residing at the southern end of Lake Turkana (Fig. 1.3).

Adaptability is key to the survival systems of indigenous groups throughout the region. This is especially true in the context of the transboundary area's wide range of environments (see Chap. 1 and below) and the periodic major stresses created by prolonged droughts, livestock and human disease epidemics, severe crop loss from pests, as well as shifting regional exchange relations and interethnic hostilities. Pressure on each group's capacity for adaptation is heightened by multiple decades of problematic government policies and especially by changes brought about by the GOE, the GOK and international finance institutions in anticipation of the Gibe III dam, along with its linked agricultural and energy transmission enterprises.

Livelihood systems of transboundary pastoral groups incorporate strategies for both *risk minimization* and *recovery from economic losses* include these general characteristics(see Figs. 4.1 and 4.2). These are outlined for the Dasanech, for example, in Carr (1977, 2012) and in much of the pastoral literature referred to above. Key among them are:

- **Maximum accumulation of capital**. Livestock are historically dominant as 'capital'—for meeting both immediate subsistence and long-term security needs.
- **High mobility of livestock herds and village settlement**. Complex and flexible seasonal movements between upland plains environments and riverine or lake zones provide ability to respond to changing environmental and social conditions.
- **Diversification of livestock types**. Cattle, small stock (goats/sheep), camels, donkeys.
- **Economic diversification to alternative production activities, including**:
    - Flood recession agriculture (the only agriculture possible in the lowermost Omo basin) on low flats along the Omo River and in modern delta lands low enough to receive annual flood waters.[2]
    - Fishing along the Omo River, in the delta region and along the northern shoreline of Lake Turkana, using primarily rafts and other minimal equipment.

Strong and pervasive social reciprocity relationships whereby material, labor and other forms of social exchange provide for precautionary as well as recovery measures. These relationships are generally rooted in age set based authority systems as well as clan/lineage and affinal ('in law') relationships.

(Footnote 1 continued)
Anthropology (1986). The Nyangatom agropastoralists have been studied in-depth by Tornay (1979, 1981), Tornay et al. (1997), Mark and Tornay (1992), Savary (2003) and Schroder (2003), with recent cultural investigation by Bassi (2011). Dasanech socioeconomy and human ecology has been the subject of early work by this writer (e.g., Carr 1977), while Almagor (1978, 1992) has produced major valuable ethnographic analysis. A critical cultural history of the Ethiopia-Sudan lands is provided by Donham and James (1986). Sobania (2011) has recently written about 'Dassanech' history, and Bassi (2011) has extended cultural history understanding by research with the Dasanech and Nyangatom). The Hamar people residing in the higher lands east of the Dasanech and the Nyangatom and closely interacting with others in the region, are not considered in this volume. Important accounts of the Hamar, with implications for the region, however, include those by Lydall and Strecker (1979) and Strecker (1988). While literature concerning the Turkana is extensive, most of it pertains to the southern and central Turkana, rather than the northern extent of Turkana territory, where the population is most vulnerable to the effects of the Omo River basin development. The works by Gulliver (1950, 1955) Gwynne (1969, 1977), Lamphear (1992), McCabe (2004), Skoggard and Adem (2010), Smucker (2006) and Fielding (2001) and Hogg (1982) are particularly useful for the purposes of understanding the Turkana in the transboundary region. The history and present crisis in the Ilemi Triangle—a contested area by all three nation states and an area with frequent and marked interethnic shifts in livelihood and territorial patterns, as well as the predictable center for a major upsurge in armed conflict among groups—is beyond the scope of this book and is the subject of a future cooperative effort.

[2]In large areas, particularly around Lake Turkana, high salinity conditions exist.

Fig. 4.1 Pastoral life along the Omo River west bank and Kibish River. *Top* Major Dasanech ritual (*Dimi*) on Omo River's west bank, with oxen/bull slaughter. *Bottom* Nyangatom male and female herders with small stock at deep (7 m) watering hole in bed of the seasonal Kibish River (see Fig. 1.3) during the long drought season

Fig. 4.2 Northern Turkana pastoralists in upland plains. *Top left* Turkana pastoralists with goats at working well (Oxfam funded). *Top right* Young Turkana camel herders. *Center* Small stock at hand-dug well in dry plains—near Ilemi Triangle. *Lower left* Northern Turkana pastoralist women and child. *Bottom right* Small stock at hand-dug well in Turkana (with hundreds of other livestock awaiting watering)

From this complex set of pastoral survival strategies, the most common means of pastoral household or village level *recovery* from livestock loss due to drought, disease, conflict or other hazards, for example, are:

- Temporary movement of herds into peripheral lands with available pasturage and water.
- Short-term barter of small stock (goats/sheep) with communities of the same or neighboring ethnic groups, in order to meet immediate food needs. Exchange or sale of animals, farm and fishing products to Somali, Ethiopian and Kenyan traders and merchants.
- Raiding of neighboring ethnic group herding camps and villages for livestock seizure.
- Use of social exchange networks (for loans, gifts and labor cooperation) in order to meet immediate household needs and rebuild livelihood systems.
- Temporary strong reliance on subsidiary production activities.
- Diversification to new forms of production, including recession agriculture, fishing, household commodity production and most recently wage labor (mostly in Kenya).

> Implementation of these strategies requires access to sufficient territory and resources—conditions that have drastically deteriorated for the Dasanech, Nyangatom and northern Turkana in the past half-century (Figs. 4.1 and 4.2).

➢ **Livelihood activities once secondary or only occasionally engaged in have recently become dominant in much of the lowermost Omo basin and northernmost Turkana region**. Among the Dasanech, for example, some communities have long been agropastoral, practicing substantial flood recession agriculture in low lying, seasonally flooded areas along the Omo River or in the Omo delta. *There is insufficient rainfall in the lowermost Omo basin for rainfed agriculture, so that flood recession agriculture is all that is possible in the Omo region.* The Kenyan government and a few aid agencies have funded small rainfall dependent agricultural projects in the dry plans of the northern Turkana segment (including near the lake), but the vast majority of these have failed.

A small segment of Dasanech have long fished and hunted crocodile, as well (Fig. 4.3). Most Dasanech, however, sustained traditional pastoral life—until recently, that is, when major pressures on livestock herding necessitated vast numbers of households shifting to substantial, even major emphasis flood recession agriculture in the riverine zone. By the same token, agropastoralists, and even some pastoralists, have taken up fishing in the lowermost river channels and northern waters of Lake Turkana. Thousands of Dasanech fishers are now locked into intense competition with Turkana over remaining fish stocks in the lake. This conflict is undoubtedly intensified by the Ethiopian government's encouragement and subsidization of commercial fishing operations in Kenyan lake waters—operations that have far greater fishing capacity due to their large and motorized boats.

In northern Turkana, a parallel set of pressures on pastoral life has also intensified in recent decades. As in Dasanech territory, inadequate rainfall in central and northern Turkana precludes rainfed agriculture and the shorelines of Lake Turkana's saline waters are also unsuitable for planting. As declining grazing conditions for livestock in the upland plains of Turkana have become dire in recent years, major numbers of northern Turkana households—often entire village—have migrated toward Lake Turkana.

As in the transboundary region more generally, massive numbers of Turkana livestock died during the extended drought between 2007 and 2010, for example, resulting in a major spike in herd losses—a process initiated years earlier. Increasing numbers of Turkana stock owners—even entire villages—began 'migrating' from the plains toward the lake's northwestern shores (Fig. 4.4). Countless thousands of animals died during these treks, due to lack of water and graze. Tens of thousands of these once pastoral households have taken up fishing, whether or not they have retained some livestock (Fig. 4.5). Hundreds of thousands of Turkana are now dependent on fishing or fishing related livelihood. They subsist only with access to sufficient Lake Turkana waters and living resources—especially fish. From this general migration pattern (Fig. 4.5), new fishing and fishing/pastoral villages have established along the northwestern shores of Lake Turkana, from the Ethiopia-Kenya border to Ferguson's Gulf, and southward.

Fig. 4.3 Riverside settlement and secondary production along the lowermost Omo River. *Top left* Dasanech boys fishing in gathering stream (*Kolon*). *Top right* Nyangatom agricultural village, with granaries, along Omo River. *Center left* Children at riverside agricultural plot. *Bottom left* Dasanech woman after burning small recession agriculture plot (before flood). *Bottom right* Dasanech villagers cutting up crocodile after successful hunt

Fig. 4.4 Northern pastoral Turkana dependency on Lake Turkana for watering and grazing. *Top left* Goats at watering. *Top right* Young goat in last stage of starvation/dehydration. *Center* Large camel herd on watering trek to the lake. *Bottom* Hundreds of small stock rushing in the last kilometer of their trek to the lake for watering

Fig. 4.5 Northern Turkana fishing villagers at Lake Turkana. *Top left* Non-motorized sailboats and villagers along the northwest shoreline. *Top right* Turkana girl at the lake. *Center* Bathing and water-getting north of Ferguson's Gulf. *Bottom left and right* Cleaning fish catch on the beach before transport to the village for drying

Indigenous Livelihoods and Survival Strategy Systems

Fig. 4.6 Indigenous village relocation (migration) to the Omo River and Lake Turkana: 1960 to present. Major livelihood dependence on lake and river is at the core of the region-wide exchange network (see Fig. 1.6)

The map of interethnic exchange relations included in Chap. 1 (Fig. 1.6) **suggests the critical nature of this system for the survival of each group**. This system of exchange relations is necessary for pastoralists, for example, both for herd-building and minimizing risks in herd management as well as for recovery from major livestock losses. For agropastoralists and pastoral/fishers too, exchange networks are critical for minimizing risk and recovery from such stressors as failure of Omo floodwaters to replenish agricultural plots and last resort grazing, decreased fish catch and failed local markets.

Despite a radical increase in agropastoralism over pastoralism among Dasanech households, and in fishing over pastoralism among many northern Turkana—changes following decades of disenfranchisement of both groups—cooperation among communities with divergent subsistence components remains. In peaceful periods between the Dasanech in the lowermost Omo basin and the Turkana around northern Lake Turkana, product (and sometimes labor) exchange is commonplace and involves pastoral, agropastoral and fishing communities.

The major food related exchange relations between just two sets of the region's groups are summarized in Tables 4.1 and 4.2. The full set of such relations is exponentially more complex, particularly when the full set of material and social transactions is considered. These include marriage-based exchange, labor cooperation, gift giving and loans, and political relations. These exchange relations are often key to economic diversification and the forgoing of new settlement and seasonal movement patterns among transboundary ethnic groups.

The tables suggest the core *food*-based exchange among three of the region's groups: the Nyangatom, the Turkana and the Dasanech—the key exchange relations among them being:

$$\text{Small stock} \Leftrightarrow \text{Grain} + \text{Small stock} \Leftrightarrow \text{Cattle}$$

As diversification from pastoral to agropastoral and fishing livelihoods has become a primary means of survival for very large numbers of the region's indigenous people in recent years, new forms are preeminent in some areas, especially:

$$\text{Fish} \Leftrightarrow \text{Small stack} + \text{Fish} \Leftrightarrow \text{Grain}$$

Pastoral Dispossession and Rising Dependence on the Omo River and Lake Turkana

> In order to carry out their highly adaptive survival strategies, including their complex system of exchange, the region's traditionally pastoral ethnic groups have depended on access to sufficient territory, including adequate pasture areas and watering points for livestock, lands for recession agriculture and most recently—fishing locales. These conditions have been drastically curtailed in recent decades, forcing the pastoralists to depend on Omo River and Lake Turkana resource for their survival.

➢ **A wide range of colonial and post-colonial policies in eastern Africa have produced major dispossession and marginalization among transboundary indigenous groups**. British colonial administrations of Kenya, Uganda, Sudan and Egypt, along with Ethiopia's monarchs—Menelik IIU and Haile Selassie—took actions that fundamentally altered the region. All of these governments engaged in border definitions, for example, that profoundly affected the region's indigenous peoples, while essentially regarding them as 'externalities.'. Major decisions were shaped without regard to the size, resource need—even the identity—of the indigenous groups concerned. This longstanding reality was described by colonial officers interviewed by this writer and in numerous written accounts in Kenyan and Sudanese archives.

Considerable documentation of these policies has emerged as part of the expanding literature for the region—most of it addressing one or another ethnic group or type of policy. Chapters 7 and 9 briefly summarize some key points from this literature, combined with information and perspective recorded from Dasanech and northern Turkana elders during investigations by SONT.

Table 4.1 Diversified food production among transboundary ethnic groups

Lower Omo Basin & Ilemi Triangle

Kara	Flood recession agriculture**/Agropastoral**
Nyangatom	Flood recession Agriculture**/Agropastoral**, Pastoral*, Fishing**
Dasanech (upland, riverine)	Flood recession agriculture**/Agropastoral**, Fishing**, Pastoral*
Suri	Agropastoral, Pastoral
Toposa	Pastoral, Agropastoral
Hamar	Pastoral, Agropastoral, Bee-Keeping

** *Year-round dependence on Omo River or Lake Turkana*
* *Seasonal dependence on Omo River or Lake Turkana*

Lake Turkana Region

Dasanech (Omo delta and east shoreline of lake)	Flood recession agriculture**, Agropastoral**, Fishing**, Pastoral*
Turkana	Pastoral*, Fishing**, Pastoral-Fishing**
Gabbra	Pastoral *
El Molo	Fishing**, Fishing-Pastoral**
Rendille	Pastoral *
Samburu	Pastoral *

** *Year-round dependence on Omo River or L. Turkana*
* *Seasonal dependence on Omo River or Lake Turkana*

Major Subsidiary Livelihood Activities (groups vary)

☐ Wild Food Gathering/Harvesting**	Utensil manufacture
☐ Beekeeping/Honey Gathering**	Fishing gear/net manufacture
☐ Boatbuilding**	Firewood/charcoal preparation*
☐ Chicken-raising	Fish marketing**
☐ Utensil and tool making	

* *Recently initiated or increased*
** *Dependent on Omo River or Lake Turkana*

Table 4.2 Major food related exchange: Turkana-Nyangatom and Dasanech-Turkana ethnic groups

Turkana- Nyangatom		Dasanech-Turkana	
From Turkana:	**From Nyangatom (Omo)**	**From Dasanech**	**From Turkana**
Animal skins/hides Live animals (mostly small stock) Cooking pans (Fish)	Sorghum, Peas, Beans, Squash, gourds	Sorghum, Peas, Beans, Squash, gourds Live small stock and cattle	Animal skins/hides Live cattle and small stock Fishing nets/gear

In the tri-nation border area, several types of colonial and post colonial policies have impacted all three major groups—the Dasanech, Nyangatom and northern Turkana, with differences in timing, intensity and outcomes of specific actions. These policies centered around:

- Redrawing of national boundaries.
- Policing agreements, often with financial subsidies to the administrating government.
- Territorial relocation or expropriation of indigenous groups and population segments within them—including for the purposes of political control, privatization of lands, settlement programs.
- Disarmament of villagers—including through forceful/brutal campaigns.
- Establishment of 'closed district' or 'no-go' zones.
- Livestock seizure—for example, as reprisals for aggressive actions, 'punishment' for incursion into declared no-go zones.
- Provision of firearms to indigenous fighting groups—for 'protection' of a particular group, and/or to establish a 'proxy' force for governmental agendas.
- Taxation—hut, population or livestock based.
- Increased militarization, with expanding security apparatus.
- Imposition of government appointed authority systems, replacing indigenous ones.

Some of these policies have been applied in one area, yet they have impacted the region more generally. For example, territorial dispossession of a particular pastoral or agropastoral group has occurred both directly, with forcible removal by a government, or indirectly, through expropriation of a neighboring group—one de facto forced to push into the transboundary region, intensifying resource competition. Such occurrences have affected the Pokot, Jie Toposa, Hamar, Gabbra, Rendille and Samburu neighboring groups. To some extent, the complex exchange network among the region's ethnic groups (Fig. 1.6) reflects such repercussions.

➣ **The Ilemi Triangle** (Figs. 1.1 and 4.7) **forms a critical part of the transboundary region's history of indigenous group marginalization and dispossession**. A disputed area between South Sudan and Kenya and one bordering Ethiopia, the Ilemi Triangle is named after a major chief (Ilemi Akwon) from the Anuak group that now resides in the Gambella region of Ethiopia/South Sudan borderlands. The Ilemi covers a vast area: 10,320 or 14,000 km^2, depending on the political boundary adopted.

The Ilemi—viewed by colonial and post colonial administrations alike as 'wasteland', 'unoccupied' (by permanent settlements) and with little if any value—nevertheless demanded certain political decisions since the early twentieth century, though these have nearly all been informal ones. The national boundaries in the tri-nation border region have remained in 'quiet' dispute—even following the Italian invasion and occupation of the region, the end of World War II, a number of

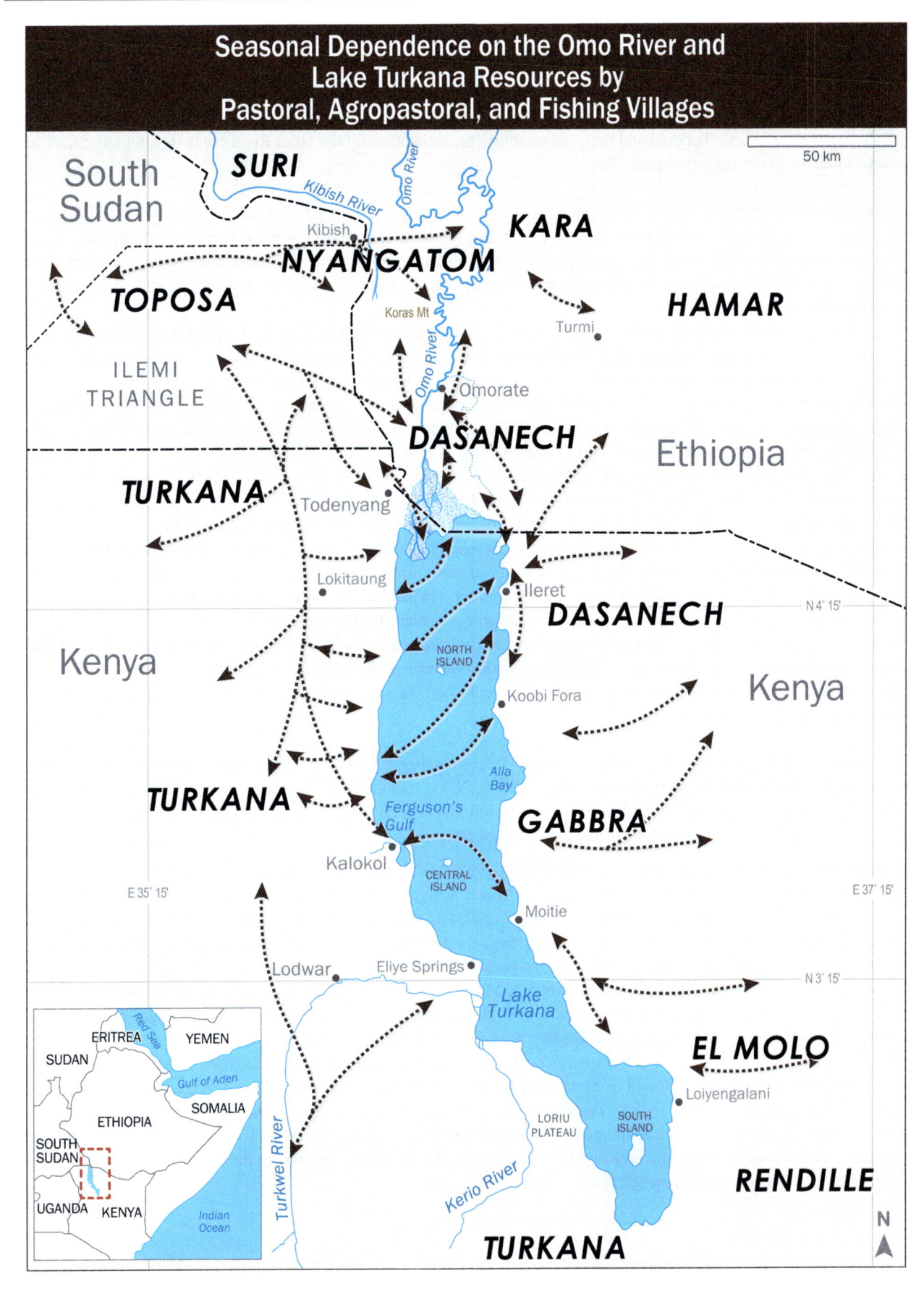

Fig. 4.7 Seasonal dependence on Omo River and Lake Turkana resources by pastoral, agropastoral, and fishing indigenous communities. East/west bank (Omo River) Dasanech and cross-lake movements by Turkana fishers have sharply increased

'modernization' efforts, and most recently, public acknowledgment by the respective governments of major oil and gas reserves there.[3] As this book goes to print, there are onconflicting reports regarding possible legal settlement between the governments of South Sudan and Kenya in this matter.

The significance of the Ilemi Triangle, and jurisdiction over it, for understanding the present status of the Dasanech, the Nyangatom and the northern Turkana and their vulnerability to decimation by the Omo River basin development (and now, extractive industry), is at least five-fold.

- In the first place, the grasslands of the region, particularly in its central and northwestern portions with rolling plains and relatively wetter foothills, are superior to nearly all those of the transboundary region. Especially during dry seasons, countless thousands of livestock from the region's different ethnic groups have flocked to the region—well prior to the military and political incursions by the region's colonial (and imperial/monarchy) powers and the shifting political arrangements among them.
- Secondly, the multiple ethnic groups in the region, while periodically engaging in conflict (with traditional weapons—not firearms, prior to the 20th century), also sustained complex cooperative and sharing relations with regard to watering locales, high value grazing areas and other issues, according to Dasanech and Turkana elders.
- Thirdly, governments have commonly exploited the indigenous groups' pressing need for access to Ilemi pastures by supplying them with arms and using them as 'proxy' forces in asserting their national territorial interests—the rhetoric of "protecting indigenous grazing rights" aside. This practice is heightened by the extensive arms trafficking within the region. The proxy character of indigenous fighters has been evident since the 1939 Italian occupation of the Ilemi and the 1941 routing of the Italians by Kenyan British forces. It continues to the present day dynamics in the region, with traditional weaponry virtually replaced by new brutality and scale of killing among ethnic groups.
- Fourthly, as numerous portions of this book suggest, the collapse of indigenous economies in the lower Omo basin and around Lake Turkana, due to Gibe III dam and linked agricultural enterprise development would catapult Dasanech, Nyangatom and Turkana into explosive new armed conflict—expanding from existing 'hot spots' (see Fig. 5.4), as all groups desperately search for rapidly disappearing resources within the Ilemi region.
- Finally, extensive oil and gas reserves in the Ilemi and the broader transboundary region, coupled with oil industry operations underway there, have intensified the boundary claims of Kenya and Sudan, in particluar. However these national claims unfold, the indigenous groups face yet worsened dispossession and marginalization, precisely when their circumstances are already at a crisis level.

Considerable scholarly accounts concerning the Ilemi Triangle/national boundary dispute have been produced in recent years, including that by Collins (2004, 2005), Mburu (2001, 2003), Lamphear (1992) and Blake (1997), as well as a large number of government and international agency reports. With some variation in specifics, these accounts are in general agreement about the historical trajectory of boundary arrangements.

A series of negotiated demarcations are chronicled from 1902/1907 through 1972, when informal agreement for Kenyan administration of the Ilemi was reaffirmed between Sudan and Kenya, along with Ethiopia's position of having no claim to the region. These boundary shifts are mapped in Mburu's (2003) account, for example. As the literature noted above details, the basic agreements among the region's governments began with Britain's 1902 survey of the border between Kenya and Ethiopia with a border referred to as the 'Maud line'—recognized in 1907, and allocating all of the Ilemi region to Sudan. This was immediately followed by the Ethiopian emperor Menelik II's claim of northern Kenyan lands to the southern end of Lake Turkana and eastward to the Indian Ocean. The British colonials responded with the Uganda-Sudan Boundary Commission agreement in 1914 that secured Sudan's access to Lake Turkana *via* the dry saltpan locale, Sanderson's Gulf (see Chap. 6). Other than that, the 1914 boundary within the Ilemi remained ambiguous. After some trading of administrative responsibility between the British of Sudan and Kenya, a 'Red line' agreement was established in 1931—five years before the Italian occupation of the region. The Red line was drawn under the rubric of 'protecting the grazing lands' of the northern Turkana. This line was once again reaffirmed between Kenya and Sudan in 1938. The effects of these administrative and policing shifts, combined with the ebb and flow of arms availability to the Dasanech, the Nyangatom and the Turkana, resulted in major conflicts among the groups.

---

[3] As Appendix A points out, the reserves were actually known to exist for decades—at least, by the oil corporations who had been exploring there, top officials of the governments and the World Bank.

Recapture of the Ilemi in 1941 by British forces, as they pushed into Ethiopia from Sudan—part of the campaign that reinstated Haile Selassie to the throne—once again led to a shift in boundaries. This time, the Kenyan administration drew a 'Blue line'—further north from the 'Red line'—thus extending its jurisdiction. In 1950, the Sudan government established a line relatively far to the west as its 'patrol line'—a move that Kenya later argued relinquished the Ilemi land. In reality, none of these boundary lines were legally binding in international terms. The Kenyan/Ethiopian border definition was reaffirmed by both parties in 1972, as was Ethiopia's confirmation that it had no claim on the Ilemi.

Meanwhile, by the 1960s, the Kenyan government designated the Ilemi a closed 'buffer zone' by agreement with the Addis government—excluding even their previously favored Turkana, as well as the Nyangatom and the Dasanech. All three groups were excluded and only small groups with their herds were able to 'poach' (as the Kenya police viewed it) the relatively rich grassland resources that they desperately needed, having experienced dispossession from their respective countries' policies and a number of years of extreme drought. Excursions into the Ilemi were extremely dangerous for the herders—risking seizure of their livestock by the Kenya police, often with brutal measures.[4]

This policy proved destructive for the Dasanech, the Nyangatom, and ultimately, the Turkana.

- The Dasanech were forced eastward into Ethiopia—effectively confining them to the area between the Kibish River and the Omo River, since their hostile relations with the Hamar prevented them from crossing the river.
- The Nyangatom were essentially split into two segments: those with villages along the Kibish River and the Omo River and those remaining near the Toposa (Fig. 1.3), slightly northwest of the heavily policed lands of exclusion. Travel and exchange between them however, continued.
- The Turkana were pushed southward into the harsher upland plains of northern Kenya.

All faced severe hardships from these changes—hardships that became crises during severe droughts and high disease incidence, even epidemics, as this writer experienced in the course of several research periods during the early 1970s.

By the late 1970s, the Kenyan government began encouraging herding and settlement in the Ilemi of Kenya's Turkana. Most observers interpreted this as a 'proxy' move to reassert Kenya's interests. The Dasanech remained excluded from lands they clearly regarded as traditionally theirs, so they perceived little recourse but to attack their neighbors. The Nyangatom, while officially barred from the Ilemi by the Kenyans, already had many households living nearby their close relatives, the Toposa (Fig. 1.3). Although under increased stress, the Nyangatom sustained much of their transhumant movement among their settlements along the Omo and Kibish Rivers, and in the Ilemi.

Ethiopian border presence remained minimal until the 1980s when the government began increasing its presence along the lower Omo River. Kenya continued to administer the Ilemi, with police posts at Kibish and several other locales. Kenyan police frequently seized large numbers of livestock belonging to the pastoralists—sometimes even shooting the herders.

> The Ilemi continued to be administered by Kenya. Until the region's oil and gas potential became an active issue in government circles (oil companies had been exploring the region for years), the contested nature of the Ilemi remained largely dormant. Oil corporations sharply increased their interest and involvement in the region during the 1980s, and this process has since accelerated. Land grabs and government issuing of leases and concessions are well underway (see Appendix A).

➢ **Largely due to these major territorial losses—compounded by prolonged droughts, shifting ethnic alliances and conflicts, and regional government policies—both people and livestock were overcrowded in numerous areas throughout the region.** Overgrazing by the increasingly crowded livestock initiated a radical decline in herd numbers—one so severe that the long-standing economic recovery strategies were no longer effective. These conditions led to ecological

---

[4]This writer did extensive ecological transecting and study in the eastern Ilemi during the early 1970s (see Chap. 7 and Carr 1977) and personally encountered a number of small groups of herders and livestock seeking safe places for their animals to graze. Relations with the Kenya police ranged from poor to atrocious, as evidenced by one Kenya police officer who stated that *"shooting Dasanech is more fun than shooting gazelle."* Kenyan police assigned to the region from the highlands commonly had antipathy for the marginalized pastoralists, and certainly they detested the extreme heat and dryness of the region.

deterioration throughout much of the dryland plains of the transboundary region, except for the restricted Ilemi lands. None of the three pastoral groups really recovered from their many years of dispossession in their respective countries and within the Ilemi. Livestock mortality continued at a high level, as village survey in both northern Turkana and Dasanech areas indicate (Chaps. 7 and 9). Herd sizes plummeted for the vast majority of households and villages.

The pastoralists' longstanding strategies for recovery outlined above have proven insufficient for a massive proportion of the northern Turkana and Dasanech in particular, so tens of thousands of them turned to the only remaining resource systems available to them—at the Omo River and along the shores of Lake Turkana. By the 1980s, the Omo delta had expanded considerably considerably, offering new resource availability for that helped local communities cope with the effects of earlier territorial losses (see Fig. 1.2).

The general patterns of migration, or relocation to the Omo River and to Lake Turkana are indicated in Fig. 4.6. Households continue to herd whatever livestock they have left—often by combining herds and cooperating in labor for stock camps, including in distant areas where still possible. Along the Omo River, most of the locales for flood recession agriculture are already spoken for as this large displaced population searches for possible planting land and for livestock grazing and watering locales. Well-established social and material exchange patterns between the Dasanech and Turkana, like those of nearby ethnic groups (Fig. 1.6), have altered with these settlement shifts and accompanying changes in mobility and economy (see Chap. 7 for changes in the traditionally complex strategies for household/village level herd management since the 1970s).

The Omo River and Lake Turkana are central to the present day seasonal shifts in survival strategies of the Dasanech, Nyangatom and northern Turkana peoples. Their general seasonal patterns are represented in Fig. 4.7—patterns for herding, planting and fishing that are conditioned by major shifts in the character of water and biological resources of the river and lake, as well as precipitation levels and other conditions in the broader region.

> The relatively recent major spike in livelihood dependency on the Omo River and Lake Turkana resources by transboundary residents leaves them vulnerable in the extreme to radical reduction of river and lake waters. The precipitous drop in river flow volume and shoreline retreat of the lake caused by the Gibe III dam and dam enabled irrigated agriculture spell unprecedented disaster for the region's indigenous population.

Environments in the Transboundary Region: From Pristine to Degraded

➢ **With the exception of the Omo riverine forest and some volcanic outcrop locales, most of the transboundary region's pristine habitats have undergone severe environmental deterioration in recent decades** (see Carr 1977, 1998; Ebei and Akuja 2008; Gil-Romera et al. 2011). As noted above, this ecological degradation results primarily from overgrazing due to overcrowding of livestock and people following the major the territorial losses noted above (also see Chaps. 7 and 9). These losses have been compounded by prolonged drought periods and other externally generated political and economic influences, including GOE expropriation of villagers' resources along the Omo River.

Only isolated pockets of natural habitat remain, including large sections of the Omo riverine forest (prior to current government cutting and clearing), certain volcanic highland locales and upland plains areas where watering places for livestock are absent or where interethnic hostilities preclude herding. Much of the unique biodiversity and ecological integrity of the region has declined to the point where environmental restoration is required. The destruction of water resource access and resources in the downstream Omo zone and around Lake Turkana, caused by Gibe III dam and irrigated agricultural development, would force villagers throughout the region to move into the remaining natural areas in the effort to avert decimation of their livelihood systems.

➢ **The transboundary zone's ecological communities are diverse compared with most other semi-arid regions of Sub-Saharan Africa**. This is largely the result of the region's complex geological, depositional and water/soil conditions (Fig. 1.9). Habitats range among those associated with the Omo River, including annually flooded riverside flats, relict river meander channels ("ox-bows"), non-flooded natural levees with variable soil moisture relations, tributaries and numerous incomplete or relict channels, and the modern delta (now about 500 km^2 in area following its recent expansion (Fig. 1.2); the

Kibish River and gathering streams; relict beach ridges. basin pans and relict floodplains; Lake Turkana with its complex lacustrine and littoral habitats, the Turkwel and Kerio Rivers and numerous small ephemeral channels; scattered volcanic highland areas, rock outcrops and tuffaceous exposures, and salt springs.

Biodiversity in the transboundary region strongly corresponds with these environmental differences. It is further enhanced by the region's position as a zone of convergence among contrasting ecological zones.[5] These characteristics are strongly reflected by the results of plant collections made by this writer in the lower Omo region, including in riverine, dryland plains and volcanic areas (Carr 1977, 1998)—species identified are listed in Appendix B.[6] Additional plant collecting has been carried out in the lower Omo and northern Turkana regions by F.H. Brown (summarized in a forthcoming article) with staff from the Kenya Herbarium in Nairobi. Raymonde Bonnefille has made extensive pollen collections in the northern Turkana/lower Omo Basin region.

Large wildlife populations have been recorded in recent decades throughout the lower Omo basin, Ilemi Triangle, and northern Turkana regions. Eland, oryx, topi, Burchell's zebra, hartebeest, lion, leopard, cheetah, elephant, bat-eared foxes, gazelle, and gerenuk have long inhabited the region's grasslands and other dryland environments. Omo riverine (or 'gallery') forest and woodland areas support a rich wildlife population including hippo, elephant, crocodile, at least three species of primates, kudu, bushbuck, waterbuck leopard, and a wide variety of bird species including fish eagle, goliath heron, and dwarf bittern. These wildlife populations have plummeted in recent years, both from habitat destruction resulting from the changes outlined above and from the pervasive use of weapons throughout the region.

The Lowermost Omo River Basin and Environs

From its source waters in the highlands of Ethiopia, the Omo River descends gradually through a series of gorges in the highlands of the northern portion of the lower basin where it is bordered by steep slopes and a mixture of riparian forest, woodland and drier plant communities. Continuing its southward flow, the river opens into broad semi-arid lowlands where it forms a strongly meandering pattern and continues to its terminus at the northern end of Lake Turkana (Fig. 1.3).

The low-lying waterside flats along the river—primarily silt berms and sand/silt bars annually flooded by the Omo—provide the conditions for vitally important flood recession agriculture by Kara, Nyangatom and Dasanech villagers. The Kara also plant in ox-bow channels when Omo floods are sufficient. Nyangatom settled along the Kibish River (Fig. 1.3) plant there when water in that seasonal watercourse is sufficient to flood their plots, and the Dasanech plant throughout much of the active delta as well as on flats along the lowermost Omo. These patterns are outlined in Chaps. 7 and 8.

Although grasslands predominate throughout this southernmost section of the lower basin (the upland areas are described in Chap. 7), the region actually supports a mosaic-like pattern of vegetation types, due to its complex soil depositional patterns and anomalous geomorphic features. Sand and silt-sand relict beaches and clay basins are punctuated by volcanic highlands, pronounced cracking patterns, salt springs and other features. The oxbow channels with seasonal floodwater inundation by the Omo River add to the complexity of plant communities.[7] Nearly all of the rangelands between the Kibish River and the Omo River (Fig. 4.7) are highly degraded from overcrowding of livestock.

The forest along the lower Omo is the last pristine riverine forest in Sub-Saharan African drylands. Forest and woodland dominate the natural levees and the immediate levee backslope along the Omo River except on strong outside river bends,

---

[5]With some notable and important exceptions, the flora and vegetation types in southwestern Ethiopia closely resemble those of the Ilemi region and eastern Africa—a relatively recently determined relationship confirming earlier predictions by the well-known British botanist in East Africa, Gillett (1955).

[6]This writer has completed plant taxonomic studies, along with collection and ecological characterization of more than two thousand plant specimens. A full set of the collection (prepared in duplicate) is deposited at KEW Gardens Herbarium in London and another at the National Museum of Kenya, in Nairobi.

[7]Arguably one of the least-described regions of eastern Africa, the impressionistic and often subjective accounts by early colonial explorers offer important clues to early subsistence patterns among the area's indigenous groups. These include descriptions by Vannutelli and Citerni (1887), D'Ossat and Millosevich (1900) and Von Hohnel (1938). Butzer (1971a, b) describes a series of expeditions that passed through the Omo delta during the early twentieth century. With the 1967 beginning of the international Omo Expedition directed by Clark Howell, Richard Leakey and Yves Coppens, concrete description of the area was undertaken in a number of scientific dimensions over a period of years. This effort included extensive geomorphology, hydrology and soil studies by Butzer, Brown, Cerling and associates as well as direct paleontological investigations. It also supported detailed riverine and upland vegetation and soil studies, as well as cultural ecological and land use investigations by this writer.

with drier (xerophytic) vegetation, from Omorate northward along the river (Figs. 1.9 and 4.7). The forest is sustained only with sufficient saturation, or infiltration of the Omo's natural levee soils from the Omo River's annual flood (generally between August and December). Overbank flooding does not occur in the forest zone of the Omo—a reality misrepresented entirely by the GOE. This writer has carried out detailed studies of the riverine forest and its development, some key features of which are summarized in Chap. 8.

The forest and woodland provide resources are of fundamental importance to the survival systems of the region's indigenous communities. In addition to the recession agriculture along waterside flats and slopes bordering the forest, transition zone vegetation from forest and woodland (see Chap. 8) provides last resort graze and browse for Nyangatom and Dasanech livestock and resources for wild plant gathering. Most species in the forest and woodland are limited to that environment and are key to the biodiversity (both floral and faunal) of the lower Omo basin.

The Omo River does not flood the vast 'floodplains,' or mudflats, lateral to the river throughout much of the lowermost basin, since these are relict (or ancient) floodplains. Moreover, rainfall in the lowermost basin is insufficient for rainfed agriculture to be practiced these relict features. The Ethiopian government's misrepresentation of these realities in its downstream impact assessment (GOE 2009b) is a key reason for the report's invalidity, since the GOE asserts not only that these lands are flooded (and thus support planting), but also that rainfed agriculture in these lateral plains could compensate for the loss of flood recession planting lands along the river, due to flow reduction by the Gibe III dam. In fact, neither survival option exists for the Dasanech and Nyangatom.

Rainfall in the region is inadequate for rainfed agriculture to be practiced in the relict floodplains, despite repetitive GOE assertions that this practice is widespread. The significance of this misrepresentation rests with indigenous agropastoralists and others not having an option of rainfed agriculture as an alternative to flood recession planting.

Since precipitation is low and erratic, ground water recharge is a critical factor in the maintenance of both riverine zone and floodplain ecosystems. The vegetation in these areas—mostly scrub-like grasses and clustered herbaceous ground vegetation—forms a highly irregular pattern, corresponding to micro-level soil and drainage variations as well as irregular distribution of rainfall and pooling during the rainy season. Vegetation cover is of major significance in preventing soil evaporation and large-scale erosion. Soil evaporation and large-scale sheet erosion.

➢ **All major dimensions of Dasanech livelihood presently depend on the sustainment of the Omo riverine/delta environment**. The roughly 500 square kilometer modern Omo delta (Fig. 1.2), detailed in Chap. 7, is flooded annually over large areas by the Omo River. Its vegetation ranges from woodland (limited in extent) to a variety of grasslands and wetlands, forming highly complex patterns in response to even small shifts in sediment deposits, nutrients and water conditions determined by the river.

> The major livelihood components of the Dasanech and Nyangatom residing along the river would be eliminated since most of the Omo River's modern delta where last option grazing, flood recession agriculture and settlements occur would be desiccated by the effects of the Gibe III dam and its linked irrigated agricultural development.

Lake Turkana and Environs

➢ **Lake Turkana is the second largest lake in Kenya and the world's largest desert lake. More than 90 % of its waters derive from Omo River inflow**. The lake is one of the most saline lakes in the Great Rift Valley and the second most saline in Africa. The lake is only borderline potable for humans, livestock, and many wildlife species.

- Both the Turkwel River and the Kerio River flow into Lake Turkana (Fig. 1.1). Inflow from the Turkwel has been radically reduced from the construction of a hydroelectric dam at Turkwel Gorge, about 150 km from the lake. Lake Turkana has no outlet since the separation from the Nile basin (Butzer 1971a).Because it is a closed-basin lake, fluctuations in the level of Lake Turkana are determined by inflow from rivers and by evaporation, which is generally accepted to be about 2330 mm/year.

- **Lake Turkana is the most saline lake in East Africa containing a normal fish fauna**. Salinity is already at a level critical for various fauna and at the extinction level for mollusks (Yuretich and Cerling 1983). Citing L.C. Beadle's early work, the authors also note that at a higher salinity, dwarfism of fish occurs. Building on studies by Hopson (1975), Wood and Talling (1988) recorded a dissolved salt content of 2440 ppm—*nearly double the safe level* in Kenya's "Guide Value." Although lake waters are technically within the range of potability for livestock, herd animals frequently refuse the saline waters, even after lengthy watering treks (Fig. 4.4). Turkana villagers report numerous health problems associated with high salinity.
- **The Omo River's annual flood produces an extensive plume of nutrients and sediment-laden freshwater and is key to the replenishment of Lake Turkana's entire ecological system, including the sustainability of its fisheries**. This nutrient and freshwater inflow can reach the central portion of the lake and stimulates fish spawning as well as fish feeding cycles in the lake. Reproductive areas of fish are mostly concentrated along the northern shoreline, in Ferguson's Gulf, Alia Bay and other shoreline areas (see Fig. 1.3 and Chap. 5).

> The major reduction of Omo River overall flow volume and inflow to Lake Turkana during the planned closure of the Gibe III dam while efforts to fill its reservoir proceed would disrupt the downstream hydrology and ecological systems of the Omo River and Lake Turkana—transforming the habitats essential to the life cycles of the more than 50 species of fish identified so far in the lake and delta waters.

- **Seasonality of the lake and access to fishing localities shifting with seasons is a key component of indigenous communities settled along Lake Turkana**. The annual flood pulse from the Omo River, combined with the prevailing winds and currents in the lake, combine to establish several major seasonal differences. The lake is exposed to strong winds during certain months, with the prevailing winds from the southeast being a critical factor in the mixing of lake waters and nutrients. Local fishers define seasons in terms of the combination of changes in winds, lake level, nutrients and fishing conditions.

➢ **Although low in fish species diversity, Lake Turkana has about 50 species, 13 of which are important in the economy of resident fishing communities**. These species range from pelagic to demersal and their distributions largely conform to the habitat factors noted above. *Tilapia* spp. and Nile perch, for example, are central to Turkana fishing economy and their survival depends on the health of the lake's ecosystem. These fish would be severely impacted, if not almost eliminated, without sustainment of their major reproductive and/or feeding habitats in shoreline or northern lake/delta areas. Fish species and Lake Turkana habitats critical to the survival of the indigenous people are identified in the section below concerning Turkana livelihood.

Wildlife at Lake Turkana has been abundant, with parks internationally recognized for their richness and uniqueness. The lake is home to Nile crocodile, hippos, turtles, and bird species numbering in the hundreds—including flamingos, cormorants, ibises, skimmers, and sandpipers, as well as highly threatened populations of turtle, Nile crocodile, and hippo populations—but remains essentially unprotected. These include populations in the UNESCO World Heritage site, Lake Turkana National Park (on the western shore), Sibiloi National Park (on the eastern shore), and island parks (both Central Island National Park and South Island Park. Most wildlife areas have suffered major degradation and habitats now face threat of destruction for a variety of reasons, especially from political and economical forces causing overcrowding of people and livestock, with resultant habitat destruction, the abundance of firearms throughout the region due to arms trafficking between Kenya and South Sudan, and commercial encroachment.

➢ **Several hundred thousand pastoral and fishing Turkana live in villages along or nearby Lake Turkana's western shoreline and are almost entirely dependent on the lake for household water, livestock watering and nearby grazing, fishing—or all of these**. As described in Chap. 8, this population is particularly concentrated around Ferguson's Gulf and Kalokol (Fig. 1.3) and northward into the Kenya-Ethiopia border. The overwhelming majority of the fishing Turkana have migrated from pastoral lands far removed from the lake, following major herd losses. Many arrive with no livestock at all. By all accounts, this migration has dramatically increased in recent years for a variety of reasons discussed below. A nearly rainless period extended for almost three years in much of the lower Omo basin and northern lake region in recent years, for example, resulting in major livestock mortality and sparking new migration.

The majority of Turkana villages along the lakeshore are oriented primarily to fishing, although some retain small numbers of livestock (small stock). Most households engage in some type(s) of secondary production, and many receive occasional international food aid. Turkana pastoralists bring tens of thousands of livestock—goats, sheep, cattle and camels from the upland plains to the lake for watering and grazing during drought months (Fig. 4.4). Recent ecological degradation of littoral and shoreline environments is a critical problem for both Turkana residents and visiting herders, as noted in Chap. 9. Fishing communities depend on the maintenance of delicate environmental and biological balances in shoreline habitats, with even small shifts affecting fish catch levels. Trading relations between fishing communities and slightly more removed pastoral/fishing communities are essential to both.

Central and northern upland environments around the central and northern portions of Lake Turkana are similar to those of the lower Omo basin and those of the eastern Ilemi in terms of rainfall. The region is semi-arid with a mean annual precipitation of less than 250 mm. Rainfall is unpredictable in the extreme, both in total amount and in distribution—conditions to which Turkana pastoralists have long been adapted (Fig. 4.2). The seasonal movement patterns indicated in Fig. 4.7 point to the high mobility of the transboundary indigenous population, including the pastoral Turkana.

> The radically deteriorated conditions for Turkana livelihood (and that of the Dasanech at the lake's northeastern shoreline) would advance to disaster level with radical withdrawal of lake waters in the extremely shallow fringes of central and northern Lake Turkana—an inevitable result even during the planned Gibe III dam filling period (whether 2 years or multiple years), let alone from continued operation of the dam and from the abstraction of river waters for the large-scale irrigated agricultural works underway.

Cross-Border Conflict and Diminishing Resources: The Ilemi Triangle Ingredient

➢ **Localized conflicts in the transboundary region have been escalating in recent years**. This is primarily due to worsening resource deterioration and poverty conditions throughout, continued dispossession fueled by government policies, prolonged droughts and disease outbreaks. Climate change, while likely to be a major factor of increased vulnerability of the local population to the impacts of the looming dam and irrigated agricultural development, has not been responsibly studied in the region. Finally, rampant arms trafficking continually expands throughout the region and contributes to the upward spiral of violence.

- Most destructive among government policies are the Ethiopian government's large-scale, forcible evictions of indigenous communities along the lowermost Omo River (as well as upstream in Mursi/Kwegu/Bodi territory) in order to make way for large-scale, irrigated commercial agricultural schemes. Major irrigation infrastructure construction brings further expropriation and destruction of resources necessary for local residents' survival, as well and creates major blockages to mobility of herd animals.
- The Kenya government's present acceptance of the Gibe III dam and irrigated agriculture—at least by the time of this writing, and the Kenyan government's active partnership in the transmission of electricity from the Gibe III to Kenya and beyond are additional major destructive policies.

Rising tensions over dwindling grazing resources and watering points has intensified long-established interethnic tensions and conflicts, major locales for which are shown in a map below. These stresses have also weakened the extensive regional system of food related exchange—a system important to the survival of all groups (Fig. 1.6). Similar impacts on social exchange have been noted for comparable pastoral regions by Abbink (2000, 2009), Teqegn (2003), and Hendrickson et al. (1998a, b).

The numerous 'hot spot' conflict localities where conflict is already intensifying and can be expected to spike even more sharply as the Gibe III dam and agricultural enterprises are allowed to operate, are indicated in a map in Chap. 5. The overwhelming response by governments and the private groups active in the region, notably including Tullow Oil, is militarization.

Literature Cited

Abbink, J. 1997. The shrinking cultural and political space of East African pastoral societies. *Nordic Journal of African studies* 6(1): 1–17.
Abbink, J. 2000. Restoring the balance: Violence and culture among the Suri of Southern Ethiopia. In *Meanings of violence: A cross cultural perspective*, Eds. G. Aijmer and J. Abbink, pp. 77–100.
Abbink, J. 2003. The yellow guns—Generations and politics in Nyangatom (Ethiopia). *Africa* 73(3): 473–474.
Abbink, J. 2009. Conflict and social change on the south-west Ethiopian frontier: An analysis of Suri society. *Journal of Eastern African Studies* 3 (1): 22–41.
Almagor, U. 1978. *Pastoral partners: Affinity and bond partnership among the Dassanetch of South West Ethiopia*, 258 pp. England: Manchester University Press.
Almagor, U. 1992. Institutionalizing a fringe periphery: Dassanetch-Amhara relations. In *The Southern Marches of imperial Ethiopia*, Eds. D.L. Donham, W. James, and Jean-Pierre de Monza, pp. 96–115.
Bassi, M. 2011. Primary identities in the lower Omo Valley: Migration, cataclysm, conflict and amalgamation, 1750–1910. *Journal of Eastern African Studies* 5(1): 129–157.
Blake, G.H. (ed). 1997. *Imperial Boundary Making: The Diary of Captain Kelly and the Sudan-Uganda Boundary Commission of 1913*. Oxford University Press.
Butzer, K.W. 1971a. Recent history of an Ethiopian Delta. Chicago: University of Chicago, Department of Geography Papers, Research Paper No. 136.
Butzer, K.W. 1971b. The Lower Omo Basin: Geology. *Fauna and Hominids of Plio-Pleistocene Formations, Naturwissenschaften* 58(1): 7.
Carr, C.J. 1977. *Pastoralism in crisis: The Dassanetch of Southwest Ethiopia*. Chicago: University of Chicago, Department of Geography Papers, 319 pp.
Carr, C.J. 1998. Patterns of vegetation along the Omo River in southwest Ethiopia. *Plant Ecology* 135(2): 135–163.
Carr, C.J. 2012. Humanitarian catastrophe and regional armed conflict brewing in the border region of Ethiopia, Kenya and South Sudan: The Proposed Gibe III dam in Ethiopia, Africa Resources Working Group, 250 pp. https://www.academia.edu/8385749/Carr_ARWG_Gibe_III_Dam_Report.
Collins, R. 2004. The Ilemi Triangle. Unpublished manuscript.
Collins, R. 2005. The Ilemi Triangle. *Annales d'Ethiopie* 20:5–12.
Donham, D., and W. James (eds.). 1986. *Southern marches of imperial Ethiopia: Essays in history and social anthropology*. Athens: Ohio University Press.
D'Ossat, G., and F. Millosevich. 1900. *Seconda Spedizione Bottego*. Roma: Studio geologico sul materiale raccolto de M. Sacchi, 212 pp.
Ebei, P., Oba, G. and T. Akuja 2008, Long-term impacts of droughts on pastoral production and trends in poverty in north-western Kenya. In *Droughts: Causes, effects and predictions*, Ed. J.M. Sanchez, pp. 103–138. New York: Nova Science Publishers.
Fielding, P. 2001. Turkana Herders of the dry Savanna (Book Review). *Quarterly Review of Biology*, vol. 76, no. 1.
Gillett, J.B. 1955. The relation between the highland floras of Ethiopia and British East Africa. *Webbia* 9: 459–466.
Gulliver, P. 1950. A preliminary survey of the Turkana compiled for the government of Kenya. University of Capetown.
Gulliver, P. 1955. *The family herds: A study of two pastoral tribes in East Africa: The Jie and Turkana*. London: Routledge and Kegan Paul.
Gwynne, M.D. 1969. *A bibliography of Turkana*. Nairobi: Royal Geographical Society South Turkana Expedition.
Gwynne, M.D. 1977. *Land use by the southern Turkana*, 16–18. Paper prepared for seminar on Pastoral Societies of Kenya: Institute of African Studies, University of Nairobi.
Gil-Romera, G., D. Turton and M. Sevilla-Callejo. 2011. Landscape change in the lower Omo valley, southwestern Ethiopia: Burning patterns and woody encroachment in the savanna. *Journal of East African Studies*, pp. 108–128.
Hendrickson, D., J. Armon, and R. Mearns. 1998a. *Conflict and vulnerability to famine: Livestock raiding in Turkana, Kenya, Drylands Programme*. International Institute for Environment and Development no. 80.
Hendrickson, D., J. Armon, and R. Mearns. 1998b. The changing nature of conflict and famine vulnerability: The case of livestock raiding in Turkana District, Kenya. *Disasters* 22(3): 185–199.
Hogg, R. 1982. Destitution and development: The Turkana of North West Kenya human nutrition. *Disasters* 6(3): 164–168.
Hopson, A.J. 1975. *Preliminary notes on the birds of the Lake Turkana area*, pp. 17.
Lamphear, J. 1992. *The scattering time: Turkana responses to colonial rule*, 308 pp. Oxford: Oxford University Press.
Lydall, J. and I. Strecker. 1979. The Hamar of Southern Ethiopia. 3 Vols. Arbeiten aus dem Institute fur Völkerkunde der Universität zu Göttingen—Bendiz. Hohenschäftlarn: Klaus Renner Verlag.
Mark, P., and S. Tornay. 1992. Les Fusils jaunes: Générations et politique en pays Nyangatom (Ethiopie). *African Studies Review* 46(1): 196–198.
Mburu, N. 2001. Military decline of the Turkana of Kenya 1900–2000. *Nordic Journal of African Studies* 10(2): 148–162.
Mburu, N. 2003. Delimitation of the elastic Ilemi Triangle: Pastoral conflicts and official indifference in the Horn of Africa. African Studies Quarterly (Sprg).
McCabe, J.T. 2004. Cattle bring us to our enemies: Turkana ecology, politics, and raiding in a disequilibrium system. Michigan: University of Michigan Press.
Savary, C. 2003. Generations and politics among the Nyangatom of Ethiopia. *Anthropos* 98(1): 278–280.
Schroder, P. 2003. Generations and politics among the Nyangatom of Ethiopia. *Anthropos* 98(1): 267.
Skoggard, I., and T.A. Adem. 2010. From raiders to rustlers: The filial disaffection of a Turkana age-set. *Ethnology* 49(4): 249–262.
Smucker, T.A. 2006. Cattle bring us to our enemies: Turkana ecology politics. *Professional Geographer* 58(2): 231–233.
Sobania, N. 2011. The formation of ethnic identity in South Omo: the Dassenech. *Journal of Eastern African Studies* 5(1): 195–210.
Strecker, I. A. 1988. The Social Practice of Symbolization: An Anthropological Analysis. Monographs on Social Anthropology, no. 60. London. Athlone Press.
Teqegn, M. 2003. The marginalization of pastoral communities in Ethiopia. *Indigenous Affairs* 4: 32–37.
Tornay, S. 1979. Armed conflicts in the lower Omo Valley, 1970–1976: Nyangatom Society. *Senri Ethnological Studies* 3: 97–117.

Tornay, S. 1980. The hamar of Southern Ethiopia, conversations in Dambaiti by Strecker, I. Homme 20(2):99–108.
Tornay, S. 1981. The Nyangatom an outline of their ecology and social organization in peoples and cultures of the Ethio-Sudan Borderlands, pp. 137–178, Ed. M.L. Bender. East Lansing, Michigan: African Studies Center, Michigan State University.
Tornay, S., K.W. Butzer, J.A. Kirchner, Fuller, G.K., Lemma, A., Haile, T., Bulto, T., Endeshaw, T., Abebe, M., Eshete, S., Petros, G.P. and D.H. Van. 1997. *The growing power of the Nyangatom*. Chicago: University of Chicago.
Turton, D. 1977. Response to drought: The Mursi of Southwestern Ethiopia. *Disasters* 1(4): 275–287.
Turton, D. 1991. Warfare, vulnerability and survival a case from Southwestern Ethiopia. *Disasters* 15(3):254–264.
Turton, D. 1995. Pastoral livelihoods in danger, cattle disease, drought, and wildlife conservation in Mursiland, S.W. Ethiopia. pp. 50. Oxfam Research Paper 12, Oxfam (UK and Ireland).
Turton, D. 2013. Aiding and abetting: UK and US complicity in Ethiopia's mass displacement. London: Think Africa Press.
Vannutelli, L. and Citerni, C. 1887. Relazione preliminare sui risulati geografici della deconda spedizione Bottego. *Society Geographical Italiana*, Series 3, 10: 320–330.
Von Hohnel, L. 1938. The Lake Rudolf region: Its discovery and subsequent exploration, 1888–1909. *Journal of Royal African Society, London* 37(16–45): 206–26.
Wood, R.B. and J.F. Talling. 1988. Chemical and algal relationships in a salinity series of Ethiopian inland waters. *Hydrobiologia* 158: 29–67.
Yuretich, R.F. and T.E. Cerling. 1983. Hydrogeochemistry of Lake Turkana, Kenya: Mass Balance and Mineral Reactions in an Alkaline Lake. Geochimica et Cosmochimica Acta, vol. 47 (Jun); 1099–1109.

Open Access This chapter is distributed under the terms of the Creative Commons Attribution-NonCommercial 2.5 International License (http://creativecommons.org/licenses/by-nc/2.5/), which permits any noncommercial use, duplication, adaptation, distribution and reproduction in any medium or format, as long as you give appropriate credit to the original author(s) and the source, provide a link to the Creative Commons license and indicate if changes were made.

The images or other third party material in this chapter are included in the work's Creative Commons license, unless indicated otherwise in the credit line; if such material is not included in the work's Creative Commons license and the respective action is not permitted by statutory regulation, users will need to obtain permission from the license holder to duplicate, adapt or reproduce the material.

Components of Catastrophe: Social and Environmental Consequences of Omo River Basin Development

5

> **Abstract**
> The Gibe III dam and its associated agricultural development would cause radical reduction of Omo River flow and inflow to Lake Turkana, as well as elimination of the Omo River annual flood—all essential to the survival of a half million residents of the lower Omo basin and the Lake Turkana region. These major changes would destroy the Omo riverine natural resource systems—eliminating 'last resort' grazing lands for livestock, flood recession agriculture and fishing habitats throughout the lowermost Omo basin. The impending destruction of indigenous survival systems is heightened by the Ethiopian government's expropriation of tens of thousands of villagers for large-scale, irrigated commercial agricultural enterprises, accompanied by political repression of communities through-out the region. Pastoralists and fishers residing near the western shoreline of Kenya's Lake Turkana also face economic collapse—primarily due to radical lake level drop causing destruction of fish habitat, lakeside grazing for livestock and potable water. As in the lower Omo basin, these conditions would produce massive scale hunger along with widespread disease. Rapid escalation of armed conflict in the cross-border region would ensue as ethnic groups battle over vanishing food sources.

Radical Reduction of River and Lake Waters by Omo Basin Development

➢ **If the Gibe III dam is completed and brought into operation, it will radically reduce the Omo River's downstream flow volume as well as the inflow to Lake Turkana, by at least 60–70 %, according to ARWG physical scientists** Fig. 5.1). This does not take into account:

- Seepage from the reservoir with a drawn-out, if not indefinite, filling period—due to the highly fractured nature of basalts and other volcanic rocks at the reservoir.
- Abstraction of Omo River waters by the planned system of GOE and commercial irrigation agricultural schemes along the lower Omo River. Utilizing estimates of the extent of irrigation, including figures from Human Rights Watch and the Oakland Institute, the consultant for the AFDB (2010) Lake Turkana assessment later estimated that abstraction of river water for the planned irrigation agricultural enterprises would cause up to a *50 % reduction* of Lake Turkana's water inflow from the river with a *lake level drop of 20 m or more* (Avery 2012).
- Failure to implement artificial flooding because of reservoir fill delay and GOE prioritization of electricity generation over the release of reservoir water for downstream indigenous economies.
- The sustainment of an artificial flood program is actually disclaimed by the GOE itself (see Chap. 6).

The drop in lake level caused by the 60–70 % Omo River flow volume loss during closure of the Gibe III reservoir and early operation of the dam would be in *addition* to rather than "within the range" of the lake's annual fluctuation. The latter is incorrectly asserted by the GOE downstream impact assessment (2009b) and the two major development bank impact assessments (EIB 2010; AFDB 2010). Annual fluctuations in lake level are about 1–1.5 m, but longer-term changes in the lake level are much larger. For example, the lake level dropped 10 m between 1975 and 1994.

A drop in the lake level of at least 10 m during the first years of reservoir filling and operation is a far more likely scenario. The maximum depth of the lake is 109-m, with a mean depth of 30-m. Its shoreline zone is markedly shallow in most of its central and northern zones. Figure 7 indicates the progressive retreat of Lake Turkana in relation to water level drop.

An ARWG physical scientist has calculated the probable drop in the level of Lake Turkana during reservoir fill and operation. The scientist's calculations relied heavily on a paper by Cerling (1986) concerning a mass-balance approach to sedimentation in recent Turkana basin history. Considerations included the following:

- The highly fractured nature of the basalt rocks in the reservoir rocks, suggesting a high level of *seepage* into the rock structure. By ignoring seepage losses, the official reports estimate a reservoir-filling period of two to three years, which is unrealistically short.
- The hydrology of the Omo River basin is misrepresented in the official reports. GOE and development bank reports suggest that a greater portion of the Omo River's discharge into Lake Turkana comes from the region below the Gibe III dam than is actually the case. In general, waters from the Hamar mountain range (Fig. 1.3) or the Omo basin lowlands do not reach the Omo River. Instead, these waters evaporate in flood basins and seasonally inundated pans.
- The *"huge evaporation losses"* from *"excessive"*, or *"uncontrolled flooding"* repeatedly asserted in the GOE's downstream impact assessment (GOE 2009b)—and referred to in the development bank impact assessments—do not in fact occur.
- Areas of sub-basins of the Omo River were measured from satellite imagery, with precipitation in each estimated from available maps and tables. The evaporation rate from the surface of Lake Turkana was calculated.

A 7-m drop is in lake level is likely during the first years of reservoir fill and operation, depending on the amount of river abstraction for irrigation, regional weather patterns and other factors. An increase in the concentration of ions in the lake from about 2330 to about 2800 mg/l is predicted.

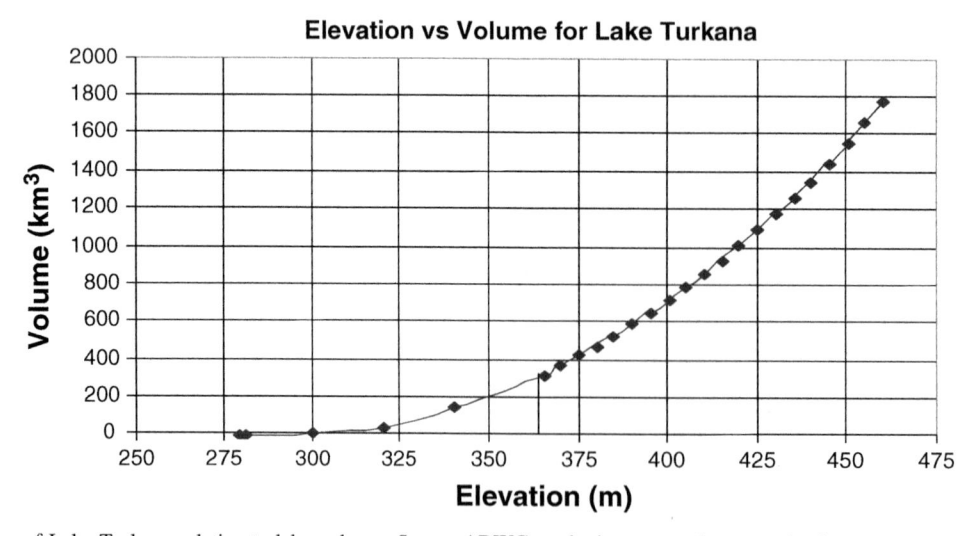

Figure 5.1 indicates the relationship between elevation and volume for Lake Turkana.[1]

Fig. 5.1 Elevation of Lake Turkana relative to lake volume. *Source* ARWG geologist, personal communication.

---

[1]This calculation by the ARWG physical scientist was included as Appendix A in Carr (2012).

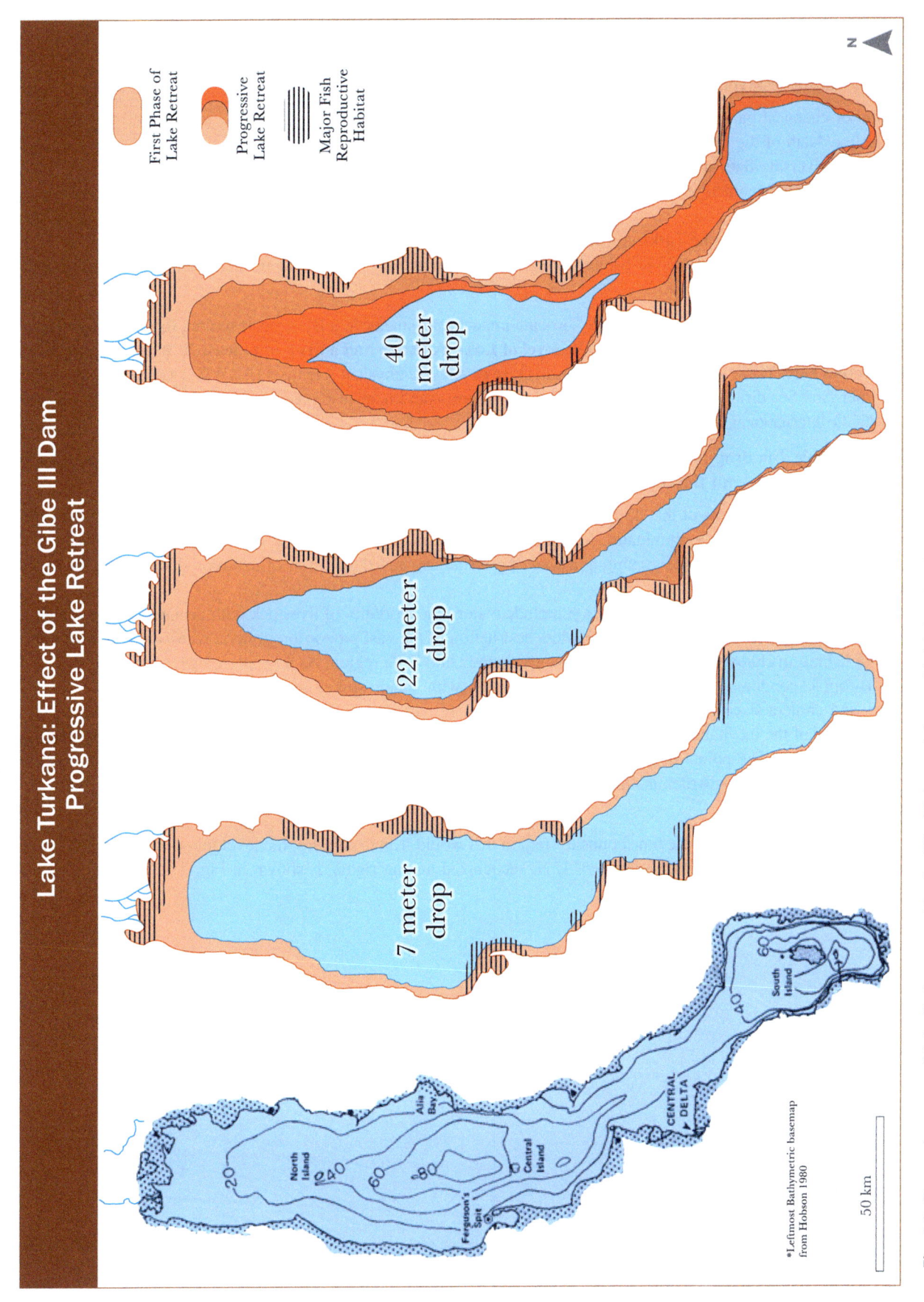

Radical Reduction of River and Lake Waters by Omo Basin Development 77

Fig. 5.2 Progressive retreat of Lake Turkana caused by the Gibe III dam and dam enabled irrigated agriculture (Base map from Hopson 1980)

Figure 5.2 indicates the progressive retreat of Lake Turkana relative to reduced lake volume, based on bathymetric measures by Hopson (1982). The maximum depth of the lake is 109 m, with a mean depth of 30 m. Its shoreline zone is markedly shallow in central and northern zones.

Even during a realistic reservoir-filling and initial dam closure period, these effects include:

- Disruption of chemical balance and biological systems—including fisheries.
- Rapid decline/death of the Omo riverine forest and woodland.
- Drying out of riverside flats and the Omo delta—including flood recession, grazing and fishing areas.
- Cessation of the Omo's annual pulse of freshwater and sediment into the lake, with salinity increase and radical reduction of critical nutrients.
- Major southward retreat of Lake Turkana with desiccation of its northernmost zone, bays and shoreline areas (Fig. 5.2).
- Destruction of major fish reproductive habitat along the river, within the modern Omo delta and in Lake Turkana's central and northern shoreline areas.

The former consultant for the AFDB's 2010 Lake Turkana impact assessment, in a later report written for the University of Oxford (Avery 2012) **calculated a 7-m drop in lake level of Lake Turkana and a drop of at least 20 m accompanying a 64 % abstraction rate**. These estimates are in line with those of ARWG physical scientists (ARWG 2009). The scenario of creating another Aral Sea disaster is raised in the 2013 Avery report—an image also presented in an article produced by the U.S. based NGO International Rivers (2013).

➢ **Even the predictable 7-m drop in lake level during the reservoir filling period and early dam operation would have catastrophic effects throughout the region**.

- The environmental dimensions of this destruction include: death of the riverine forest, desiccation of the Omo delta and northern end of Lake Turkana, elimination of flood recession agriculture along the river and throughout the delta, destruction of riverine and lakeside grazing and watering resources, and destruction of fish reproductive and life cycle habitat.
- The livelihood and associated human destruction include major new mortality of livestock with decimation of remaining herds for countless pastoral villages and households, destruction of the vast proportion of flood recession agriculture and collapse of fishing livelihood around Lake Turkana and in the Omo delta region. As a result, indigenous residents would face catastrophic conditions of malnutrition and starvation, with major spikes in disease—even epidemic levels of dysentery and cholera since Turkana levels of these diseases are recorded as some of the highest in East Africa.
- A regional view of the inevitable expanding *interethnic* armed conflict in the already heavily weaponized border zone is indicated in Fig. 5.3. The points of already existing conflict, along with predictable locales and directions of conflict expansion—identified by this writer and the SONT research team in conversations with community elders—are indicated in Fig. 5.3.[2]

➢ A summary of the above impacts in the transboundary region that would lead to catastrophic level human and environmental destruction, involving GOE and GOK violation of U.N. recognized human rights, is shown in Fig. 5.4.

---

[2]Human casualties would far fewer in the more upstream. This 'highland' portion of the lower Omo basin is relatively removed from the tri-nation border zone and has sufficient rainfall to support rainfed agriculture. Like the transboundary region, however, land expropriation and displacement are producing major socioeconomic crisis.

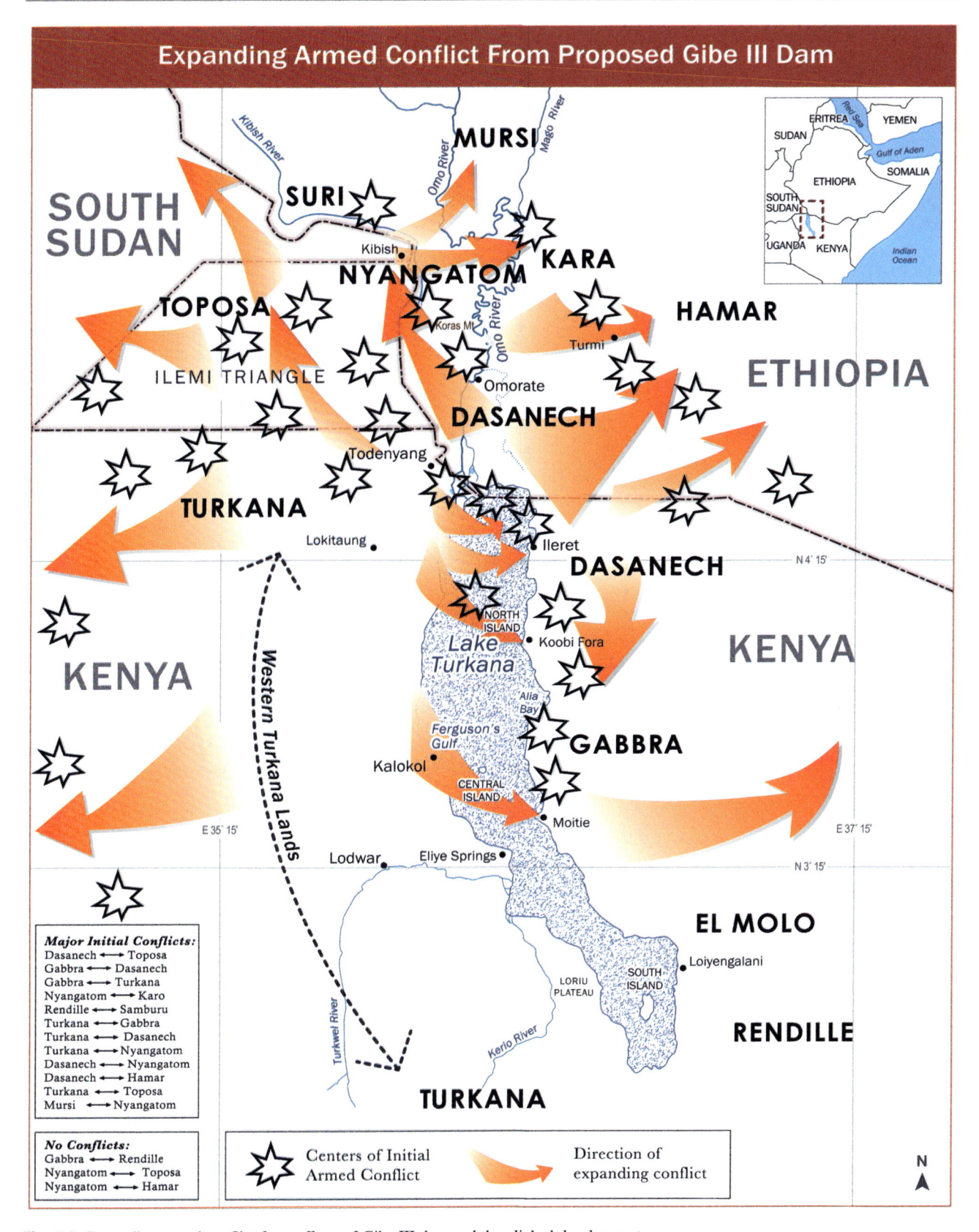

Radical Reduction of River and Lake Waters by Omo Basin Development

Fig. 5.3 Expanding armed conflict from effects of Gibe III dam and dam-linked development

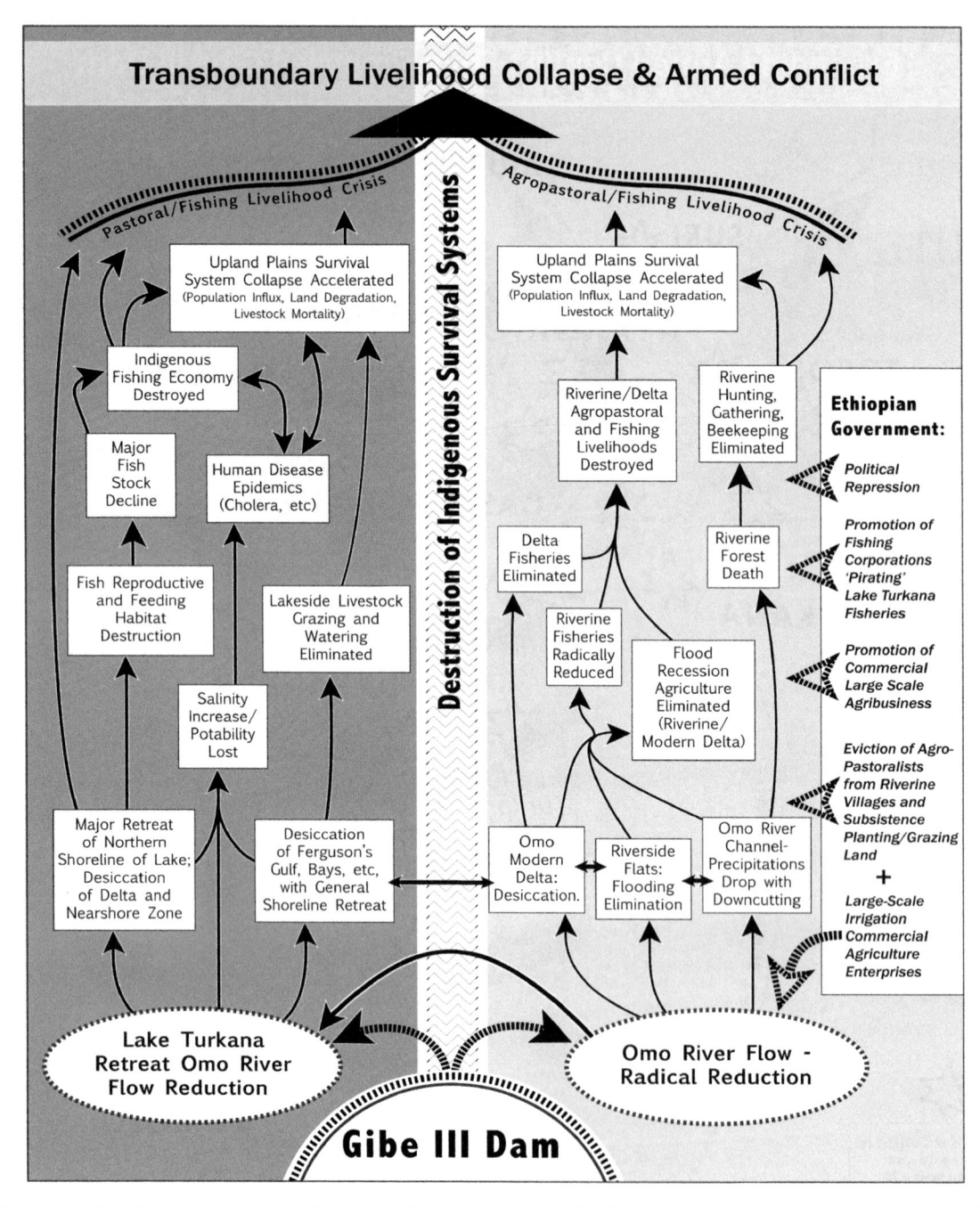

Fig. 5.4 Summary of the humanitarian catastrophe and conditions for armed conflict from the Gibe III dam and its linked large-scale irrigated agricultural development

Consequences for the Lowermost Omo River Basin

1. Omo River flow volume reduction of at least 60–70 % during the reservoir filling period and early dam operation, not including additional river water abstraction/diversion by GOE and private company irrigation agriculture; with elimination of the river's annual flood.
2. Elimination of Omo River flooding in the modern delta (about 500 km^2) and southward migration of Lake Turkana's northern shoreline by at least 8–10 km (Fig. 5.2); desiccation of the modern Omo delta and northern Lake Turkana waters.
3. Cessation of annual flooding of waterside point bars (and sand/silt spits) along the entire lower Omo River, with elimination of vegetation communities that are adapted to this flood-driven habitat replenishment.

4. Additional downcutting of the Omo River channel in adjustment to the new lower base level of the lake, with channel scouring, riverbank erosion and disruption of oxygen and nutrient levels in the river—all with major impacts on water quality in Lake Turkana.
5. Destruction of the livelihood of at least 80,000 residents—primarily Dasanech and Nyangatom—relying on flood recession agriculture along the lower Omo River and in the modern Omo delta region, along with tens of other indigenous pastoralist, fishers and others whose survival depends on trade for grain and other agricultural product from farms at the Omo.
6. Elimination of fish reproductive and critical life cycle habitat throughout the modern Omo delta and northern Lake Turkana waters, as well as in Omo riverine environments impacted by major disruption of nutrient, sediment and oxygen content.
7. Destruction of Dasanech and Nyangatom artisanal fishing—the major livelihood for many of the poorest communities.
8. Desiccation of last resort grazing for livestock along the river in response to extended drought periods, intense overgrazing and other hardship conditions—causing major spikes in livestock mortality and new dependence on the cultivated areas at the same time that they are being eliminated.
9. Radical decrease in floodwater quantity and moisture *residence time* (the duration of riverine soil moisture retention) in Omo River natural levee soils, causing destruction of the riverine forest—the last pristine such forest in semi-arid Africa.
10. Destruction of riverine forest/woodland-based secondary food production for local residents, including wild food gathering, beekeeping, hunting and other activities essential to the survival of many of the most impoverished indigenous communities in the lowermost basin.
11. Elimination of ground cover (mostly grasses and forbs) along riverbanks, on natural levee back slopes and in active delta environments—vegetation that requires sustained 'residence time' of substantial soil moisture during the river's annual flood.
12. Lowering of the current limited but critical ground water recharge in the dryland plains lateral to the lower Omo River channel, causing grassland deterioration, increased susceptibility to overgrazing by livestock and accelerated desertification.
13. Major 'outmigration' by tens of thousands of households in response to their livelihoods collapsing in the Omo riverine and northern lake region—a process newly impacting neighboring pastoralists who are themselves struggling with similar radical changes.
14. Sharp increase in malnutrition throughout the region, with conditions of starvation taking hold in both riverine and regional dry land plains areas. Estimates by U.N. and non-governmental aid organizations for the region *presently* hover around 40 %, not including years of extreme stress, and sometimes famine.
15. Major spike in disease incidence and conditions promoting widespread increases in malaria, dysentery and other diseases, facilitated by large areas of stagnant water along the river, in canals and backup channels or pools, with severe threat of cholera and other diseases.
16. Major escalation and initiation of armed conflict among ethnic groups, particularly among the Dasanech, Nyangatom, and Turkana in the Ilemi, Kibish River, Koras Mountain. and Omo delta regions, as well as with the Hamar to the east (Figs. 1.3 and 5.3).

Consequences for the Lake Turkana Region

1. Initial lake level drop of at least 2–3 m during reservoir filling and early operation, with continuing lake retreat to at least 7 m within 5 years. This would create:
   - An extended reservoir-filling period due to seepage through fractured volcanics in the reservoir locale.
   - Inadequacy of the GOE's planned artificial flood and low likelihood of its implementation.
   - Reduced inflow from Ethiopia's major diversion of the river for irrigation agriculture.
2. Southward migration of the northern shoreline of Lake Turkana by about 8–10 km further into Kenya during the earliest phase of the reservoir-fill period (Fig. 5.2) with continued shoreline regression likely under reduced inflow conditions. This ARWG prediction is in agreement with the EIB's (2010) report noting that such a retreat is likely.
3. Desiccation of the modern Omo River delta at the northern end of the lake as well as the entire northern shoreline zone (see Chaps. 7 and 9) and all shallow areas along the western and eastern shorelines.
4. Elimination of the Omo River's annual flood with its pulse of freshwater, sediment and nutrients—contributing to destruction of major fish reproductive habitats in the modern delta, Ferguson's Gulf, Alia Bay and other shoreline areas (Fig. 5.2).

5. Salinity increase in lake waters that are already borderline in potability (see above). Lake Turkana is one of the most saline large lakes in Africa. Increased salinity threatens its availability for consumption by hundreds of thousands of local residents, as well as by livestock and wildlife watering.
6. Collapse of fish stocks critical for indigenous fishing communities due to the combination of lake withdrawal and destruction of reproductive and feeding habitats.
7. Destruction and likely elimination of the northern Turkana's artisanal fishing economy—their major last resort means of survival.
8. Collapse of the exchange system fishing, agropastoral, agricultural and pastoral communities, leading to the destruction of food related reciprocity relations in the region.
9. Elimination of most existing shoreline vegetation essential for last resort grazing/browsing by livestock—sparking a radical increase in livestock disease.
10. Loss of remaining grassland vegetation and increased (often irreversible) establishment on unpalatable invasive species, with lowered water table in dryland plains adjacent to Lake Turkana. (These species include *Prosopis juliflora* and other species toxic to livestock.)
11. Destruction of major portions of the northern Turkana pastoral economy due to elimination of water and grazing resources, forcing major new migration from upland plains to the lake, just as fishing there is collapsing.
12. Southern movement of villages along Lake Turkana's western shoreline—conforming to longer term pattern (Fig. 4.6), as fishing and fishing/pastoral Turkana desperately search for relief aid or any last option available in towns (for example, Kalokol and even Lodwar) or at Ferguson's Gulf. Likely swelling of villages and population around Ferguson's Gulf, although the Gulf would be rapidly drying.
13. Epidemic-level disease outbreaks at Lake Turkana, including cholera—already well known in the region, especially around Kalokol/Ferguson's Gulf (Figs. 1.4 and 5.2) and other densely populated locales. Susceptibility to new levels of malaria, schistosomiasis and cholera due to major new areas of stagnant water. All of these health crises would be heightened by the extreme malnutrition conditions.
14. Region-wide extreme malnutrition and starvation conditions among the Turkana, with large populations attempting to migrate to towns or internally displaced persons (IDP) camps.
15. Major interethnic armed conflicts, exacerbated by the widespread availability of arms and the upward spiral of indigenous survival system collapse in the region west of Lake Turkana and in the lower Omo basin (Figs. 5.3 and 5.4).

Consequences for the Ilemi Triangle and the Broader Region

1. Major increase in overgrazing and ecological deterioration throughout the dryland plains of the Ilemi Triangle, due to a large influx of people fleeing hunger conditions along the Omo River and around Lake Turkana (Fig. 5.5).
2. Immediate escalation of existing patterns of interethnic armed conflict due to overcrowding by ethnic groups and competition for vanishing pasturage and water. Traditional patterns of interethnic conflict remain, but with greater inclusiveness as stresses on all groups within the region increase. For the Dasanech, conflicts are primarily with their Nyangatom, Hamar, and Turkana neighbors, but also with the Toposa in the Ilemi region and the Gabbra along the eastern shores of Lake Turkana. For the northern Turkana, hostilities most frequently erupt with the Dasanech and Nyangatom. For Turkana fishers camping along the lake's eastern shoreline, hostilities with the Gabbra also occur.
3. Rapid geographic spreading of armed conflict in the cross-border region and internally within the three nations concerned—an inevitable outcome in view of rampant arms availability in the region. AK-47 rifles and other types of weaponry are ever expanding through Kenyan arms merchants and multiple sources—military and otherwise—in the region. Many of the region's residents and outside observers regard the national governments as tacitly supporting the arming of "their own" ethnic groups—as 'proxy forces' in the contested Ilemi region, where oil and gas exploration is also active (Appendix A).
4. Increased militarization in the region by both the Kenyan and Ethiopian governments. This process accelerates as governments react to interethnic conflicts and as they, together with domestic and international investors, advance plans for 'development,' with de facto assertion of control over the region's resources.
5. Cross-border livelihood collapse and conflict (Fig. 5.4).

Fig. 5.5 Ilemi triangle border area with Ethiopia. Lightly grazed Ilemi grasslands on relict beach ridge to the *left*; heavily grazed area with pastoral villages (present and abandoned) to the *right*

These developments include expropriation of indigenous lands for major scale diversion of Omo waters for irrigated commercial agricultural schemes with major infrastructure construction including canals, irrigation systems, roads, and bridge-building. For the region more broadly, they include oil and gas exploration—in Lake Turkana and throughout much of the vast surrounding land areas where petroleum deposits are known or predicted, and land privatization. Within Ethiopia, these development policies are closely tied to the GOE's policies of political repression and violation of U.N. defined human rights—the right to water central among them. Within Kenya, the government's failure to protect its indigenous population from the effects of the Gibe III dam and its attendant agricultural development also violates a major indigenous population's human right to water. Although the matter of severe political repression in northern Turkana remains ambiguous, widespread fear of government reprisals is evident.

> Political repression and a climate of increasing hostility accompany militarization in such remote pastoral regions—a reality strongly in evidence in major pastoral areas in the Horn and East Africa. All indications are that increased militarization by the Kenyan and Ethiopian governments will only further the destabilization of the region rather than to promote 'peace,' as the governments and a host of international aid organizations continually assert in public statements.

Literature Cited

Africa Resources Working Group (ARWG). 2009. A Commentary on the environmental, socioeconomic and human rights impacts of the proposed Gibe III Dam in the lower Omo River basin of Southwest Ethiopia. http://www.arwg-gibe.

African Development Bank (AFDB). (2010, November) S. Avery, Assessment of Hydrological Impacts of Ethiopia's Omo Basin on Kenya's Lake Turkana Water Levels. Final report. 146 pp.

Avery, S.T. 2012. Lake Turkana and the lower Omo: Hydrological impacts of Gibe III and lower Omo irrigation development, vols. 1 and 2, African Studies Centre, University of Oxford. 92 pp.

Avery, S.T. 2013. What future for Lake Turkana? The impact of hydropower and irrigation development on the world's large desert lake. African Studies Centre, University of Oxford.

Carr, C.J. 2012. Humanitarian catastrophe and regional armed conflict brewing in the border region of Ethiopia, Kenya and South Sudan: the proposed Gibe III Dam in Ethiopia, Africa Resources Working Group (ARWG) (Dec 2012), 250 pp. https://www.academia.edu/8385749/Carr_ARWG_Gibe_III_Dam_Report.

Cerling, T.E. 1986. A mass-balance approach to basin sedimentation: constraints on the recent history of the turkana basin. *Palaeogeography, Palaeoclimatology, Palaeoecology* 54: 63–86.

European Investment Bank (EIB). 2010. Sogreah consultants, independent review and studies regarding the environmental and social impact assessments for the Gibe III hydropower project, Final Report, Mar 2010, 183 pp.

Government of Ethiopia (GOE), Ethiopian Electric Power Corporation (EEPCO). 2009b. Agriconsulting S.P.A., mid-day international consulting, level 1 design, environmental and social impact assessment, additional study of downstream impacts. Report No. 300 ENV RAG 003B.

Hopson, A.J. 1982. Lake Turkana, a report on the findings of the lake Turkana project, 1972–1975, Vols. 1–6, funded by the Government of Kenya and the Ministry of Overseas Development, Overseas Development Administration.

International Rivers. 2013. Gibe III's impacts on Lake Turkana. East Africa's "Aral Sea" in the making? 10 Jan 2013. www.internationalrivers.org.

Open Access This chapter is distributed under the terms of the Creative Commons Attribution-NonCommercial 2.5 International License (http://creativecommons.org/licenses/by-nc/2.5/), which permits any noncommercial use, duplication, adaptation, distribution and reproduction in any medium or format, as long as you give appropriate credit to the original author(s) and the source, provide a link to the Creative Commons license and indicate if changes were made.

The images or other third party material in this chapter are included in the work's Creative Commons license, unless indicated otherwise in the credit line; if such material is not included in the work's Creative Commons license and the respective action is not permitted by statutory regulation, users will need to obtain permission from the license holder to duplicate, adapt or reproduce the material.

The Rush to Rationalize: Public Policies and Impact Assessments

They came to our village and us that there would be a wonderful new development that would bring us schools and more fish and shillings and a better life. We said that we heard about the dam and that the lake would go down and that we cannot lose our fish who mostly are at the shore. We don't want the dam. They said that we were wrong and that the dam would be good for us. Then they went away. We do not know what they did after that.

[Turkana man in shoreline village following AFDB consultants' visit]

> **Abstract**
> The common interests of the Ethiopian government, international development banks and global consulting firms in promoting, implementing and legitimizing the Gibe III dam and its associated development are starkly apparent in environmental and socioeconomic studies and impact assessments. None of them address the actual impact area of the Gibe III dam project—namely, one including human and environmental effects of the project in the tri-nation Ethiopia/Kenya/South Sudan transboundary region. The Ethiopian government (GOE) downstream impact assessment is invalidated by its major omissions, misrepresentations, and fabrications. These failings include false assertions of 'disastrous' Omo River floods 'requiring' river regulation floods which do not occur; misrepresentations of Omo basin environmental hydrology and socioeconomy; and exclusion of the impacts of Gibe III dependent, large-scale irrigated agriculture. Global consulting industry assessments of the dam commissioned by the European Investment Bank and African Development Bank (AFDB) do not significantly challenge GOE failings. Instead they offer primarily 'suggestions' for future consideration, rather than the identification of analysis that must be conducted before approval of any impact assessment. They pave the way for the World Bank and AFDB to violate their own human rights protocols by funding an infrastructure to allow Ethiopia to export hydrodam generated electricity.

Launching the Gibe III Dam—And a System of Bias

Any environmental consequence has to be recognized early and taken into account in project design. (GOE 2009b)

> This statement by the Ethiopian government is directly at odds with the its apparent predetermined decision that contracting and initiation of construction of the Gibe III dam could disregard the funding procedures and safeguard requirements of international aid organizations, that the Gibe III dam would be built whatever the cost to the livelihoods of the nation's own indigenous population, and that any impact assessments produced during construction would basically endorse, or legitimize the project.

➢ **The most common statements of 'justification' for the more than $1.7 billion Gibe III project that have been issued have been the "need for electricity" by the Ethiopian population—including its poorest elements—and the 'urgency' for regulation of an allegedly "uncontrolled" and "disastrous" Omo River.** These have been repeatedly issued by the GOE (including by the Prime Minister's office, EEPCO officials and Ministers) and by development agencies involved in various stages of the project. Both statements are based on false premises. In the first instance, a high proportion of Gibe III power generation has been slated for export since the inception of planning. Moreover, the cost of electricity would inevitably be far too high for a vast proportion of the population. In the second, no such 'disastrous' floods exist. These issues are detailed below.

- **As with the Gibe I and Gibe II projects, the GOE awarded a no-bid, 'turn-key' contract for Gibe III construction to its long-term global associate, Salini Costruttori—exempting the firm from any oversight.** The GOE (with Salini) contracted two Italian-based global consulting firms for environmental and socioeconomic impact assessments. Both reports were produced in 2009. They are described below as so fundamentally flawed that they must be considered unacceptable—or invalid.
- **The Ethiopian government began its construction of the Gibe III dam construction at the beginning of 2007—more than two years prior to environmental or socioeconomic impact assessment.** This action was in clear violation of its own laws and procedures as well as those of the international development banks from which it was to seek funding. Chapter 2 outlined the political and economic context for river basin development in prior decades within the country, including as a lead-up to the Gibe III project.

The GOE/Salini initiation of construction and after-the-fact impact assessments were done without oversight by its alleged regulatory agency, the Environmental Protection Authority (EPA)—a new agency formed under advice from international aid agencies. Directly responsible to the Prime Minister's office, under whose auspices the project had been initiated, the Authority was incapable of an 'independent' evaluation. The EPA later stated that it had 'approved' the project and repeated the Executive Office's insistence that the Gibe III dam is a matter of "national priority" and "urgency".

➢ **GOE violations of required procedures notwithstanding, the GOE requested funding from the European Investment Bank, the World Bank and the African Development Bank for its approximately USD 1.7 billion project.** Despite their long-term and comprehensive support of such developments (see Chap. 2), the GOE's violation of their required procedures for consideration of funding placed the banks in a difficult position. The situation was particularly problematic since the Gibe III dam was designed as an important first phase of plans for an East Africa-wide energy network and industrial development program.

The banks continued with their means of supporting the dam, however, in multiple ways:

- Support for GOE infrastructure projects.
- Funding of Ethiopian executive and administrative agencies, including EEPCO which is directly in charge of the project.
- 'Direct funding' for "protection of basic services"—a highly fungible type of funding that is directed to the lower Omo basin, among other areas of Ethiopia.
- Funding for construction of a major transmission line (primarily by the World Bank and the AFDB) for electricity export from Ethiopia to Kenya as part of the initiation of a multi-billion dollar 'energy highway' system for eastern Africa, the EAPP. Despite declaimers The Gibe III would clearly contribute to this power export system.

> The transmission line—in handling electricity from the overwhelmingly largest source of power in the Ethiopian region—the Gibe III dam—has *cumulative effects* with the hydrodam and therefore must be included in any impact assessment, including of the transmission line itself. No such comprehensive assessment was produced by the development banks, nor by the GOE.

➢ **EEPCO released the reports from both contracted firms in 2009. The agency released an ESIA for the dam 'project area' in January of that year: "Gibe III Hydroelectric Project, Environmental and Social Impact Assessment"** (GOE 2009a). This report was prepared by the Italy based global consulting firms, Centro Electrotecnico Sperimentale Italiano S.p. A (CESI)/Mid-Day International Consulting Engineers—companies with longstanding connections to both the GOE and Salini. 'Scoping' stipulations in CESI's contract, however, limited this environmental and socioeconomic impact assessment (ESIA) to the area proximal to the dam and the 'impound area' for the reservoir. This GOE assessment effectively dismissed the possibility of serious downstream environmental and social concerns as a consequence of the Gibe III dam.

A few of the GOE's caveat statements in its assessment are sufficient to demonstrate its disregard for the downriver impacts of the dam—certainly for the hundreds of thousands of indigenous people residing there, as well as their environs:

- *There will be 'negligible impact' on livelihood bases of the population.*
- *There are no tribal people in the project area whose traditional lifestyles could become compromised through the implementation of the proposed hydropower project.*
- *No adverse direct or indirect impacts are anticipated in respect of sensitive habitats.*

A second ESIA for the 'downstream zone' from the Gibe III dam was released by the GOE later in 2009. This assessment is one with extensive major omissions, misrepresentations and fabrications, the dimensions of which are indicated in Fig. 6.1. Another global consulting firm—Agriconsulting of Italy, in association with MDI Consulting Engineers—prepared this assessment, entitled *Gibe III Hydroelectric Project: Environmental Impact Assessment—Additional Study on Downstream Impact* (GOE 2009b).[1]

The omissions, misrepresentations and fabrications indicated in the figure apply to both baseline and 'empirical' information presented and are detailed in the sections below. They include, for example, the exclusions or gross distortion of:

- The transboundary character of the Gibe III's impact system, including not only the lower Omo River basin but Kenya's Lake Turkana region, and the contested Ilemi Triangle Region with the extreme southeastern portion of South Sudan.
- Major seismic issues in the Gibe III dam region, including plausible catastrophic destruction from earthquake and landslide events.
- Fundamental hydrologic characteristics of the region, especially the Omo River's flow volume and Omo River inflow to Lake Turkana—and the impacts of their radical reduction from the effects of the Gibe III dam and the large-scale irrigated agricultural systems planned, the characteristics of multiple watercourses in the lower basin, and the nature of Omo River flooding patterns.
- Hydrologic patterns of Lake Turkana and critical freshwater inflow from the Omo River.
- Livelihood systems of pastoral, agropastoral and fishing ethnic groups within the lower Omo River basin and around central and northern Lake Turkana: their customary resource tenure patterns, present economic status, and vulnerability to radical destruction of their survival systems from river and lake reduction by the Gibe III dam.

The Myth of Flood 'Disasters' as Rationale for Megadam Development

> The crisis confronting hundreds of thousands of indigenous peoples struggling to survive along the Omo River and the shorelines of Lake Turkana is too little flooding, not excessive flooding.

➢ **The rush to rationalize the proposed Gibe III dam was starkly evident in the Ethiopian government's reports of 'catastrophic losses of human life and property (livestock, particularly) from an alleged Omo River flood "disaster" in August of 2006.** The GOE's numerous reports and requests for international financial and material aid 'for relief efforts' repeatedly declared that at least 350–1000 people had drowned (depending on the specific report) and at least 3000–4000 livestock had been swept away. Hundreds of press releases and declarations from the GOE's information permeated African and international media.

The GOE has continually asserted that the "catastrophic losses" of 2006 were just the most recent of such events from the Omo River's alleged "excessive" and "destructive" flooding with repeated "major losses" of human life and livestock. Based on this fallacious assumption and the complicity of international development agencies and investing corporations, the GOE proceeded with the clear assumption that the Omo River must be regulated, and that this would occur through Africa's largest hydrodam, the Gibe III. As stated above, construction of Gibe III infrastructure was already underway in 2006, without any impact assessment.

---

[1] Unless otherwise specified, references to the GOE's environmental and socioeconomic impact assessment (ESIA) refer to the GOE's downstream impact assessment (GOE 2009b).

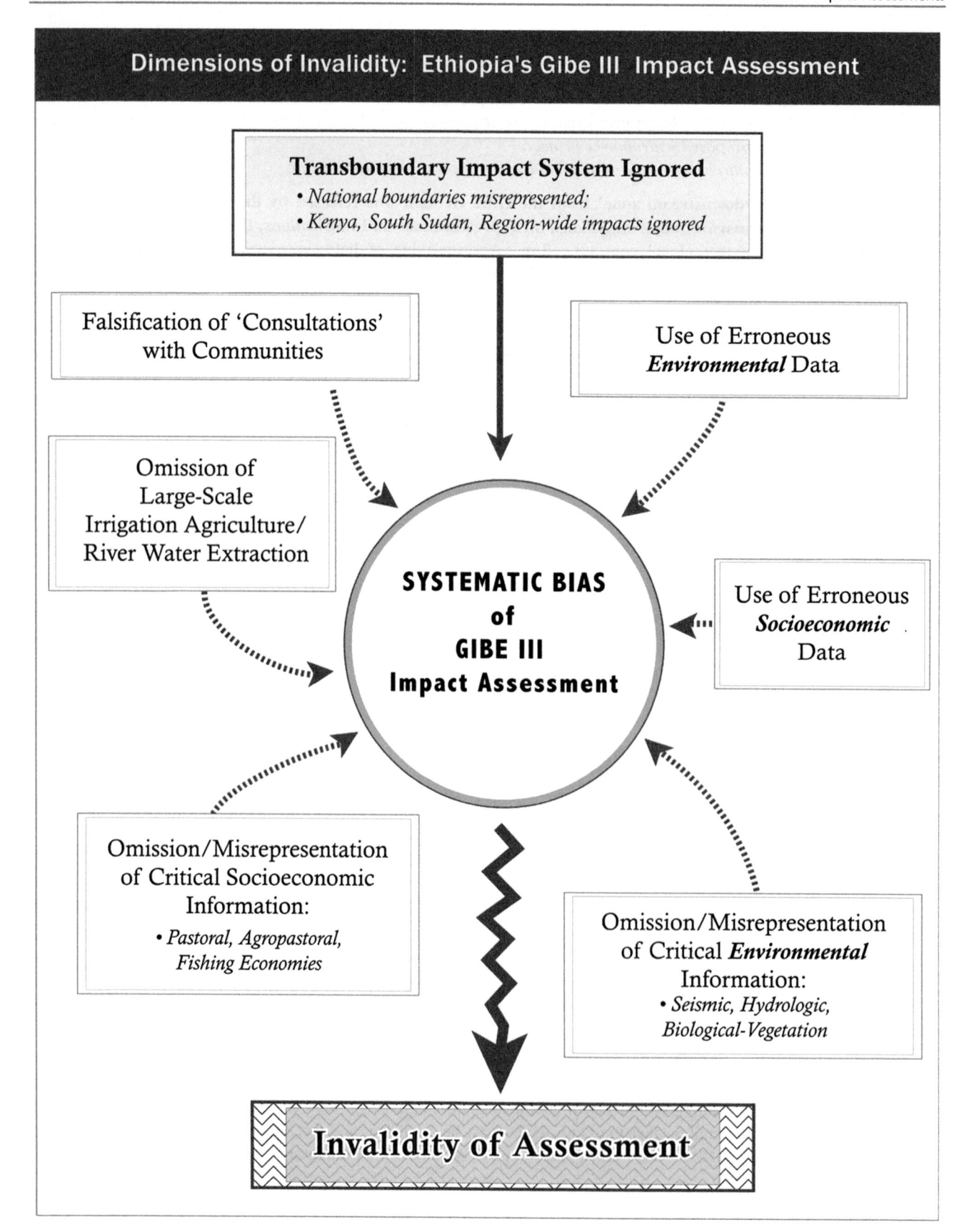

Fig. 6.1 Dimensions of invalidity in the Ethiopian government's downstream impact assessment of the Gibe III dam

The GOE's false assertion of disastrous flooding caused by the Omo River is accompanied by its equally false assertion of 'excessive evaporation' occurring broadly in the region from alleged standing water following overbank flooding in broad plains adjacent to the river—with 'lost waters' that would be recovered following Gibe III dam construction. Two related assertions repeatedly made by the GOE are false.

- Firstly, regulation of the Omo River is necessary, making the Gibe III dam an imperative. In fact, no such disastrous floods have occurred, so there is no 'imperative' for major dam construction.
- Secondly, the Gibe III dam "will provide recovery" of "lost waters" from excessive flooding—'compensating' for the radical reduction of Omo River flow volume caused by the dam. In fact, only relatively small isolated pockets of standing water occur, including from seasonal rainy periods.[2]

The GOE maintained strict control over the reporting of the alleged flood disaster and information available to representatives from a host of national governments, international relief and development agencies, and private foundations personnel visiting the region. Visitors were situated in Omorate (where the police station and some other structures in the 20-year-old frontier town were flooded)—tens of kilometers upstream from the alleged devastation—where they were instructed by the GOE in the 'gravity of the crisis' and the 'urgent need' for substantial funds and supplies from international donors. For some, fly-overs were arranged for an aerial view of what were, in fact, swollen waters in the active Omo delta region and the Omo River's terminus at Lake Turkana.[3] Inquiring visitors were told that the absence of visible dead bodies could be explained by the 'fact' that they had floated down the river and into Kenya's Lake Turkana. No explanation was offered for the absence of reports of such dead bodies or carcasses around Lake Turkana.

Hundreds of media reports repeated the same GOE 'news'—with no tolerance of dissenting views. In fact, when an Ethiopian non-governmental organization based in the Omo delta region publicly stated the absence of human life loss, it was quickly removed from the region and its agricultural aid work was discontinued.

> The GOE's misrepresentation of the Omo's annual flood rests on its false assumption that massive flood waters spread throughout the plains extending east and west from the river, and upstream from the modern Omo delta.

➤ **There was no calamitous flood in 2006—nor have there been 'repeated disastrous floods,' as overbank flooding by the Omo River occurs only within the active (modern) delta and along the delta's northernmost limit.** This fact is repeatedly misrepresented by the GOE in its general statements as well as its environmental impact assessments. The absence of such flooding is clear from both abundant satellite data and accounts by indigenous community leaders throughout the so-called 'flood disaster zone'.

Omo flooding does not occur in what are in fact relict, or ancient, floodplains lateral to the river upstream from the active delta. ARWG physical scientists estimate these relict floodplains to be at least 10,000 years old. Dasanech elders also reported that according to their oral history, these lands have not flooded in recent memory. General field observation and detailed field-based studies of soil and vegetation characteristics in these broad relict floodplains—including by this writer and other ARWG researchers—have yielded no evidence of overbank flooding.[4]

Large portions of the Omo delta have emergent vegetation that appears in satellite photos during the highest portion of the flood. The 'submerged island' described in multiple Ethiopian press releases is an area that was emergent only in recent years, as the delta has expanded into Kenya's Lake Turkana. Its elevation is about 363 m, so it would normally be flooded

---

[2]The flooding apparent in the highly saline depression of Sanderson's Gulf (see Chap. 7 maps) indicated in the USDA satellite photos—is most likely from an ephemeral watercourse, the Kibish River—(see Fig. 1.1). Alternatively, a backup of water from the Omo's terminus could have produced these waters.

[3]The Ethiopian head of state, the late Meles Zenawi, made a high profile 'stop-over' in the delta region, underscoring the GOE's reports of 'calamity' in the region.

[4]Overbank flooding is a distinctly different phenomenon from small breaks in the natural levee at a few points, allowing movement of Omo waters at flood stage into small locales here and there behind the levee. This writer has spent weeks in these flats conducting soil and vegetation transects, as well as conducted helicopter reconnaissance throughout the region.

when the lake reaches that elevation. Rather than spreading beyond the active Omo delta, the flood extended only into the northern edges of the active delta and the basin-like depression west of the Omo River, known as Sanderson's Gulf (referred to as *berar* by the Dasanech). The U.S. Department of Agriculture satellite photos (Fig. 6.2) substantiate this assertion, as do NASA photos from the same period.

Even the actively flooded Omo delta did not undergo the destruction claimed by the GOE. The *"raging waters"* described by the GOE to the media, for example, are fully contradicted by both Turkana and Dasanech fishers' accounts. Dasanech and Turkana elders interviewed by this writer and other SONT researchers gave accounts of the 2006 flood and those of prior years that are contrary to those of the GOE. SONT interviewed community elders from multiple locales in the alleged 'flood crisis' zone, including: (i) the interior of the Omo delta, (ii) the west bank of the lowermost Omo River, (iii) Todenyang/Lowarengak villages at the extreme northwestern Lake Turkana shoreline, (iv) the Ileret region at the northeastern extreme of the lake, and (v) Ferguson's Gulf (Fig. 1.3).

The Turkana and Dasanech elders consistently described annual Omo floods—and the 2006 flood in particular—in the following terms:

- The August 2006 Omo River flood was extraordinarily large, but not destructive of human life and livestock as the GOE had portrayed it. The impacts described by indigenous residents in the Omo delta and along the northernmost shores of Lake Turkana (excluding those in the service of the government) consistently described the flood as "very large" but "not destructive"—in fact, useful for their multiplicity of economic activities in ensuing months.[5] The flood was not given a special name by the Dasanech or the Turkana: a clear indication that they did not regard the flood as a 'crisis'.
- Large proportions of both ethnic groups have recently been forced by their radically declining pastoral lives to relocate to the river or the lake, where they have brought their remaining livestock for 'last option' grazing and watering and where they have often taken up recession agriculture (along the Omo) or fishing, or both. They share major dependence on the annual Omo River flood whose inflow of freshwater sustains Lake Turkana and the fishery upon which they depend for their survival. The Dasanech term for 'big Omo' flood—*war gudo'ha'*—is entirely positive for them.

> The crisis that all riverside and northern lakeshore communities face is *too little flooding*, not excessive flooding. They regard the Ethiopian government to have brought crisis to their area—not the Omo River.

Table 6.1 summarizes the flood history as reported to SONT by 25 Turkana and Dasanech elders from villages in the modern delta region and along Lake Turkana's northwestern shoreline, who universally described the loss of human life and livestock during the 2006 flood as extremely small—ranging from 4 to 10 people lost and 5–90 cattle lost. Individual cases of humans or cattle lives lost were well known among villagers,—underscoring these events as "special" or "news" to villagers. When told of the GOE's estimates of lost people and livestock, all informants were either angry or scoffed at such accounts.

The 2006 flood waters were not even strong enough at the mouth of the Omo River to cause many fishers to cease fishing or suffer damage to their nets and other equipment during the approximately ten day high flood waters. Villagers near the river's terminus at the lake—for example, at Lopelele—describe continued active fishing by both Turkana and Dasanech at the time. They also described the Ethio-Fisheries Corporation as remaining in the main channel for fishing during the flood. In other words, all three groups of fishers continued their activity throughout the flood's duration, while agriculturalists and pastoralists moved laterally from the river to await its return to a lower level.

- Crops were submerged, but this was universally described as an exception from the predominant experience over many years of Omo flooding, with success in recession agriculture when they were fortunate enough to receive 'enough' water.

---

[5]Individuals typically functioning in service to the government include church missionaries, both the Hiwot mission in Ethiopia and the Catholic mission at Todenyang, Kenya. Both are 'partnered' or at least complicit with the GOE's policies toward the indigenous communities and the establishment of the dam enabled irrigated agriculture along the river. During Sont investigations, both missions were charging local residents for payment of services—windmill use in the case of the Hiwot church and repair of church-installed wells (on traditional Turkana land), in the case of the Catholic diocese.

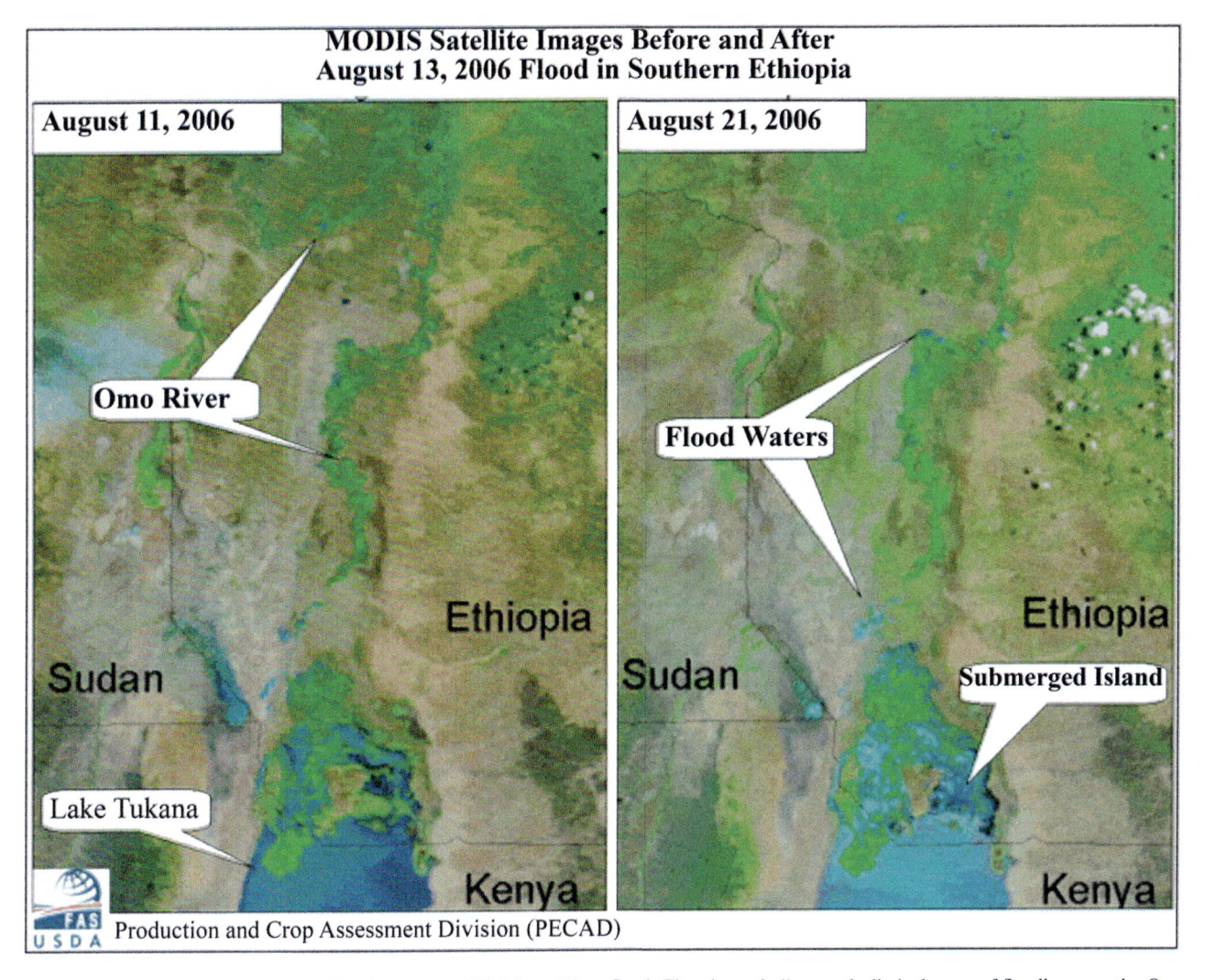

Fig. 6.2 Satellite images before and after the August 2006 Omo River flood. The photos indicate only limited areas of flooding near the Omo River and Lake Turkana, not massive scale flooding as described by the GOE. *Source* U.S. Department of Agriculture (USDA)

- One common account among residents was similar to that of one well-known Dasanech man considered to have a problem, because of "two dead cows—one black adult cow and one calf." The health clinics at Loyere and Toltale along the Omo also took on water, with some boxes of medicines, biscuits and other materials carried away. "That is all," local elders stated.
- Residents reported that the GOE made a powerful show of force in military terms during the flood, particularly when Ethiopia's Prime Minister visited the area. Locals described the arrival of "many engine boats," moving only a "few" people who wanted to be transported. Villagers generally knew to move away from the river with their livestock during floods. No reliable estimates exist for the population that actually moved, but elders estimate that no more than 6000–8000 people were involved.

Villagers with social ties to villages along the Omo's west bank moved there—especially to Attala, Kalama, Kipercheria, and Bokom village areas.[6] Other Dasanech in the delta went to the east bank of the river, including to Afor (Afewerk) and Kapusie—locales south of Omorate where international observers were based. A small number of these new settlements

---

[6]Many of these Dasanech had to flee Atalago the following year, due to hostilities with the Turkana.

Table 6.1 Recent Omo River flood history from a SONT 2009 survey of village elders

2009 SURVEY OF WESTERN OMO DELTA & NORTHWESTERN TURKANA VILLAGERS REGARDING RECENT FLOOD HISTORY (25 RESPONDENTS)*

YEAR	FLOODS 'SERIOUSNESS'	DELTA COVERAGE	AMT.DELTA/WEST	CATTLE LOSS	GOATS/ SHEEP
1968	BIG floods	Covered all the delta	West channels were full	Small losses	No reports
1986	Normal floods	Full deltas	„	„	„
1987	Little floods	Little to the deltas which had already started to dry	Gave room to Dasa…to move forward to the deltas	no	no
1988	Low inflow	Little coverage	West channels a half full/deltas	„	„
1993	No impact	Small flow into delta	West channels were full; others lacking	„	„
1994	Small inflow	„	„	„	„
1995	Too little.	Didn't reach most of delta	Very small	„	„
1996	The deltas got enough.	full	Normal	„	„
1997	„	„	„	„	„
1998	Floods that up rooted reeds that covered the whole part of the lakeshores.	„	„	Unknown number	„
1999	Small floods.	Extremely small	Very limited	no	no
2000	„	„	„	„	„
2001	moderate	Moderate coverage	Covered the deltas to west channels.	„	„
2002	Very very small	Small inflow inside the river.	Very small	„	„
2003	Very small	Little out but most the river	Very small to the channels in the west.	„	„
2004	„	„	„	„	„
2005	„	„	„	„	„
2006	Lasted 6-9 days lasted five to eight days.	Mixed with rains which caused additional water in streams to enlarge, so *people moved aside.*	Full water to both west channels and deltas. Fishing continued, villagers in delta moved to shores. *Several people drowned trying to retrieve animals.*	50-90 lost; 'stuck in mud' along river;	No reports, but same as cattle.
2007	Big floods	full	„	no	No reports
2008	No impact at all.	Normal coverage	No coverage	„	no

* **Terminology used by SONT interviewers and by respondents are retained in this table.**

Elders interviewed were from Dasanech villages within the Omo delta and Turkana villages along Lake Turkana's northwestern shoreline. 2009 survey of western omo delta and northwestern Turkana villagers regarding recent flood history (25 respondents) (Terminology used by SONT interviewers and by respondents are retained in this table)

were perpetuated after the flood. Some of the newly settled villagers were incorporated as 'paid' labor on government farms, while others attempted to survive on their own or migrated back into the delta region or off to the east bank of the Omo. Those Dasanech who moved out of the active delta region as the Omo waters rose planned to return to their replenished lands when the waters subsided, as usual. However, many were blocked by the Ethiopian government from doing so.

Invalidity of the Ethiopian Government's Downstream Impact Assessment

➢ **A small selection of prominent statements in the GOE's downstream impact assessment** (GOE 2009b) **illustrates the major failures of the downstream assessment.** These issues referenced here are addressed below.

(i) **Seismic Risk.**
GOE Statements:

• *While the Gibe III dam is located in Ethiopia, in the vicinity (about 70 km) of the eastern branch of the East African rift system the entire area interested in the project, according to the Level 1 Design Geological Report, doesn't show any evidences of present existing seismic activity.*
• *There are no confirmed occurrences of geothermal activity in dam area and because of its distance from the major Ethiopia seismic centres, located in the rift valley; any tectonical [sic] event will have negligible effects on the project area.*

False. Historical records and current geophysical information freely available in the public domain do not support these statements. As Chap. 3 details, there is a 20 % chance of a 7 or 8 earthquake within the Gibe III dam region within the next fifty years, and there has been significant seismic activity within the correct geographic unit of concern, as established by international standards. Moreover, there is clear evidence that earthquakes of a magnitude greater than 5, accompanied by a long series of foreshocks and aftershocks, have been related to reservoir impounding.

To repeat the statement by an ARWG geologist researching volcanics in river basins for several decades:

In the case of Vaiont, Italy, accelerated mass movement displaced water in the reservoir, the water overtopped the dam, and a 226-foot wall of water drowned some 2600–3000 people downstream along the Piave River. Such an event in the Omo would obviously affect a great many more people, most likely tens or hundreds of thousands of people.

(ii) **'Elimination of Drought Risk'.**
GOE Statements:

• *The main beneficial impacts of the Gibe III reservoir operation on the downstream hydrological regime are, therefore, as follows: … Reduction of the extended drought periods (as the 1986–1987 ones). The [Gibe III] plant will allow the complete regulation of the river flows reducing the highest peak floods and avoiding extended drought periods…*
• *[Artificial floods] nearby Lake Turkana will be similar therefore to the monthly average flows of August/September during the dry years thus avoiding any critical drought event.*
• *Floods that occurred in 2006 (with return period of less than 10 years) caused destructive effects on human and animal life, private assets and public infrastructure in the river delta, while the extended drought[s] period 1986–1987 originated a famine crisis for humans and wildlife. The [Gibe III] plant will allow the complete regulation of the river flows reducing the highest peak floods and avoiding extended drought periods by means of: …The reservoir live capacity of 12,300 mm^3 (Comparable to the mean annual inflow volume of 13,800 mm^3)…*
• *Increase in the Omo River's flow volume during the dry season (by dam engineers, following the construction of the dam), will alleviate the "severe drought" and associated overgrazing conditions in the lower basin.*

False. Droughts are the result of lack of rainfall, not river flow volume. There is no causal relationship between volume of river flow during the dry (or wet) season, on the one hand, and drought conditions within the extensive rangelands of the lower Omo basin on the other. The EIA's assertion that there would be a carefully regulated flow that would alleviate grazing conditions is also unfounded. Overgrazing results from a complex of changes, including territorial restriction of indigenous groups, causing overcrowding of livestock. The reverse is overwhelmingly likely to occur: precipitous drop in river flow would radically reduce available pasturage and browse near the river and force major overgrazing and even worse hunger conditions throughout the region. The following chapters detail these issues.

(iii) **Kenya's Lake Turkana Level.**
GOE statement:

The Gibe III dam will stabilize, not lower, the present level of Lake Turkana. No appreciable drop in lake level or constitution will result from the dam's construction.

False. Dam completion, reservoir filling and early dam operation, would produce a substantial drop in the level of Lake Turkana—by ARWG scientists' estimate, at least 7 m within 5 years—or about a 60 % reduction of flow volume; higher estimates extend to anywhere from 10 to more than 20 m. A 2–4 m drop in the lake level, calculated purely from the volume of water required for reservoir fill, is improbable, in view of:

- The high frequency of fissures in the basalts at the reservoir, with major potential seepage from the reservoir (which would take decades or longer to reach Lake Turkana) and
- Major abstraction of water for planned large-scale irrigated commercial agriculture and industrial development.

(iv) **'Excessive' Flooding and Evaporation.**

The GOE's ESIA makes more than fifty references to alleged repetitively occurring 'destructive floods', including the supposedly 'disastrous' flood of 2006. Its statements include:

• *Large and sudden floods (peak flows up to 5200 m³/s, return period 30 years, Gibe III site). Floods occurred in 2006 (with return period of less than ten years) caused destructive effects on human and animal life, private assets and public infrastructure in the river delta while the extended droughts* [sic] *period 1986–1987 originated a famine crisis for humans and wildlife.*

Huge evaporation losses as a consequence of excessive, uncontrolled flooding [of the Omo river] further contribute to the current recession of Lake Turkana.

• Omo River overflow into the extensive adjacent floodplains, resulting in excessive evaporation, has caused the recent drop in the level of Lake Turkana ...and deprivation of vital waters for recessional cultivation.
• The Gibe III dam will provide for 'recovery' of these waters, thus compensating for much of the river flow volume decrease during reservoir filling.
• The long-term benefits [of the Gibe III dam] *will include the reduction of the unproductive evaporation losses taking place in the floodplains after the floods retreat. These losses largely exceed the expected total evaporation from the proposed Gibe-3 reservoir.*
• The Omo River, in fact, provides over 80 % of the lake water resources. By retaining the unproductive runoff surplus of the extreme high-flow years in the reservoir, and releasing it in the extreme low-flow years, the dam will secure an overall increased inflow to the Lake.

False. There is no 'excessive flooding' or 'excessive evaporation' in the 'floodplains,' as these broad mudflats are relict, not active floodplains—thus, there are no such 'benefits' of the Gibe III dam. There is no such cause of the recent drop in the level of Lake Turkana, nor is there any 'withholding' of waters from cultivated lands. Flooding has never been controlled, yet Lake Turkana has risen and fallen significantly in the past fifty years; the floodplain has played no role in the lowering of Lake Turkana. The sharp drop in lake levels in recent years results from reduced precipitation in the large upper Omo basin (more than 200,000 km²). As detailed in the next section, there has been only one major flood within the past fifty years—in 2006—and this flood was not a disaster for humans or livestock.

(v) **Indigenous 'Rationality' of Resource Use.**
GOE statement:

In fact, technicalities and relevant characteristics, potentials and constraints of practiced Agriculture, Animal Husbandry and Grazing, and Fisheries have evidentiated [sic]—***a largely backward and primitive concept of land use, in which natural resources are being utilized "at will" without much regard being paid to ensure any sustainability*** *to the utilization process, in most cases simply exploiting the resource to the best of individual capabilities, given the numerous formidable constraints in terms of adverse environmental conditions in which human beings are forced to dwell.* [Emphasis added]

False. The indigenous communities in the lower Omo basin have evolved highly adaptive and effective strategies for coping with some of the most difficult environmental conditions on the earth. These strategies became increasingly more difficult to implement, as powerful colonial and post-colonial economic and political forces caused dispossession and marginalization within the region. The above statement by the GOE reveals its extreme level of ignorance of the indigenous socioeconomic systems where "at will" use of resources and "disregard of sustainability to the utilization process" are completely antithetical to their centuries old traditional resource understanding and practice.

(vi) **Riverine Forest and Forest Indigenous Livelihood.**
The riverine forest in the lowermost portion of the Omo River is a critical settlement area and the source of livelihood for thousands of indigenous peoples—primarily the Nyangatom, Kara and Dasanech (see Chap. 8). The GOE document falsely represents the forest as: **(i)** "*not of unique value, since* [it is] *the same as the highland riverine forest*," whereas the lowland riverine forest is entirely distinct from the higher altitude forest system, and it is therefore significant in Ethiopia's natural diversity; **(ii)** "*stressed by overbank flooding by the Omo River*," when no such flooding occurs upstream from the modern (active) Omo delta and its northern perimeter (Figs. 1 and 3), and **(iii)** benefitting from Gibe III regulation of the river. For example, it states:

The herbaceous stratum will also benefit from more limited flooding, increasing its photosynthesis activity and establishing itself on the river banks down to the river stabilized water level, thus contributing to limit riverbank erosion and filtering sediments reaching the river from the interior.

False. In reality, the opposite of all three assertions is true for this last pristine riverine forest in semi-arid Africa, with its uniquely undisturbed vegetation and rich fauna, which would be destroyed by the Gibe III dam and associated agricultural development. There is no evidence for overbank flooding, let alone 'prolonged submersion' of the forest. The riverine forest is described earlier to include a complex flora and ecology described by this writer (Carr 1977, 1998, 2012) and large fauna including leopard, elephant, buffalo, kuku, monitor lizards, at least three primate species and remarkably rich bird life. Instead the forest and woodland vegetation is adapted to and depends on soil water retention during the Omo River's annual flood for its survival. Far from benefitting, the forest would be destroyed by the elimination of annual Omo high waters during the 'flood' period since the shallow rooted forest and woodland trees are wholly dependent on soil water retention (termed 'residence time') in natural levee soils for their survival (see Chap. 8). With the destruction of this forest environment, the livelihoods of thousands of riverine dwelling indigenous residents would be dismantled.

Table 6.2 summarizes key failures of the GOE downstream assessment (GOE 2009b) for the region most impacted by the Gibe III.

Table 6.2 Dimensions of invalidity of the Ethiopian government's environmental and social impact assessment (GOE 2009b)

	INVALIDITY OF ETHIOPIAN GOVERNMENT ASSESSMENT OF GIBE III DAM IMPACTS ON TRANSBOUNDARY INDIGENOUS PEOPLES & ENVIRONMENTS	
	ASSESSMENT COMPONENT	**MAJOR ETHIOPIAN GOVERNMENT (GOE) ASSESSMENT FAILURES: (Omission, Misrepresentation, Fabrication)**
1	Transboundary impact system assessment: (Ethiopia, Kenya, Ilemi/South Sudan).	*Omission, Misrepresentation* Full-scale disregard for transboundary social and environmental impacts of the Gibe III dam and associated irrigation agricultural enterprises. • Omission of impacts of Omo River radical flow reduction on more than 100,000 indigenous (including Nyangatom, Kara and Dasanech) livelihood systems along the lowermost Omo and within the modern Omo delta, as well as on the 300,000 indigenous population living along Kenya's Lake Turkana in the transboundary zone. All face massive scale hunger, or starvation. • Disregard for Kenyan sovereignty over Lake Turkana's northern shoreline zone and a significant portion of the Omo delta; expansion of the delta since the 1970s brought the Omo River terminus and much of the modern delta within Kenyan national borders. • Omission of impacts on pastoralists of northern Kenyan and the Ilemi Triangle/South Sudan, due to loss of river-based resources essential to their survival — particularly during prolonged drought periods. • Omission of above changes causing major increase in cross-border interethnic armed conflict and destabilization, as pastoral, agropastoral and fishing groups compete for disappearing riverine, lake and adjoining grassland resources.
2	Seismic Risk at the Gibe III dam: Earthquake, landslide, dam collapse threat.	*Omission, Misrepresentation* • Misrepresentation of appropriate geographic unit for seismic analysis (identified by U.N., Ethiopian, USGS and BGS seismic offices). False claims of "no serious threat," despite a 20 percent probability of a 7 or 8 (Mercalli) intensity earthquake in the Gibe III dam region within the next 50 years, and plausible dam disaster occurrence with landslide events. • Omission of available data (e.g., literature, satellite images) indicating frequent major seismic events in the Gibe III dam locality's geologically defined province. Dismissal of landslide danger, including potential danger of threat to the dam's integrity. ARWG geologists point to the steep rock faces of the reservoir walls in the gorge, additional landslide potential from saturation of the volcanic rocks, and seismically induced instability. • Misrepresentation of sediment buildup danger as "not affecting the life capacity of the reservoir before 130 years," despite counter assessment that sediment accumulation capacity would reached within a few years, with dam failure.
3	Baseline data for Omo River downstream flow volume & Omo River inflow to Lake Turkana.	*Omission, Misrepresentation* • Failure to establish baseline data/measurements for Omo River downstream flow volume, with projections of 'mean flow,' instead of direct measurement of Omo flow in the lowermost basin and the river's inflow to Lake Turkana. • Invalid projections of existing river flow volume for both downstream zone and inflow to Lake Turkana — based on incorrect/misrepresented data regarding precipitation and geomorphic characteristics of lower Omo basin, including: *(i)* Incorrectly identified precipitation regime for rainfall projections between the dam region and the lower Omo basin); *(ii)* False representation of watercourses as Omo River 'tributaries' and thus 'alternative' sources of inflow prior to the main Omo channel, whereas these channels are not tributaries, as they dissipate prior to intersecting the Omo River (or Lake Turkana). *(iii)* Misrepresentation of the ephemeral Kibish River (Fig. 1.3) as flowing into Lake Turkana, whereas the Kibish dissipates in a large depression (Sanderson's Gulf) northwest of Lake Turkana.
4	Impacts on Inflow from Omo River to Lake Turkana: reduction from Gibe III.	*Omission, Misrepresentation, Fabrication* • Misrepresentation of major components of Omo River inflow to Lake Turkana, based on above failures. • Fabrication of extensive and repeated Omo flooding, with 'lost' waters otherwise available to sustain Lake Turkana — whereas, in fact, these vast lands referred to are relict (ancient) floodplains not flooded for thousands of years. • Fabrication (based on the above false assertion) of 'excessive evaporation' and vast floodwaters that are 'recoverable' by river regulation (the Gibe III dam) — allegedly contributing to the sustainment of Lake Turkana. Since there are no such waters, there would be no such 'recovery' or sustainment.
5	Excessive vs. insufficient Omo River flooding re: human life, livestock, and habitat.	*Fabrication* • Fabrication of frequent 'uncontrolled' flood —' 'highly destructive' of human life and livestock. By GOE accounts, 360 to 900 people drowned, and more than 3000 livestock were destroyed in the 2006 flood. To the contrary, local elders reported very little loss of human life through drowning (fewer than 10 people and fewer than 100 livestock, mostly those mired in mud). Satellite data supports local accounts of the 2006 flood. No such destructive floods as described by the GOE are experienced by the region's indigenous groups. • Residents throughout the lowermost Omo basin/modern Omo delta and northern Lake Turkana shoreline region insist that the worst problem they face with the Omo River is *too little flooding*, rather than excessive flooding.

(continued)

Table 6.2 (continued)

6	Projected Gibe III reservoir 'filling' period and factors impacts on downstream and Lake Turkana people and resources.	***Omission, Misrepresentation*** ● GOE exclusion of readily available geological literature and field-based information indicating highly fractured basalts and other volcanic rocks at the Gibe III dam/reservoir site — indicating a high probability of major water seepage from the reservoir into fractured rocks, with a major extension of 'filling' period, with sustained radically reduced Omo flow. ● Misrepresentation of water movement in the region - re: seepage water 'returning' to Lake Turkana "in any case" — whereas the alleged 'migration' of seepage waters would require decades, if not a century or more.
7	Baseline data for indigenous livelihood systems (pastoral, agro-pastoral and fishing)	***Omission, Misrepresentation, Fabrication*** ● Omission of available information and investigation of the livelihood dynamics of indigenous groups in the lower Omo basin; total omission of information regarding indigenous residents around Kenya's Lake Turkana.[1] The failed dimensions of assessment include: *(i)* The dynamics of pastoral, agropastoral and fishing livelihood systems. *(ii)* The natural resource dependencies of these economies on Omo River and Lake Turkana waters (livestock raising, recession agriculture and fishing). *(iii)* Recent changes in livelihood systems and the causes for these changes. *(iv)* Current status (including malnutrition and disease) and key stresses on survival systems. *(v)* Exclusion of the content of available literature concerning the ethnic groups and economies in the lower Omo basin - including Turton, Carr, Almagor, Tornay and Abbink, among others. (Some of this literature is referenced in the GOE's bibliography but its content is essentially excluded from consideration.) *(vi)* Overall *extreme vulnerability* of indigenous economies to the major resource losses predictable for the Gibe III dam. ● Omission of plans for GOE eviction of tens of thousands of indigenous residents along the lower Omo River in order to establish private and government irrigated commercial agriculture and major physical infrastructure (e.g., irrigation canals, commercial facilities). No account of the impacts of widespread expropriation, evictions and major infrastructure construction on indigenous survival systems within the lower Omo basin. ● Omission of the impacts of large-scale irrigation agricultural development: to indefinitely prolong the radical reduction of Omo River flow volume and Omo River inflow to Lake Turkana, in turn causing major scale Lake Turkana retreat and destruction of its fisheries, along with Omo River channel incision and cessation of annual flooding of grazing and recession agricultural lands. ● Fabricated Ethiopian government demographical data and other socioeconomic information incorporated into a report.[2] ● Fabrication of quantitative data for ethnic groups (demographic, livestock populations, other livelihood particulars). Quantitative projections from landsat/satellite materials with little if any bearing on actual conditions. ● Fabrication of 'planned' agricultural and social development 'for' indigenous communities whereas, in fact, they are being evicted by the thousands from riverine lands.
8	Flood recession Agriculture: impact from the radical river flow reduction	***Omission, Misrepresentation*** ● **Omission of all information regarding the direct dependence of lower Omo basin indigenous communities on flood recession agriculture** — as well as information regarding the dependence of tens of thousands of other indigenous peoples on regional exchange relations involving agricultural produce, particularly as the pastoral economies of all ethnic groups continue to decline. ● Failure to describe accurately the locations, cropping patterns, water requirements and other specifics of flood recession agricultural production, both along the Omo River's waterside flats and throughout much of the active Omo delta. ● As a consequence of the above failures, inability to establish any assessment of the impacts of radical Omo River flow reduction caused by the Gibe III dam and dam enabled irrigation on this mainstay food source. Omission of flow reduction issue, northern shoreline retreat of Lake Turkana into Kenya and desiccation of the lowermost Omo flats and delta — thus eliminating flood recession agriculture in the lowermost Omo basin (and riverine zone last resort grazing for livestock). ● Fabrication of the practice of rainfed agriculture occurrence in the Lower Basin (this practice is not possible to have observed because it does not occur, due to limited rainfall).
9	Artisanal fishing in River and delta: Impacts of Omo River flow volume reduction	***Omission, Misrepresentation*** ● **Failure to detail and evaluate the importance of fishing - the last resort food source for tens of thousands of indigenous people in the lower Omo basin -primarily (but not only) among the Dasanech.** Consequently, failure to establish a baseline for considering the impacts of the Gibe III on this pervasive and increasing form of livelihood. ● Exclusion of bathymetric and other data (available, but shown by ARWG, the EIB and others to indicate that the Gibe III would cause southward migration of Lake Turkana's northern shoreline — at least a retreat of 10 kilometers, but likely substantially more due to major Omo water diversion for the GOE-promoted irrigated commercial agriculture). Critical fish reproductive habitat throughout the northern lake extremity and delta locales would be destroyed, with elimination of the fishing livelihood as a means of survival throughout the lower Omo basin (with parallel destruction of fish dependent communities around Lake Turkana). ● Omission of the Ethiopian government subsidized private fishing corporations active in the lowermost Omo River, including the modern delta waters. Major piracy of Kenyan fisheries in Lake Turkana by the well-equipped and aggressive commercial Ethiopian vessels — with the extraction of resources that are essential to both Dasanech and Turkana fishers and local residents. ● Disregard for artisanal fishing in Lake Turkana - the mainstay of survival for upwards of 200,000 indigenous Kenyans settled around Kenya's lake, as well as those depending on exchange with them (at least an additional 100,000 people).

(continued)

Table 6.2 (continued)

10	**Riverine forest & forest-based survival activities**	***Omission, Misrepresentation, Fabrication*** ● Omission of available data regarding Omo riverine/forest ecology[1] (e.g., Carr 1998) and comparative riverine environments where forest destruction has followed large hydrodam construction. ● No apparent field-based investigation of lower Omo basin riverine forests. Erroneous description of the riverine and woodland vegetation along the lowermost Omo River. ● Disregard for the need to consider the Omo riverine forest structure and dependence on the river's flood regime - thus its vulnerability to destruction from Gibe III development, as has occurred in all other Sub-Saharan Africa semi-arid environments with riverine forest development. ● Fabrication of overbank floods in riverine forests: no field evidence for such exists. In order to be sustained, riverine forests require particular 'residence times' of water retention in natural levee soils during the Omo River's annual flood. ● No account of livelihood activities in riverine forest (food gathering, hunting, and bee-keeping) critical to the survival of the most impoverished indigenous communities (Nyangatom, Dasanech and Kara) and tens of thousands of additional pastoralists who depend on riverine environments for their own and their herds' survival during prolonged drought. All would be eliminated by radical river flow volume reduction from the Gibe III dam. ● Omission of the GOE's major eviction of indigenous communities along the Omo River for clearing of the forest, construction of canal and irrigation systems and the establishment of large-scale commercial irrigation farms.
11	**Artificial flood: efficacy of GOE 'assurances' to sustain downstream ecological and indigenous economies**	***Omission, Misrepresentation, Fabrication*** ● Misrepresentation of GOE's 'planned' artificial flood program as calculated from adequate measures of downstream river flow volume and lake inflow, when no such measurements have been taken. ● Misrepresentation of GOE 'planned' (but elsewhere disclaimed) artificial flooding — as "sufficient" to sustain downstream ecological systems as well as to mitigate problems that "might" occur for indigenous economies as a consequence of sharply reduced river flow volume. Calculations used by the GOE to bolster its assertion of effective mitigation are shown to be erroneous by the EIB, the Africa Resources Working Group (ARWG) and others. ● Omission of the fundamental conflict between the GOE's key objective — generating hydroelectric power at the Gibe III dam and that of releasing water to sustain downstream indigenous communities and their environments. ● Fabrication of a GOE plan to undertake controlled flooding along with agricultural development, for the 'benefit' of indigenous communities, whereas a GOE program of widespread eviction and privatization of indigenous lands, is well underway.
12	**L. Turkana shoreline & littoral/nearshore zone impacts (physical, biological) of the Gibe III dam.**	***Omission, Misrepresentation*** Omitted from any impact consideration, since GOE denies any significant impact will occur for the level of Lake Turkana from the Gibe III dam.
13	**Armed conflict and crisis in the transboundary region.**	***Omission*** While the GOE is fully engaged in trying to quell small scale but mounting armed conflict among ethnic groups in the broader region, and neither supports or permits various peace initiatives in the region, the GOE assessment indicates no cognizance whatsoever of the rising armed conflict, - conflict that would move to explosive proportions in a far broader region if the Gibe III is completed, since hundreds of thousands of indigenous people would be facing the devastation of their communities from the loss of the Omo River. The 2006 flood, rather than spreading beyond the active Omo delta, extended only into the northern edges of the active delta and the basin-like depression west of the Omo River, known as Sanderson's Gulf (referred to as *berar* by the Dasanech). The U.S. Department of Agriculture satellite photos (Fig. 6.2) substantiate this assertion, as do NASA photos from the same period. *Dasanech and Turkana male villagers strongly assert that — when faced with the deaths of their women, children, and youngsters to hunger and disease — they will fight for their survival.*

Environmental and social impact assessment (ESIA)—additional study on downstream impact

[a]Villagers (Dasanech) throughout the west bank communities along the Omo River and within both middle and upper portions of the active Omo delta reported to SONT researchers that no individuals from the government or other agency had interviewed or questioned them about their resident populations, settlement patterns, livelihood activities (livestock raising, recession agriculture, fishing activities), or their major resource needs. Respondents described only police and government personnel taking action to move them or evict them altogether (with 'employment' of only a few individuals), or government/'outsider' presence to investigate an instance of armed conflict, most frequently with the Turkana (or Nyangatom)

[b]One non-indigenous member of the governmental office stated to this writer that he himself had filled out government questionnaire forms "for" the local Dasanech population

[c]For details of the lower Omo basin riverine forest/woodland floristics and vegetation (see Carr 1998); some summary characteristics are presented below

The False Promise of an Artificial Flood 'Solution'

➢ The GOE's stated plan to begin releasing reservoir waters annually as "artificial flood" (also termed "controlled flood" or "simulated flood") forms the centerpiece of its alleged 'mitigation' program following the operation of the Gibe III dam (GOE 2009b), which the government acknowledges wouls cause a 60–70 % reduction of Omo River flow volume. The gap between the rhetoric of the GOE's publically announced program and the reality of its actual intentions and the economic and political constraints on such a program, however, is extreme.

There is an effectively unblemished record of failure to implement artificial flood programs in major hydrodam development within Sub-Saharan Africa, despite government and development agency assurances. These failures include the Pongolapoort—one of the largest dams in South Africa, the Cahora Bassa dam in Mozambique, the Iteshi Teshi dam in Zambia, and the Manatali dam in Senegal. The Manatali Dam in Senegal is often cited as an exception to the rule of failed artificial flood programs. But even it was short-lived and ended in failure.

Assurances by governments and development agencies of planned artificial flood programs are not difficult to interpret, since large dams have produced such an abysmal record of downstream problems for indigenous communities and non-indigenous small-holders, as well as the environments upon which their livelihoods depend. Such promises of mitigation programs have served as a cornerstone for the argument that although large dams are "flawed, they remain necessary" for economic development within Africa and elsewhere in developing countries.

There are clear economic and reasons for the record of failure of these programs.

- **For one thing, there is a clear contradiction between management of large dams for the maximization of power generation, on the one hand, and management geared to sustaining downstream livelihoods and riverine environments, on the other.** This is evident enough in the case of Senegalese floodplain livelihoods that were sustained by the river's annual flood for centuries—until international finance and the Senegalese government constructed the Manatali. A report to the World Bank described the severe impacts of the Manatali dam on the regional ecology, agricultural production, fisheries and public health in the downstream zone. The report, by Thomas and DiFrancesco (2009) points out that inundating the floodplain below the dam:

...does require that some amount of power deliveries and revenues be foregone, as well as storage for irrigation purposes in subsequent drier years.' ... '[Artificial] **flood releases appear to have been provided only on those rare occasions when they do not result in a reduction in power revenues or irrigation deliveries**. *They have occurred only when the flood water could not be retained in storage in order to avoid the risks of overtopping the reservoir and creating a safety hazard.* [Emphasis added]

A newly formed agency, the OMVS (*Organization pour la Mise en Valeurdu Fleuve Senegal*), was charged with implementing a downstream artificial flood. In order to provide the intended 2000 m^3/s release of floodwaters (an amount *double* that assured by the GOE for the GIBE III project). The Thomas and DiFrancesco report to the World also stated:

...the necessity to release water from the sluice gate and bypass the generator, requiring the OMVS members to forego some amount of potential hydropower generation and revenues. Thus, there remains a serious economic constraint in order to operationalize the Charter's commitment to optimal annual artificial flood releases...As a consequence of these enduring trade-offs, the Water Charter, and the Operational Manual **do not** [sic] **guarantee that the artificial flood will be provided** *in the magnitude and frequency that the floodplain needs...This decision is also likely to be influenced by economic and political considerations.*
[Emphasis added]

The massive scale capital outlays—typically, loans to be repaid, with interest and fees—necessitate maximizing financial return from dam operations. In Africa—especially in dryland regions—this objective is hardly met by prioritizing the subsistence needs of pastoralists, agropastoralists and small-scale fishers in the downstream zone, compared with providing electricity for high energy demand industrial enterprises, whether agricultural, extractive or manufacturing.

- Another key constraint on implementing an artificial flood program is that rainfall and its related conditions (erosion, land use, etc.) pertains to the fact that catchment area upstream from major dams are major determinants of reservoir water potential. Since these conditions fluctuate widely, and in the case of Ethiopia. are largely unpredictable, there is no possible 'guarantee' of reservoir water release to serve the survival needs of downstream peoples and their environments. Of many such instances, the Cahora Bassa dam situation provides a good example, as noted by Richard Beilfuss, hydrologist and present CEO of the International Crane Foundation:

One of the big problems with the management of Cahora Bassa dam is that each year they need to create capacity to store a possible large flood that could overtop the dam and so in order to do so they often have to release waters during dry season. Especially in wet years, they have to release waters downstream to make storage capacity available in the dam and in those dry season floods that have really been a terrible problem for farmers and fishers downstream because they can come at any time of the year. They can come in August or October and they can come and wash out crops along the lower Zambezi. It's quite a big problem, so not only is there the loss of floods during the normal time of peak flooding - there's also the potential for floods at any time of year.

➢ **There is no evidence to suggest that the Ethiopian government would be the first African nation to implement an artificial flood program for a large dam—quite the contrary**. GOE spokespeople themselves—including the Prime Minister, EEPCO executives and the General Manager of Ethiopia's Environmental Protection Authority—have asserted the priority for electrical power generation in the "*national interest.*" Even by the most optimistic scenarios set forth by the GOE, industry consultants and development agencies, such a program would not be feasible for a substantial number of years, following reservoir 'fill.' But there are direct GOE disclaimers, as well.

There are multiple disclaimers by the GOE of even its intention (or ability) to implement an extended artificial flooding program in its downstream impact assessment (GOE 2009b). These statements are ignored in the 2010 EIB and the AFDB assessments.

They include the following:

- *In the event that the annual release is not being implemented as scheduled, some possible adverse impact may occur as described below.*
- *Further desirable instruments have been considered and may be applied to both integrate the above main measures, compensate in case planned artificial floods are partially withheld due to unforeseen circumstances, and as desirable confidence-building actions to strengthen support of local authorities and populations.*
- *Some of the potential interventions meant to offset possible negative impacts stemming from possible difficulties in fully implementing designed controlled floods as planned.*
- *This may be necessary in case disruption of artificial floods may no longer fully contribute, by adequate submersion, to the rejuvenation of grasses.*
- *In case the annual release does not fully succeed as scheduled, some other compensation measures could be implemented including…conflict prevention and resolution training programs.*

Beyond the electrical power allocated for marketing within the national grid system of Ethiopia, the GOE is bound by contracts for power production for export, a system of contracting fully supported by the international development banks. As detailed in Chap. 10, the banks and other key aid agencies are in fact engaged in funding the 'energy highway' system for energy transfer. Moreover, The 2010 EIB's (conservative) estimate of a USD 10 million cost *per annum* for an artificial flood program conflicts with Ethiopia's contracts for export of power and its debt repayment obligations.

Even if the artificial flooding plan assured by the GOE were to be actually implemented, it would be *wholly* inadequate to provide the flooding necessary for recession agricultural plots, riverside grazing habitat and sustainment of the riverine forest—nor would it prevent or compensate for radical retreat of Lake Turkana (see Chap. 5 and Fig. 5.2). These multi-dimensions of destruction would ensue from the *outset* of impoundment—a reality asserted in the 2009 ARWG report and in Carr (2012).

Finally, GOE assurances that its plan for an artificial flood program would satisfy downstream environmental and socioeconomic needs are based on a number of invalid assumptions and calculations. GOE assertions that: (i) its projections are 'analytical,' (ii) "*suitable environmental flow and controlled floods" have been arranged for*, and (iii) "*the proposed river hydraulic model and outcomes of the Environmental Monitoring Plan will permit* [sic] *to optimize the required controlled flow.*" All of these assurances are baseless.

The EIB stated in its 'Independent Review' that *"the technical baseline for the estimation of the flood level is extremely limited and further hydrological and hydraulic investigations are necessary..."*. A similar point was made by the report when it stated, *"It is actually not possible to evaluate the effectiveness of such a mitigation measure without a scientific detailed survey which will determine what is required to obtain successful results from flood recession cultivation."*

The GOE suggests a "suitable environmental flow" from which controlled flow calculations have been designed. According to EEPCO, *"The proposed river hydraulic model and outcomes of the Environmental Monitoring Plan will permit [it] to optimize the required controlled flow."*

The value proposed is entirely inadequate for sustaining traditional artificial flood systems along the river's waterside flats and within the modern delta for the indigenous communities' continued survival within the lowermost basin, as the EIB consultant points out and ARWG physical scientists concur. The necessarily strongly increased figure would be would have to be based on data not yet even collected or analyzed.

- **Whatever artificial floodwaters the GOE would release from the reservoir would be highly compromised if not entirely consumed by the large-scale irrigated agricultural enterprises downstream.** Since the GOE downstream assessment omits any account of these enterprises it is not possible to accurately assess the abstraction of waters that would be required for the commercial operations but it is clear that this would be a massive removal of essential flow. The EIB inadvertently makes this point in labeling the GOE's planned releases from the dam wholly inadequate—even though the bank assessment omits consideration of abstraction for the irrigated systems. It is striking that the EIB report, like the 2010 AFDB assessment, ignores the GOE's own disclaimers of a sustained artificial flood program (see below).
- **Finally, artificial flooding even minimally sufficient in order to sustain recession agriculture, grazing and fisheries locales downstream would likely be highly destructive for expensive and fixed irrigation works along the lower Omo River.** Hundreds of thousands of hectares of large-scale irrigation systems are involved along the lower Omo River are planned by the GOE—many of which are already under construction (see above chapters and Chap. 7). This point was first raised in the main text of the 2010 EIB assessment. This argument against substantial controlled flood releases is partially responsible for the failure of implementation of 'promised' controlled floods in much of Africa.

> In sum, the bare survival of hundreds of thousands of indigenous people would be destroyed—a crisis only worsened by dam-dependent, irrigated agricultural enterprise abstraction of remaining Omo River flow. This major reality is acknowledged by the EIB (2010) impact assessment, but only obscurely in the report's text while omitted in its Summary and Conclusion. These points are emphasized in the discussion of development bank assessments below.

Multilateral Development Banks and the 'Complicity Treadmill'

➢ **At its core, the relationship between multilateral development banks and the global consulting industry is one of reciprocity**. In return for basic compliance with development bank interests and parameters for impact assessment, global industry consultants are rewarded with lucrative contracts and favorable status in future bidding processes. This enables the development banks to maintain relationships with 'reliable' contractors for subsequent projects and simultaneously satisfy their operational requirements for 'open bidding' and 'arms length' assessment. Meanwhile, the global consulting industry firms and individuals concerned can maintain their professional standing as 'independent' analysts—even among many non-governmental organizations critics and the concerned public.

This reciprocal relationship amounts to a '*contract treadmill*' that can persist for decades. Chapter 1 describes much of the basis for this treadmill emerging from the progression of institutions and political relations involved in river basin development within Ethiopia.[7] The connection between development banks, global industry consulting firms and individuals involves both explicit and implicit agreements, or 'understandings,' regarding boundaries and approaches for impact

---

[7]Similar reciprocal relationships exist between executive offices of (African) states and major bilateral aid agencies, on the one hand, and global consulting firms and individuals, on the other.

assessment. Contract Terms of Reference, or 'Scoping' and 'Bounding' specifications, set parameters such as geographic extent and emphasis on—or de facto exclusion of—particular environmental and socioeconomic impact dimensions in assessments. Implicit 'understandings' are equally important in shaping of assessment content and conclusions, particularly in capital-intensive projects such as river basin developments.

> Together, these explicit and implicit directives lead to assessments by global consulting industry firms and individuals that provide 'complicity with comment'—lending credibility to the assessment process without fundamentally questioning the projects in question.

Development bank and other policymakers rely on summary sections of assessments, including Executive Summary, Conclusions, and Recommendations. It is commonplace that policy-makers and their staff policy officers and their staff only peruse main texts of assessments and key reports in order to confirm their apparent 'empirical' character or for some highly specific purpose. Consequently, brief mention or 'notation' of major problems embedded in the main text of assessments, are seldom substantively considered. Even problems mentioned in the summary and conclusion sections are easily passed over unless their significance is prominently stated.

Global consulting industry representatives commonly downplay impact problems potentially severe enough to merit delay or halt of the project at hand. Even the most destructive environmental and socioeconomic impacts predictable for a project are frequently raised only as 'suggestions' for 'further study' or inclusion in alleged mitigation or monitoring efforts. Finally, consultants acknowledge major impacts inconspicuously in the often highly technical text of assessments but omit or barely mention in Conclusion or Executive Summary section.

These different means of minimizing negative impacts are pervasive in the GOE, EIB and AFDB assessments of the Gibe III dam—as is the avoidance of the contentious issue of dam-enabled major irrigated agricultural enterprises planned, information about which was fully accessible to the consultants.[8]

> **The European Investment Bank and African Development Bank, while considering requests from GOE for Gibe III project funding, each contracted long-established global industry consultants for 'independent' impact assessments of the dam**. The EIB extended a contract for a review (EIB 2010) of the GOE's environmental and socioeconomic downstream assessment—relating *only* to Ethiopia. The contract was awarded to the France-based global firm, Sogreah. For its part, the AFDB limited (by its Terms of Reference) its impact assessment to Kenya's Lake Turkana region, even though it had previously funded the most recent Master Plan for the Omo basin (Woodroofe and Associates 1996). The AFDB contracted for two separate impact assessments for the Lake Turkana region—the most notable of which was for the dam's hydrological impacts on the lake, including its water level and fisheries (AFDB 2010).[9] This contract was awarded to a former senior manager and executive for 26 years with the Nairobi-based global firm, Gibb Africa.[10] The second assessment was for the dam's socioeconomic impacts (AFDB 2009).

There was no apparent link between the AFDB's environmental (Lake Turkana) and socioeconomic assessments, nor between these assessments, on the one hand, and lake level drop caused by the Gibe III dam and linked developments, on the other. The hydrological assessment (AFDB 2010) is by far the one referred to by government and development agencies. However, the socioeconomic assessment has been largely ignored in policy statements and documents—likely for reasons outlined below.

Both Sogreah and Gibb Africa (formerly, Sir Alexander Gibb and Partners) had been engaged in Ethiopian river basin development since the Awash Valley developments in the 1960s. As outlined in Chap. 2, both firms had received contracts from international development bank and major bilateral agencies for river basin and related water resource development projects that totaled hundreds of millions of dollars from major aid agencies.

---

[8] As the earlier sections of this chapter detail, 'downplaying' also occurs as straightforward falsification and misrepresentation—both of which are described above to be pervasive in the GOE's assessment of downstream impacts (GOE 2009b).
[9] Unless otherwise specified, the AFDB assessment referred to below is the 2010 AFDB assessment of Gibe III impacts on the hydrological conditions of Lake Turkana, rather than the 2009 AFDB socioeconomic assessment.
[10] Shortly before signing a contract with the AFDB for the hydrological assessment of the Gibe III dam, this consultant joined a new consulting group of experienced individuals: Nairobi-based Water Resource Associates.

> The fragmentation of the Gibe III dam assessments into separate country approaches basically defined away, by simple Terms of Reference, consideration of the *worst* impact zone in terms of human lives at risk, namely, the transboundary region.

➢ **There are a number of shared failings of the EIB (2009) and AFDB (2010) assessments**. Beyond the longstanding contract history with development banks and other key aid agencies and the obvious limitations by Scoping agreement with their client institutions, many of these failings can be traced to an uncritical view of GOE information and perspective or to the reality that these assessments were primarily 'desk studies' with brief field trips, rather than substantial field-based investigations.[11] The key problems include:

- **Failure to cite inadequate impact assessment parameters** as defined in their own Terms of Reference (or Scoping/Bounding instructions) agreed upon with their respective clients—the EIB and AFDB.
- **Failure to analyze the transboundary region**, where the most intensive destruction of human life and environment by the Gibe III dam and associated development would be inflicted.
- **Discounting of the major seismic threat** documented for the Gibe III dam geological province, despite available data indicating a 20 % chance of a 7–8 magnitude earthquake within fifty years, as detailed in Chap. 3. The EIB and AFDB assessments do not challenge the GOE's statements (nor those by the company constructing the Gibe III dam—the Salini corporation) that no evidence exists for a significant seismic threat to the Gibe III dam. Such statements are indefensible by any scientific reckoning.

The EIB consultant's response to the ARWG's 2009 reporting of data in support of serious seismic threat in the dam region —information expanded in this writer's subsequent report (Carr 2012)—was simply stated as:

"Some risk cannot be avoided".

The 2010 AFDB assessment referenced the EIB's stance, with this added 'suggestion':

It would be sensible for the EIA studies to evaluate the consequence of a dam-break situation, especially as the dam is being constructed in a seismically active zone and will store a massive volume of water equal to a depth of two meters on Lake Turkana.

- **Failure to challenge the Ethiopian government's false assertions of 'disasters and 'uncontrolled flooding' by the Omo River, and particularly the GOE's falsification of catastrophic loss of human life and livestock**. As described earlier in this chapter, no such phenomena have occurred.[12] This major flaw in the assessments is closely related to the their failure to address the GOE's false assertions of "excessive evaporation" from overbank flooding by the Omo into vast floodplains—evaporation of waters that the GOE maintained "contribute to the current recession of Lake Turkana, and waters that would be "recovered" following the Gibe III dam, thus providing major water to Lake Turkana. In fact, no such broad flooding of these lands occurs since they are ancient floodplains, as noted earlier.
- **Acceptance of GOE and Salini (the Gibe III's construction firm) assertions of a 'brief' reservoir filling period, without risk of water seepage leakage from the reservoir** (GOE 2009b).[13] Without reference to available literature contrary, the EIB consultant further contends that even if leakage from the reservoir were to occur, it would '*migrate back into the Omo basin*' and flow to Lake Turkana. Despite the obvious major importance of the issue for the hydrologic impacts of the dam on Lake Turkana, the 2010 AFDB assessment concurs, stating:

---

[11]Like the 1996 AFDB-financed Master Plan for the whole Omo basin and the AFDB-supported 1990 Master Plan for Ethiopia's river basins in total, the 2010 EIB and AFDB assessments are primarily document studies, with limited field-based investigation The EIB, for example, reported 11 days field investigation between the Gibe III dam site and the lowermost end of the Omo basin.
[12]The 2010 AFDB consultant later questioned GOE claims of such destruction in two publication for the African Studies Centre at Oxford (Avery 2012, 2013). This point had been made by the ARWG in 2009 and again in = Carr 2012.
[13]The GOE (2009a) notes that food aid will be required for just "*two years*", when no flood will occur. No mention is made that adequate floods for recession agricultural plots would cease.

In 2009, the reservoir area had not yet been studied due to the challenges of access (Salini et al., 300 GEO RSP 002A, 2007). However, studies have since been reported, and Salini remain of the view that there are no appreciable losses (Pers. Comm. Studio Pietrangeli, 2010). This view has not been disputed by others conducting reviews, such as Sogreah (Sogreah, 2010), and if there are any losses, the topography dictates that the losses will feed back into the Basin and will not be lost.

There is ample geological literature and observation in the Gibe III region (see Chap. 3) documenting the highly fractured volcanic rocks, including basalts—conditions for strong leakage from the reservoir. Furthermore, statements of "no loss" of even leaked waters because of movement back into the Omo basin are meaningless— certainly from the standpoint of human life and ecological conditions in the lowermost basin and Lake Turkana. Seepage waters would require multiple decades— perhaps centuries, according to ARWG geologists—to migrate the hundreds of kilometers to the lowermost basin and Lake Turkana. This 're-balancing' would obviously have no impact on the human devastation already having occurred.

- **Agreement with the GOE's (2009b) false statement that a lowering of Lake Turkana's level by 1.8 m "during the reservoir filling" period" would be "insignificant" since this decrease in lake level is** *"within the annual fluctuation of the lake."*[14] In fact, the drop in lake level would be in *addition* to, not within, whatever annual fluctuation occurs. Even the first two meters of lake level drop—with its corresponding cessation of Omo River flooding of vast areas along the lowermost river and within the modern delta—would have disastrous impacts on the human communities already facing major malnutrition and health crises.

➢ It is possible to cite specific EIB and AFDB (2010) assessment patterns that reflect the widespread practice of complicity on the part of the global consulting industry (GCI) with their client government and development agency institutions. This complicity takes some particularly common forms, partially noted above.[15] They include:

- Agreeing to (sometimes even drafting) Terms of Reference (or Scoping) for assessments—terms that exclude consideration of possible or even likely major human and environmental impacts: for example, impacts on resource accessibility, local survival systems and preexisting livelihood stresses, as well as the critical matter of *cumulative effects* of actions—in the case of the Gibe III dam, cumulative effects of the dam, irrigated commercial agriculture and the hydroelectricity export transmission system.
- Offering 'critical' remarks regarding impacts of projects, but as *suggestions* for (optional) 'future study,' as 'considerations' for monitoring or mitigation of completed projects or other *optional* courses of action.
- Noting a destructive impact of major proportion without identifying its significance or the necessity of resolving the matter *prior to* inception or continuance of the project in question.
- Identifying serious, even life threatening or potentially disastrous impacts—in detailed or highly technical text within reports or assessments, rather than in clear form in Conclusions. Summary or Recommendation sections.

Table 6.3 summarizes these and related patterns in the 2010 EIB and 2010 AFDB reports.

The EIB assessment presents a combination of content ranging from all-encompassing positive statements regarding the 'benefits' of the Gibe III dam to the highly specific criticisms of GOE (2009b) methodologies. This statement in the report's summary, for example, exemplifies the overall positive evaluation of the project by the EIB consultant:

The ESIR Consultant considers the Gibe III project ... to initiate the economic development of the Lower Omo, one of the least developed regions of Ethiopia. It is recommended that any financial support to Gibe III development is closely linked to the simultaneous socio-economic development of the Lower Omo region, in order to maximize the benefits from the river flow regulation.

---

[14]"The first filling of the Gibe III reservoir will have a limited incremental impact on the lake's natural water level fluctuations. It is expected that there will be a lowering of the water level by about 1.8 m. This value is close to the annual fluctuations of water level in the lake, which is around 1.5 m, and well within the inter-annual fluctuations observed in the past (EIB 2010)."
[15]Between the two AFDB assessments, this failing is most evident in the 2009 socioeconomic assessment.

Table 6.3 European Investment Bank and African Development Bank impact assessments

Bank Impact Report	I Suggested mitigation or monitoring measures	II Suggested studies or reviews of issues or potential problems	III Identification of definite problem without analysis of significance	IV Identification of critical problem without assertion of urgency re: possible major destruction	V Identification of critical issue threatening major destruction — needing resolution *prior* to project continuance[1]
EIB	21	16	17	8	0
AFDB	3	19	10	2	0

Information is based on summary sections of both reports (Executive Summary, conclusions, recommendations) in 2010 EIB and AFDB assessments

[a] In the case of the Gibe III dam impact assessments by the GOE, EIB and AFDB, all such efforts were initiated only well after project

The AFDB's assessment of 2009, with Terms of Reference to identify the socioeconomic impacts of the Gibe III dam in the Lake Turkana region, is rife with analytical misinformation, inconsistencies. This assessment fails to undertake a substantive investigation of even the most basic aspects of pastoral, mixed pastoral/fishing and fishing livelihoods in the Turkana region, let alone the potential impacts of the Gibe III dam—or its linked irrigated agricultural enterprises—on these systems. The report blatantly asserts the Gibe III's overall "benefits" to residents in the lake region.

The EIB report identifies at least eight major problems predictable from the Gibe III dam development, including the virtual absence of baseline data for Omo River flow volume and river inflow to Lake Turkana, inadequate consultation with local communities, and inadequate 'compensation' for residents of the lower Omo basin. However, the report stops short of asserting that any of these are key issues necessitate resolution prior to continued dam building and operation (Table 6.3).

At least seventeen additional problems of significance are identified, though inconspicuously, in the EIB report. Most significant among them are the following.

- Absence of a sound management plan and the failure to substantively evaluate 'compensation' for the region's resident population.
- Fundamental conflict between power generation at the Gibe III dam and the high opportunity cost of releases of water for controlled flooding downstream.
- Likely occurrence of serious disease outbreaks downstream from the dam.
- Drying out of lands within the Omo delta and low riverside flats, where flood recession agriculture would be eliminated (see below).
- An 8–10-km retreat of Lake Turkana's northern shoreline—as a result of the radical reduction of Omo inflow to the lake (Fig. 5.2).
- Possible violent conflict among ethnic groups within the Ilemi region.

The EIB report fails to draw the obvious implications of such problems for the overall impact of the Gibe III project, even when some of them clearly spell disaster for the resident population and the region's environmental integrity. The main exception to this pertains to the report's recognition of the loss of recession agriculture, due to effects of the dam.

> *In the Lower Omo plain, the construction of the Gibe III dam will have significant and direct impacts on the reduction of cultivated land on the riverbanks and flooded areas (flood recession agriculture)...Approximately 67,600 people depend to some extent on flood recession agriculture.* (EIB 2010)

Even these statements are fully inconsistent with other assertions in the report. For example, the EIB apparently relies on GOE 'estimates' and assertions of 'limited indigenous reliance' on flood recession agriculture—therefore, limited vulnerability of the indigenous to livelihood systems to destruction from the Gibe III dam. The EIB quotes the GOE's false statement, shown below, without comment:

> **Only** *for Karo people where livestock is less significant might recession agriculture constitute the main source of livelihood.* [Emphasis added.]

In contradictory fashion, the EIB report references GOE estimates of 20,000 households within the delta that are practicing flood recession agriculture ('and cropping'), yet states elsewhere:

> *This means that 100,000 people could be affected in the delta alone.*

For Dasanech communities along the Omo River, the loss of flood recession agriculture would have disastrous consequences. Yet the EIB report repeatedly asserts that any loss of livelihood from elimination of flood recession agriculture could be "compensated" or "offset" by rainfed agriculture—a patently false statement. Even brief conversation with Dasanech residents or ecological ground reconnaissance in the lowermost basin and delta region would reveal—firstly, that there is no 'rainfed cropping' in the region (in contrast with the Mursi region upstream), and secondly, that multiple thousands of households depend on flood recession agriculture for their survival because their pastoral economy has been dismantled in recent years.[16]

A dominant feature of the EIB assessment's summary section is its inclusion of at least sixteen 'suggestions' for further study or review—few if any of which have been carried out since the report was released. Some of these suggestions pertain to impacts identified above; others are major issues that most critics would view as essential to resolve prior to conceptualizing the dam, and certainly before actually constructing and operating the dam. The EIB report suggests various studies studies, including of Omo River flow volume and inflow to Lake Turkana, the technical feasibility of a controlled flood program, livelihood development plans for the indigenous population, the socioeconomic and ecological character of the delta, indigenous recession agricultural systems and riverine zone biodiversity. *None of these are suggested as essential to the planning or implementation of massive development in the basin, when the reality is that virtually all of them are.*

Offsetting any such problems noted by the EIB are its assertions of 'positive' Gibe III dam outcomes for local residents and for the ecology of the lower basin (Table 6.3). The report offers at least twenty-one positive statements about the Gibe III project and its objectives, as well as GOE intentions and efforts to address local problems.[17] *The vast majority of these statements are unfounded, and the likelihood of GOE adoption of the numerous mitigation 'suggestions' is understood to be miniscule.*

The EIB assessment repeats GOE's assertions that its planned irrigation and canal works in the lowermost Omo basin will be for "social development" projects—mainly, small-scale irrigated agriculture for pastoralists and distribution of social services (GOE 2009b). The opposite was already occurring, however, even by the end of 2009 and early 2010—namely, the expropriation of major resource and village areas along the river for the establishment of irrigated, *large-scale government* as well as private farm enterprises. Desperation was already widespread, especially among the Dasanech.

---

[16] As detailed in Chap. 7, Dasanech villagers throughout the region consistently stated to South Omo/North Turkana (SONT) researchers that no GOE or other individuals had been to their villages for information about household members and population, floods, recession agriculture or any other activities.

[17] Although the GOE's violations of EIB funding procedural requirements may well have raised a roadblock for the bank to fund the dam project directly (a parallel situation with the AFDB and the World Bank), there are different interpretations as to why EIB funding did not actually materialize. Whatever the decision of the EIB would have been, the Ethiopians reached an agreement with the Chinese for just under USD 500 million for construction (see Chap. 1).

By comparison with the EIB, the AFDB has been far central in the rationalization of the Gibe III development, in several regards. These include its financial and technical support for::

(i) River basin development master plans and feasibility studies of dam and irrigated agricultural development in the Gibe-Omo basin, from 1990 through 1996.
(ii) Impact assessments for the Gibe III project (2009/2010)—reports largely complicit with the development.
(iii) Financial support and legitimation of the major transmission line for exporting electricity Ethiopia to Kenya and the broader eastern Africa region—a line including Gibe III electricity by all GOE accounts, despite AFDB and World Bank disclaimers.

The development bank's socioeconomic impact assessment (AFDB 2009) lacks an even minimally acceptable livelihood description and analysis; various failings of the report are dealt with in Chap. 9. The AFDB socioeconomic consultants apparently viewed their task as promoting the Gibe III dam project, as their own descriptions suggest. However, the assessment does briefly note Turkana community opposition to the dam because of its likely destruction of the lake and its fishery—and even to threats of direct action by local leaders. Nevertheless, the consultants refer to their role as one of emphasizing project 'benefits' to local residents.

The AFDB 2010 hydrological assessment is the primary one referred to by GOE, development bank and other officials. It presents a largely complicit with the project. As indicated in Table 6.3, it generally states concerns as 'suggestions' for studies or reviews, or as notations for possible monitoring or post-operational mitigation. The report's major reservation about the Gibe III dam project, stated only hypothetically, pertains to 'possible' irrigation schemes upstream. (As noted earlier, the AFDB consultant produced two later reports for the African Studies Centre at the University of Oxford (Avery 2012, 2013) that took a far more critical view of the impacts of irrigated commercial agricultural development on Lake Turkana—predicting, in fact, likely major destruction of the lake.

The 2010 hydrological assessment does identify some significant problems with the GOE's impact assessment of the downstream Omo riverine zone within Ethiopia (GOE 2009b), such as in its statement, *"Positive impacts on the lake's hydrology have been claimed, but there was no basis for these claims."* Other problems with the GOE's assessment that are identified in the AFDB report include: the absence of Omo flow volume and lake inflow baseline data—for which the AFDB review lays out an alternative method of calculation; predictable Omo River incision and water table drop, sediment capture and buildup at the dam, and impact of flood regime changes on fish breeding in the lake.[18]

Beyond the deficiencies identified above as in common for the AFDB and EIB reports, other major problems exist in the 2010 AFDB hydrological assessment.

(i) **Rather than recognizing the transboundary character of the Omo River itself, with its terminus within Kenya's national borders** (Figs. 1.2 and 1.3), **the AFDB report repeatedly asserts the river to be only within Ethiopia**. For example, the Executive Summary states:
The lake is almost entirely within Kenya, whereas the Omo River is entirely within Ethiopia.
Rather than acknowledge the transboundary nature of the Omo River itself, the AFDB report asserts only that the effects of reduced flow (unspecified as caused by the dam or irrigation) would impact Kenya, so that "management of the Omo Basin and lake water resources presents transboundary challenges"—presumably with management techniques after dam operations have been launched.

---

[18] A prediction of Lake Turkana's likely drop in level during and following the Gibe III reservoir-fill, by an ARWG physical scientist long familiar with the regioni, is presented in Chap. 5 and subsequent chapters.

> The matter of the Omo's terminus with Kenya, making it an *international* river, is obviously of major political significance in any economic development affecting the river's delivery of water to neighboring countries and must be negotiated with those countries' governments—*prior* to project approval or inception. This point reflects both international principles and Ethiopian regulations. The AFDB assessment skirts any explicit recognition of this reality.[19]

(ii) **Nearly all of the concerns raised in the 2010 AFDB assessment summary sections take the form of 'suggestions' for further study or review, mitigation or monitoring, as** (Table 6.3) **indicates.** Many of the 19 studies or reviews that are suggested in AFDB summary sections involve issues that most analysts would consider fundamental to even a minimally adequate impact analysis. These require resolution *prior to* the project's undertaking. Because the dam was already under construction for nearly three years by the time the AFDB assessment was produced, such suggestions clearly remain discretionary on the part of project developers and they presumably were meant to be undertaken *during or after* project completion and operation (see Table 6.3). For example, studies or reviews are suggested for bathymetric studies of the lake, hydrologic measurement, flooding patterns of the river, irrigation utilization of the river, fisheries status of the lake, calculation of the 'economic value' of the lake, socioeconomic studies, and 'integrated basin environmental and socioeconomic assessment,' and dam 'breakage.'

(iii) **The AFDB 2010 assessment fails to deal with the predictable 8–10-km retreat of Lake Turkana's northern shoreline during Gibe III reservoir filling period (EIB 2010) and the consequences of this retreat for major fish reproductive habitats there**. Such radical change spells disaster for hundreds of thousands of Dasanech and Turkana residents who depend on the northern lake's resources for their survival. The absence of the AFDB report's consideration of such a retreat is particularly disturbing in view of its recognition of the importance of the Omo's annual flood pulse of fresh water, sediment and nutrient input to the northern end of the lake.

(iv) **The AFDB assessment erroneously and repeatedly describes the extensive irrigation agriculture development along the Omo River as a separate phenomenon from the Gibe III dam**. The fundamental dependence of the irrigated farms on the Gibe III dam's regulation of the Omo River is essentially denied, or downplayed, as evidenced by these statements:

> *"This hydrological study has demonstrated that with the potential abstractions that might be implemented, the lake could drop up to 20 m.* **This is not attributable to Gibe III...**" [Emphasis added.] "*Irrigation abstraction is not a project component of the Gibe III project, as the dam is developed* **solely** *to generate power, but indirectly, the regulated flow sequence from the dam is expected to stimulate* **small-scale** *irrigation*". [Emphasis added.]

While the latter statement is technically accurate, it omits the reality that the dam's hydroelectric potential and the Omo Valley's possible irrigation agriculture, including large-scale commercial operations—were long considered integrally by development interests.

The 2010 EIB impact report references GOE information (see below) regarding irrigation plans in its consideration of irrigation feasibility potential in the lower basin, and report includes a GOE map indicating major irrigation areas planned (see chap. 7). The GOE has openly solicited international agribusiness and related industrial investment in the Omo basin prior to and during preparation of the EIB and AFDB assessments.

Since feasibility studies and planning (even some construction) of large-scale irrigation infrastructure were undertaken well before the AFDB assessment and were, in fact, referred to by its author, planning for both the hydroelectric dam and its linked irrigated agricultural enterprises was certainly known to the AFDB consultant.

---

[19]The AFDB does quote, but without comment, the 1996 Master Plan for the Gibe-Omo river basin: "…This means that in the international context a bilateral agreement should be reached between the two countries (Kenya and Ethiopia) before either country changes the natural flow of the river…Any major change in the river's regime as, for instance, by the construction of a dam for the development of hydro-power, or, more significantly, by the development of large-scale irrigation in the south of the basin, would be almost certain to raise issues internationally…".

The development banks had actually supported various aspects of planning for river regulation and large-scale irrigated agriculture in Ethiopian river basins for decades. In the case of the AFDB, the bank had directly funded two desk studies of major policy importance. One of these was the 1990 Master Plan for all Ethiopian river basins, prepared by the global consulting firm, WAPCOS (see Chap. 2). The other—contracted to Woodroofe and Associates a few years later for USD 6.1 million—was the 1996 Master Plan for the Gibe-Omo River Basin, which estimated both the basin's potential hydrodam electricity and irrigation agricultural development.

The GOE's 2009 downstream impact assessment formulated three types of irrigation:

- Small-scale 'settlement' agriculture,
- Small-scale commercial irrigated farms, and
- Large-scale irrigation farms—i.e., over 7200 ha (cotton and sugarcane plantations).

Two additional statements within the AFDB 2010 impact assessment are sufficient to illustrate the report's failure to link the Gibe III dam and major irrigation systems and to directly address the seriousness of the combined dam-irrigation effects on the indigenous population.

"If irrigation development proceeds as planned in the Omo Basin, the lake will diminish, as will biomass and fisheries. **Whether this is of consequence should be the subject of a separate study and consultations** *with the Kenyan Government and stakeholders, and should be based on a proper economic evaluation of Lake Turkana and its resources."* [Emphasis added.]

"The biggest impacts can be expected to arise from Ethiopian Government plans for large scale irrigation within the Basin. These have not been studied in this Report, but they must be taken into account at some point, as it will be futile to disregard developments in the Basin as a whole as they affect mitigation measures. As developments are inevitable to cope with rising population pressure and food security needs, a balanced view needs to be agreed upon between Kenya and Ethiopia, through detailed studies and dialogue, on what environmental impact is acceptable, and on what mitigation measures can be adopted, and how they will be managed. This process has already begun." [Emphasis added.]

In reality, major scale irrigated agricultural enterprises planned or under construction downstream from the Gibe III dam are clearly dam enabled enterprises, not 'unrelated' phenomena. Among numerous linkages are these closely related ones:

- Large irrigation systems, commonly for monocrops and high chemical use—require predictability of water and calibration of cropping phases—predictability that can only be provided by regulation of the river flow volume and its periodicity.
- Irrigation infrastructure cannot accommodate wide fluctuations of the river's natural state.

Linkages between river regulation by major dam construction and large-scale irrigated agricultural enterprises are well-documented in developmental policy and economic and scientific literature concerning drylands Africa. They have been in play throughout the decades of 'modern' Ethiopian river basin development, beginning with that nation's first irrigated agricultural schemes and major (at the time) dam development—in the Awash Valley.

Since large-scale irrigated agricultural development along the lower Omo River is dependent on the river's regulation by the Gibe III dam, these two developments have a **causative connection**. This matter is discussed in Chap. 10, along with the critical matter of human rights violations being committed.

Literature Cited

African Development Bank (AFDB). 2009. A. S. Kaijage, N. M. Nyagah, Final Draft Report, Socio-economic analysis and public consultation of Lake Turkana communities in Northern Kenya, Tunis.

African Development Bank (AFDB). 2010. S. Avery, Assessment of hydrological impacts of Ethiopia's Omo Basin on Kenya's Lake Turkana water levels, Final Report. 146 pp.

Africa Resources Working Group (ARWG). 2009. A commentary on the environmental, socioeconomic and human rights impacts of the proposed Gibe III Dam in the lower Omo River Basin of Southwest Ethiopia. http://www.arwg-gibe.

Avery, S. 2012. *Lake Turkana and the lower Omo: hydrological impacts of Gibe III and lower Omo irrigation development*, Vols. 1–2. African Studies Centre, University of Oxford.

Avery, S. 2013. *What future for Lake Turkana? The impact of hydropower and irrigation development on the world's large desert lake*. African Studies Centre, University of Oxford.

Carr, C.J. 1977. *Pastoralism in crisis: the Dassanetch of Southwest Ethiopia*. Chicago: University of Chicago Department of Geography Papers. (.90), 319 pp.

Carr, C.J. 1998. Patterns of vegetation along the Omo river in southwest Ethiopia. *Plant Ecology* 135(2): 135–163.

Carr, C.J. 2012. Humanitarian catastrophe and regional armed conflict brewing in the border region of Ethiopia, Kenya and South Sudan: The Proposed Gibe III Dam in Ethiopia, Africa Resources Working Group (ARWG), 250 pp. https://www.academia.edu/8385749/Carr_ARWG_Gibe_III_Dam_Report.

Government of Ethiopia (GOE). 2009a. Ethiopian electric power corporation (EEPCO), CESI, mid-day international consulting engineers (MDI), Gibe III hydroelectric project, environmental and social impact assessment, Report No. 300 ENV RC 002C Plan.

Government of Ethiopia (GOE). 2009b. Ethiopian electric power corporation (EEPCO), agriconsulting S.P.A., mid-day international consulting, level 1 design, environmental and social impact assessment, additional study of downstream impacts. Report No. 300 ENV RAG 003B.

Government of Ethiopia (GOE)., and Ethiopian Electric Power Corporation (EEPCO). 2010. Environment and social issues related to Gibe III hydroelectric project, GIB HEP Office. http://ethiopianembassy.org/pdf/giveiihydroelectricproject.pdf.

Kinde, S., and S. Engeda. 2010. Fixing Gibe II—engineer's perspective. http://www.digitaladdis.com/sk/Fixing_Gilgel_Gibe_II.pdf.

Thomas, G.A., and K. DiFrancesco. 2009. The heritage institute for the world bank. Final report, rapid evaluation of the potential for reoptimizing hydropower systems in Africa.

USDA, United States Department of Agriculture. 2006. MODIS satellite imagery August 11, 2006; August 21, 2006.

Woodroofe, R., and Associates, with Mascott Ltd. 1996. Omo-Gibe river basin integrated development master plan study. Final Report. I–XV Vols.

Open Access This chapter is distributed under the terms of the Creative Commons Attribution-NonCommercial 2.5 International License (http://creativecommons.org/licenses/by-nc/2.5/), which permits any noncommercial use, duplication, adaptation, distribution and reproduction in any medium or format, as long as you give appropriate credit to the original author(s) and the source, provide a link to the Creative Commons license and indicate if changes were made.

The images or other third party material in this chapter are included in the work's Creative Commons license, unless indicated otherwise in the credit line; if such material is not included in the work's Creative Commons license and the respective action is not permitted by statutory regulation, users will need to obtain permission from the license holder to duplicate, adapt or reproduce the material.

The Dasanech of the Lowermost Omo Basin: From Adaptation to Development Debacle

Abstract

The decline of Dasanech pastoral economy in recent decades, due to increasing marginalization by powerful external political and economic forces, has forced the majority of Dasanech to move to areas along the Omo River and its active delta or around the northeastern shores of Kenya's Lake Turkana. Radical reduction of river flow volume, lake retreat and elimination of the river's annual flood brought about by the Gibe III dam, together with dam enabled irrigation agricultural enterprises, would destroy the key components of Dasanech livelihood. Most flood recession agriculture would be eliminated, along with 'last resort' livestock grazing lands, forest resources and fish reproductive habitats in the lowermost Omo and Lake Turkana northern shoreline. Even if the highly unlikely and in any case inadequate artificial flood program promised by the Ethiopian government were implemented, Dasanech survival systems would have already been decimated. The looming crisis of region-wide hunger and mortality is intensified by the Ethiopian government's eviction and expropriation of thousands of Dasanech villagers for large-scale irrigated commercial agriculture. Political repression and a culture of fear prevail. As the crisis unfolds, Dasanech communities, faced with vanishing means of survival, would inevitably contribute to rapid escalation of cross-border, interethnic armed conflict.

Dasanech Pastoral Decline: Roots and Responses

➢ Throughout the first half of the twentieth century, the Dasanech pastoralists sustained a system of wide-ranging movements throughout the region's dryland plains, as did other indigenous groups in the transboundary region. These movements ranged across different types of habitats (Carr 1977; Bassi 2011), facilitating diversified food production and complex patterns of risk minimization, as described in Chap. 4.

By the second half of the century, the combination of direct territorial restriction by the Kenyan and Ethiopian governments, conflict with other disenfranchised pastoral groups in the region and other pressures effectively confined the Dasanech to the plains west of the Omo River and east of the Kibish River for many years. The Dasanech were forced to relinquish most of these areas:

- The upland plains of the Ilemi Triangle—formerly, a 'buffer zone' created by agreement between the Ethiopian monarchy and the Kenyan colonial administration.

- Grasslands in the Kenya/Ethiopia border area to the northwest of Lake Turkana and southeast of the Ilemi Triangle—especially critical during drought periods.[1]
- Woodlands and grasslands along the Kibish River, Koras Mountain and much of the pasture lands between Koras Mountain and the Omo River (Fig. 1.3)—lands previously shared with the Nyangatom but increasingly fought over as resource degradation and loss of land access have taken hold in the region.
- Semi-arid plains and relatively wetter foothills east of the lowermost Omo River, due to hostilities with the Hamar group to the east (Fig. 1.3).

These territorial restrictions caused severe overcrowding of Dasanech livestock, with a loss of access to critical resources during prolonged drought and other stress periods. Centuries old risk minimization and recovery strategies of the Dasanech were no longer effective. With no relief available to them, the Dasanech were subjected to continued deterioration of their lands, and the herds of individual pastoral families plummeted.

While Dasanech pastoralism persisted in the second half of the twentieth century (Fig. 7.1), their livelihood was severely threatened to the point where major adjustments were necessary for their survival.

➢ Widespread ecological degradation was evident throughout much of their remaining territory by the late 1960s and early 1970s. Instead of the uniquely complex, mosaic-like complex of habitats and vegetation, vast areas became susceptible to invader species of plants—a process that has continued to the present. This writer studied the structure and floristics of 'natural' versus disturbed grassland communities in major plains habitats, including including relict sandy beach ridges, black cracking (margallitic) clay basins and ancient floodplains adjacent to the Omo River (Carr 1977). There are some common features of ecological deterioration in these different communities:
- Sharp reduction of total vegetation cover with the creation of significant bare areas.
- Invasion and spread by numerous 'disturbance indicator' plant species—mostly unpalatable species.
- Increasing erosion (both water and wind driven) with the loss of topsoil. Much of this degradation is irreversible in practical terms, particularly once sheet erosion and certain species invasion occur (most recently, *Acacia nubica*, *A. horrida* and *Prosopis juliflora*—the last of these, an introduced species, has spread throughout the region). Vast areas are now with negligible plant cover and susceptible to intensive wind and water erosion.

This writer (Carr 1977) described 'phases' of this deterioration for much of the lowermost Omo basin based on construction of ecological (vegetation and soil) transects along a gradient of grazing pressure in the above geomorphic units (Figs. 7.2 and 7.3).[2]

Most Dasanech pastoral lands are now so severely degraded that their recovery potential is in serious question, even under the best of circumstances.

➢ **Dasanech elders uniformly describe the degradation of their grasslands as the most important change in their survival efforts, at least until the recent aggressive policies by the GOE along the Omo River**. The statement of one Dasanech elder on the west bank, recorded in 2011, is illustrative of the narratives throughout Dasanech lands.

When I was a young man, our land was big. Now we don't live in those lands, as the governments have taken them from us, and now they let others into our lands. Once we had the lands of good grass, but now we have no grass except for short times when the rains come, and even then the grass goes away quickly. Before, there were so many wild animals roaming the land: topi, oryx, wildebeest, lion, cheetah, foxes and many more. Now most of them are gone. Have they gone north? Or west? They have been chased away by the loss of grass that is killing our cattle too, and killed by poachers and by those who have gotten many guns from the war.

---

[1]The British administrator from Kenya (Mr. Whitehouse) who played a key role in the decision to remove the Dasanech (termed the "Marille" by the Kenyan government) from this region described to this writer that no account was taken of the Dasanech population or neighbors' survival needs in boundary determination.

[2]Grassland and other vegetation types west of the Omo River are determined by a combination of factors including ancient sediment depositional patterns (for example, alluvial and fluvial—with soils ranging from silty clay relict floodplains through margallitic, or black cracking soils to sandy beach ridges) and land use pressures.

Fig. 7.1 Dasanech herders and livestock. *Top* Cattle at pastoral village in highly overgrazed area. *Center left* Women slaughtering goats in pastoral village. *Center right* Young male herder at stock camp. *Bottom left* Mid-day milking. *Bottom right* small stock watering at Omo River

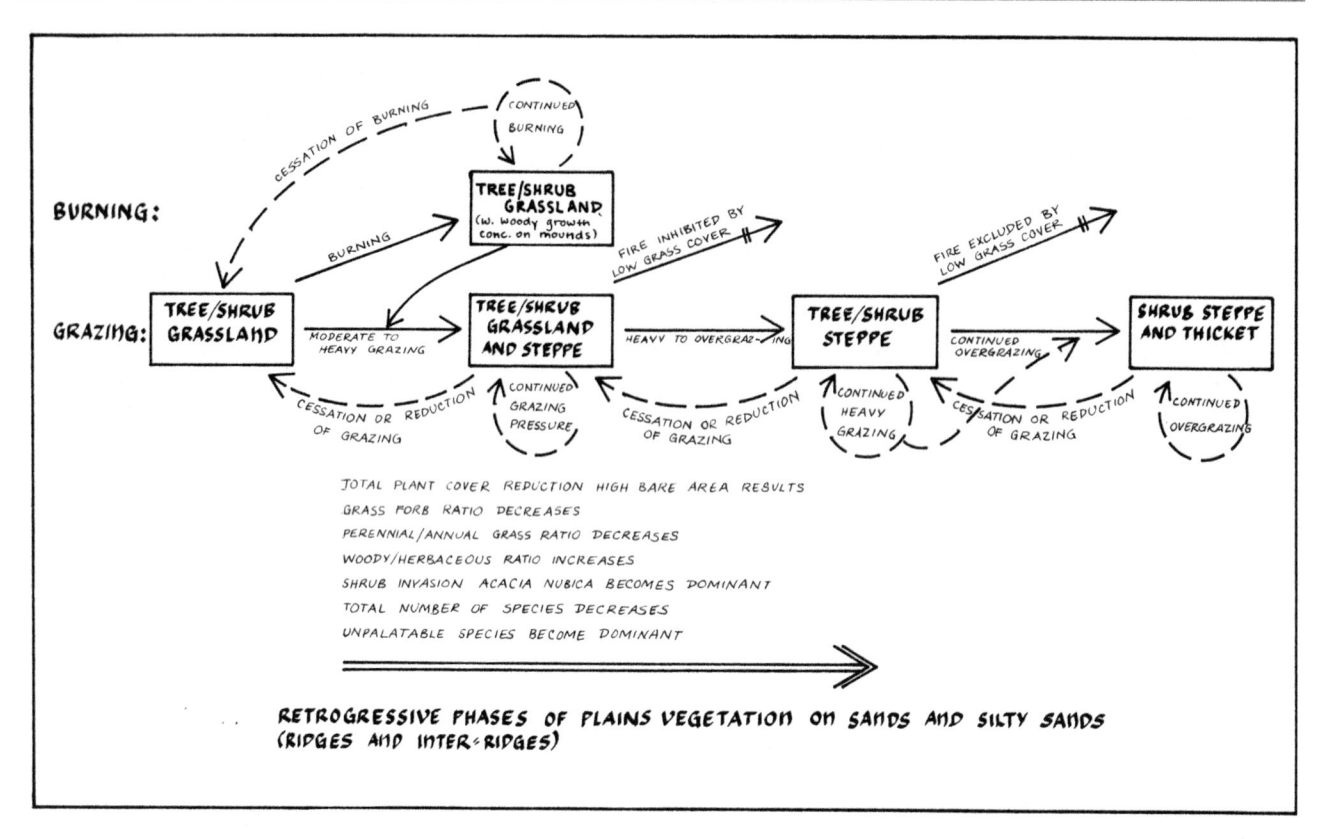

Fig. 7.2 Pasture deterioration phases in upland plains. Dominant and comparable soil type in Dasanech region of the lower Omo basin (sandy-silt soils on relict beach/interridge areas)

Pastoral Dasanech households (Fig. 7.4) **attempted to recover from unprecedented resource losses by any means possible**. Among other adjustments, they altered their seasonal herding patterns, lessened the mobility of villages—relying instead on highly mobile stock camps, utilized all types of social cooperation and exchange relationships, and engaged in raiding of neighboring groups deemed hostile at the time. This writer studied these complex patterns during the 1970s, with herd sizes, seasonal movements and production activities recorded for six different village areas, indicated in Fig. 7.5 (Carr 1977).

The differentiation of Dasanech social segments is reflected in the 'nodes' of pastoral settlement: for example, the Rendelli segment is distinguishable from Inkabela and Oro segments. These patterns were identified by this writer (Carr 1977) and are summarized in Fig. 7.5. At the time, most Dasanech were pastoralists residing in the upland plains—taking their livestock to the Omo River primarily during the dry seasons and especially during prolonged drought periods. Those Dasanech households engaged in flood recession agriculture along the Omo River went there primarily during periods of intensive farming labor, although some Dasanech were well-established there in semi-permanent villages. A significant proportion of villages along the river were of the Eleli segment of the Dasanech, for example. Some of the poorest Dasanech had already begun fishing in the lowermost delta and along Lake Turkana's northern shoreline in the delta.

A series of prolonged drought periods in the region during the 1970s and 1980s, accompanied by increased livestock raiding between the Dasanech, and their neighbors—especially the Turkana and Nyangatom, greatly worsened Dasanech coping efforts. Herd numbers plummeted for the vast majority of households. Kenyan officers administrating the Ilemi Triangle began permitting Turkana pastoralists back into Ilemi lands by the 1980s, so even the longstanding 'illegal' but persistent use of the Ilemi by the Dasanech became sporadic, at best. This situation continues to the present day.

Fig. 7.3 Phases of ecological decline in lower Omo basin pastoral lands. *Top left* Ilemi Triangle 'healthy' grassland—well-drained silty/sand relict beach ridge. *Top right* young herders with cattle in seasonally inundated cracking clay grassland basin within the Ilemi Triangle with relict beach ridge behind. *Center photos* two intermediate overgrazed conditions in silty/sand soils with discontinuous cover and plant 'invader' species. *Bottom photos* Severely overgrazed conditions; *left*—unpalatable vegetation (e.g., *Cadaba rotundifolia*, *Euphorbia nubica*) severely malnourished cattle

The urgency for additional grazing lands for Dasanech livestock in these three different environments emerged from a combination of factors, including:

- Worsening environmental degradation of the dryland plains, due to overgrazing, with increased livestock mortality and herd losses.
- Increasing hostilities with neighboring groups, particularly the Turkana and Nyangatom.
- Continued exclusion from the Ilemi Triangle (and contiguous lands dominated by the Turkana).

Fig. 7.4 Dasanech pastoral villagers and activities. *Top left* Dasanech at major (*dimi*) ritual. *Top right* Woman at noon-time milking in village. *Bottom left* Hut-building with riverine shrub branches and animal skins. *Bottom right* Men slaughtering ox in plains village

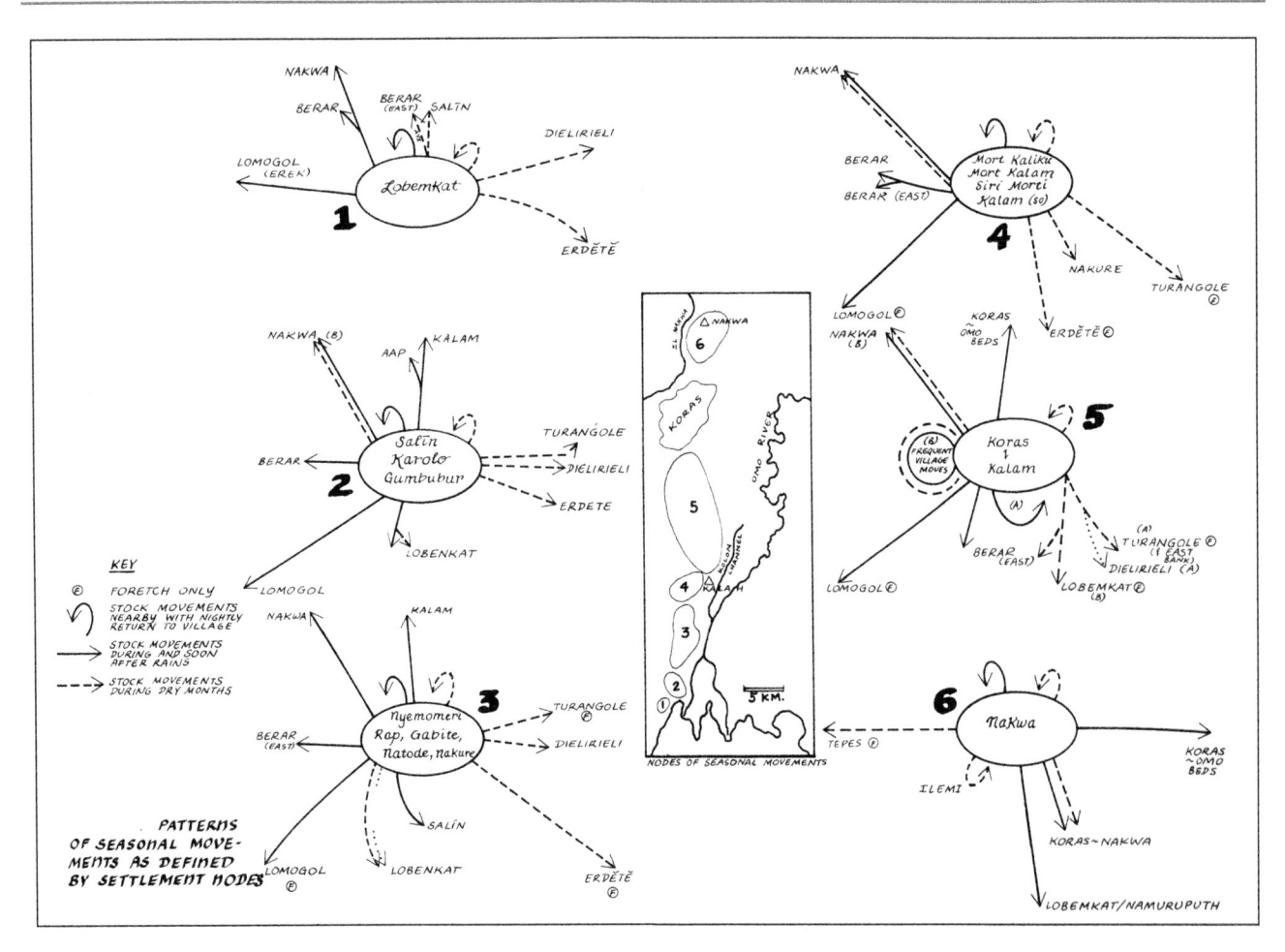

Fig. 7.5 Seasonal movement patterns of six Dasanech settlement areas west of the Omo River. *Source* Carr (1977)

Conflicts between Dasanech and Nyangatom communities intensified, as both groups competed for water and grazing at the seasonally flowing Kibish River, where they have long done extensive well digging in the dry months, in lands around Koras Mt. and eastward to the Omo River (Fig. 1.1). They also competed for settlement, wild food gathering and hunting locales along the Omo River. Similar conflict relations existed with the northern Turkana—primarily over access to the Ilemi region and the grasslands of the Ethiopia-Kenya border region northwest of Lake Turkana. Venturing into these lands for the Dasanech was at great risk—both because of possible Turkana and Nyangatom attack and because of seizure of their livestock and sometimes shootings by Kenyan police.

Adapting from Upland Pastoral Life to Diversified Economy at the River

Faced with rapidly diminishing herds and environmental degradation throughout their pasturelands, by the early 1980s, Dasanech pastoralists had no choice but to rely on the Omo riverine zone and lake environments, both for last resort grazing and economic diversification to recession agriculture. The only real options for them were locales (i) within the delta, which was actively expanding (see Chap. 1), (ii) along the river upstream from the delta to the southern extent of the Nyangatom—initially, along the west bank (due to the danger of attack from the Hamar to the east), and (iii) around the northeastern shoreline of Lake Turkana to Ileret, Kenya—and southward along the lake (Fig. 1.3). At first, most Dasanech settlements in these area were seasonal, but over time many of them remained throughout the year.

The overall migration pattern is shown in Fig. 7.6. Some Dasanech pastoralists remained highly mobile, even with greatly diminished herds. The more mobile pastoral households continued sending stock camps to the eastern Ilemi Triangle from

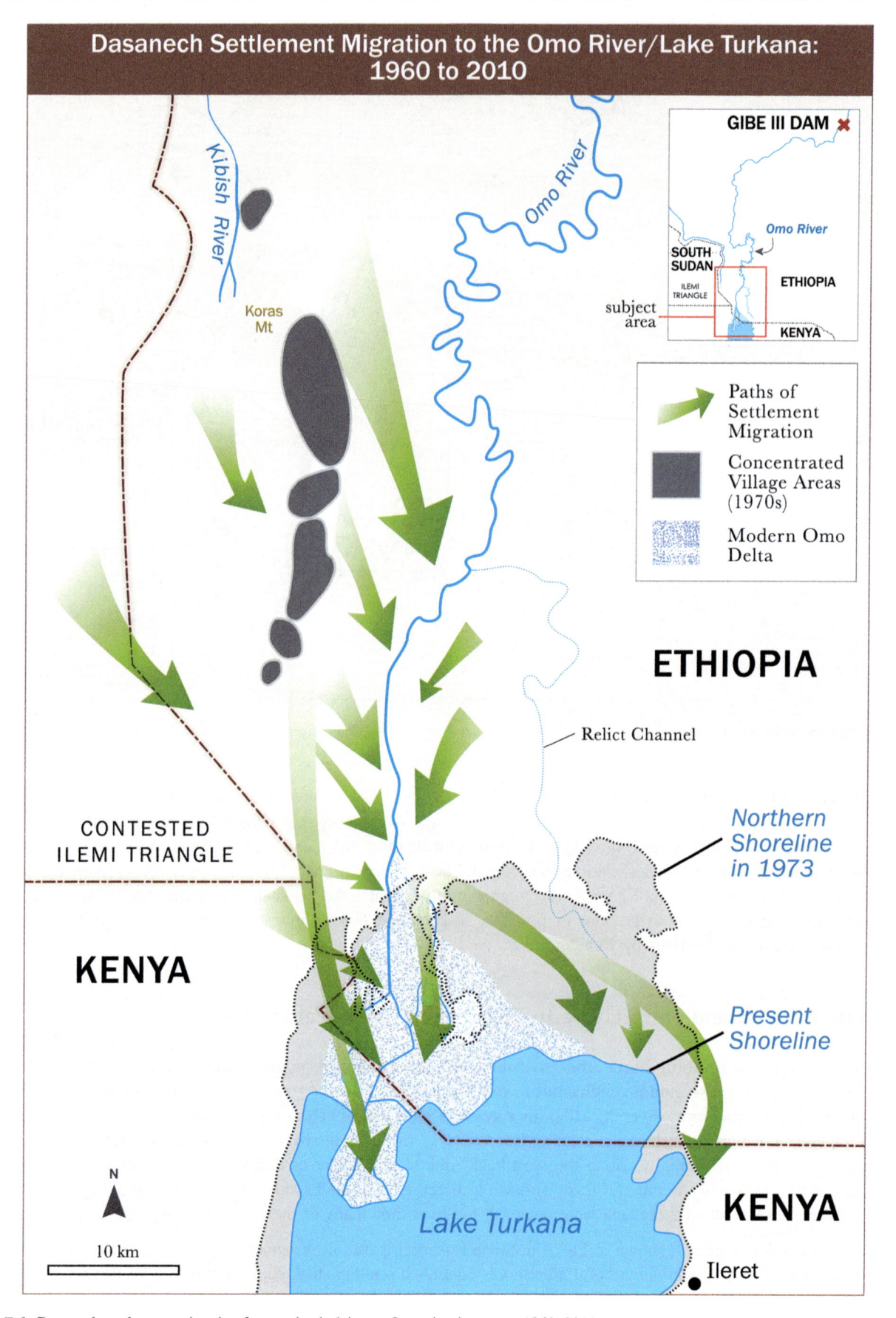

Fig. 7.6 Dasanech settlement migration from upland plains to Omo riverine zone: 1960–2011

which they were officially excluded, as well as around Koras Mountain and the Kenyan-Ethiopian border areas—including the coveted grazing lands of *Meyen* and *Labur*. (Their use of the Ilemi lands, however, was with strong risk of having their livestock seized by police and herding in the border lands and around Koras Mountain carried equal danger of attack by the Turkana and Nyangatom (a reciprocal threat that remains and intensifies in times of stress).

> In spite of these efforts to access their traditional lands, Dasanech pastoralists continued to suffer major cattle and small stock losses. They became increasingly dependent on the Omo River zone for both water and pasturage. Even pastoral villages remaining several kilometers away from the Omo River had at least some family members engaged in food production in the riverine zone—primarily flood recession agriculture and fishing.

Several new possibilities for Dasanech livelihood activities emerged during the 1980s, even though these have been wholly insufficient to compensate for their massive economic decline.

- The Omo delta began a period of expansion southward, with its terminus extending well into Kenya's Lake Turkana (Fig. 1.2). This new delta area has amounted to more than 500 km^2 for possible new settlement as well as livestock grazing, flood recession agriculture and fishing.
- Local Kenyan officials began permitting the Dasanech to return to the arid lands around the northeastern shoreline of Lake Turkana where the group had once resided. Large numbers of Dasanech, particularly from one cultural segment, responded to this opportunity. In fact, many of them continued southward from Ileret to their fluctuating border with the Gabbra (Fig. 1.3). Relations between the Dasanech and Gabbra are generally hostile although they were once largely peaceful, with resource sharing in many locales. Hostilities have intensified as available pasturage has disappeared.

Settlement shifts by most Dasanech from their degraded upland plains to the Omo River and Lake Turkana environments directly reflects the failing economic conditions for the group as a whole.[3] Many Dasanech who settled west of the Omo began moving back to the riverine zone on the east bank of the Omo River—lands where they had resided decades earlier. The danger of attack by the Hamar to the east effectively confined their villages and herds close the river.

A variety of village and household livelihood patterns have emerged, but with overall economic decline and shift in food production activities (Fig. 7.7). Despite the Dasanech's long-term cultural dislike of fishing, an increasing number of the poorest households have had no choice but to begin fishing in the channels of the lowermost Omo River or the northernmost waters of Lake Turkana. The number of fishing households has increased substantially since the 1980s, for several reasons. first, the plains environment and livestock herds have continued to deteriorate; second, the traditional lands of the Ilemi were no longer available; and third, the Ethiopian government has evicted thousands of villagers from their riverine lands—forcing most of them into the already crowded Omo delta where planting land is scarce. A wide variety of subsidiary but essential livelihood activities are carried out in both riverine and upland areas (Fig. 7.8).

➢ **Tens of thousands of Dasanech now reside in three major areas along the lowermost Omo River—the west bank, the east bank and the active delta region.** There are no reliable demographic data for these Dasanech, despite the obvious importance of such information for a detailed assessment of the human impacts that would be caused by the Gibe III dam. SONT researchers identified major Dasanech village areas that are variously seasonal or of year-round duration. Village areas recorded and the large zone of GOE expropriation of villagers from their settlements and riverside lands are indicated in Fig. 7.9.

GOE population estimates vary from 40,000 to more than 200,000. The Ethiopian government's census results were clearly projections rather than actual village based counts—projections likely generated in towns well removed from Dasanech settlements. Villagers throughout the region emphatically state that neither government nor other individuals have

---

[3] Although strong differences in wealth have long existed, the resulting precipitous herd decline affected the Dasanech as a whole, since those stock owners fortunate enough to retain larger herds were obliged to distribute at least some of their wealth to varying combinations of clan, age-set and affine (in-law) relations. These social structural relations are best described by Almagor (1978).

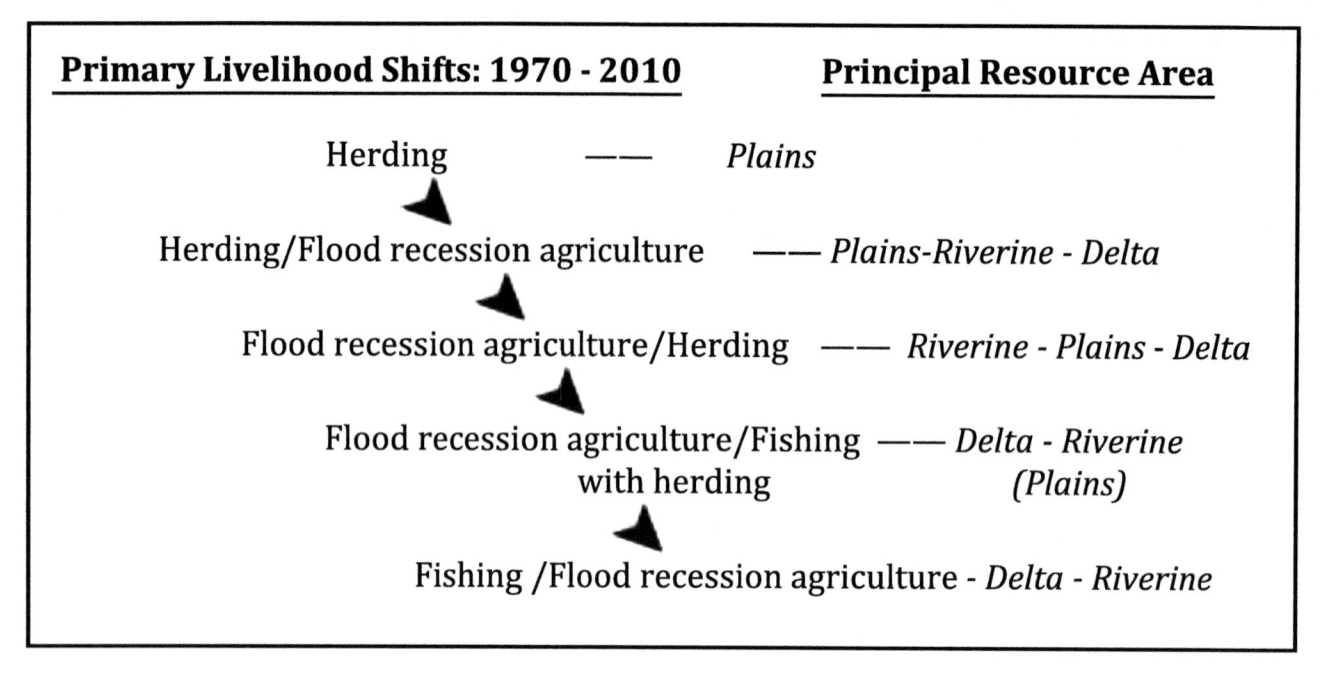

```
Primary Livelihood Shifts: 1970 - 2010        Principal Resource Area

              Herding                  ——      Plains
                 ▲
    Herding/Flood recession agriculture   ——  Plains-Riverine - Delta
                 ▲
    Flood recession agriculture/Herding   ——  Riverine - Plains - Delta
                 ▲
    Flood recession agriculture/Fishing   ——  Delta - Riverine
              with herding                        (Plains)
                 ▲
    Fishing /Flood recession agriculture - Delta - Riverine
```

Fig. 7.7 Major Dasanech livelihood decline from upland pastoral economy (west side of the Omo River)

visited to record census information.[4] Staff members of non-governmental organizations engaged in periodic relief operations amongst the Dasanech expressed their frustration at having to rely on official government estimates.

> Any figure between these extremes points to the major numbers of Dasanech whose lives are threatened by the Gibe III dam and dam enabled irrigated agricultural development.

It is likely that the European Investment Bank's impact report (EIB 2010) relied on GOE population figures, given the EIB consultant's statement that the duration of its field investigation of the entire area from the Gibe III dam to Lake Turkana was only ten days. Such a period is wholly inadequate for even a minimally acceptable population sampling among the Dasanech. Estimates based on satellite imagery without extensive ground-based investigation would be entirely inaccurate, since Dasanech village complexes fluctuate in location—sometimes extremely rapidly—in response to shifts in pasturage, water, disease, immediate security and other conditions. Moreover, villages visible in satellite images are often abandoned ones, or reoccupied by another ethnic group—especially common in contested areas. There are also major Dasanech settlement shifts between Kenya and Ethiopia, reflecting rapidly changing environmental and socioeconomic conditions.

SONT researchers could not conduct systematic population counts, due to GOE restrictions and widespread fear of government reprisals by villagers. Dasanech village areas were therefore identified only through ground reconnaissance, and primarily during the dry season (Table 7.1). Small scattered villages were not recorded.

The combination of information from all reliable sources, including SONT's own field-based investigation in selected village areas and several non-governmental officials interviewed, suggests **a minimum Dasanech population of 60,000 to 70,000**—an approximate figure, at best. Whatever the accurate figure for the Dasanech population, it is clear that there are tens of thousands of them residing along and nearby the Omo River with their livelihood dependent on sustainment of and access to the Omo River.

---

[4]These villager statements were corroborated by a senior government officer in Omorate (see Fig. 1.3 and maps below), who affirmed to this writer that no *ground* census had been carried out.

Fig. 7.8 Dasanech woman making dugout canoe from a tree trunk in a small clearing in the Omo riverine forest

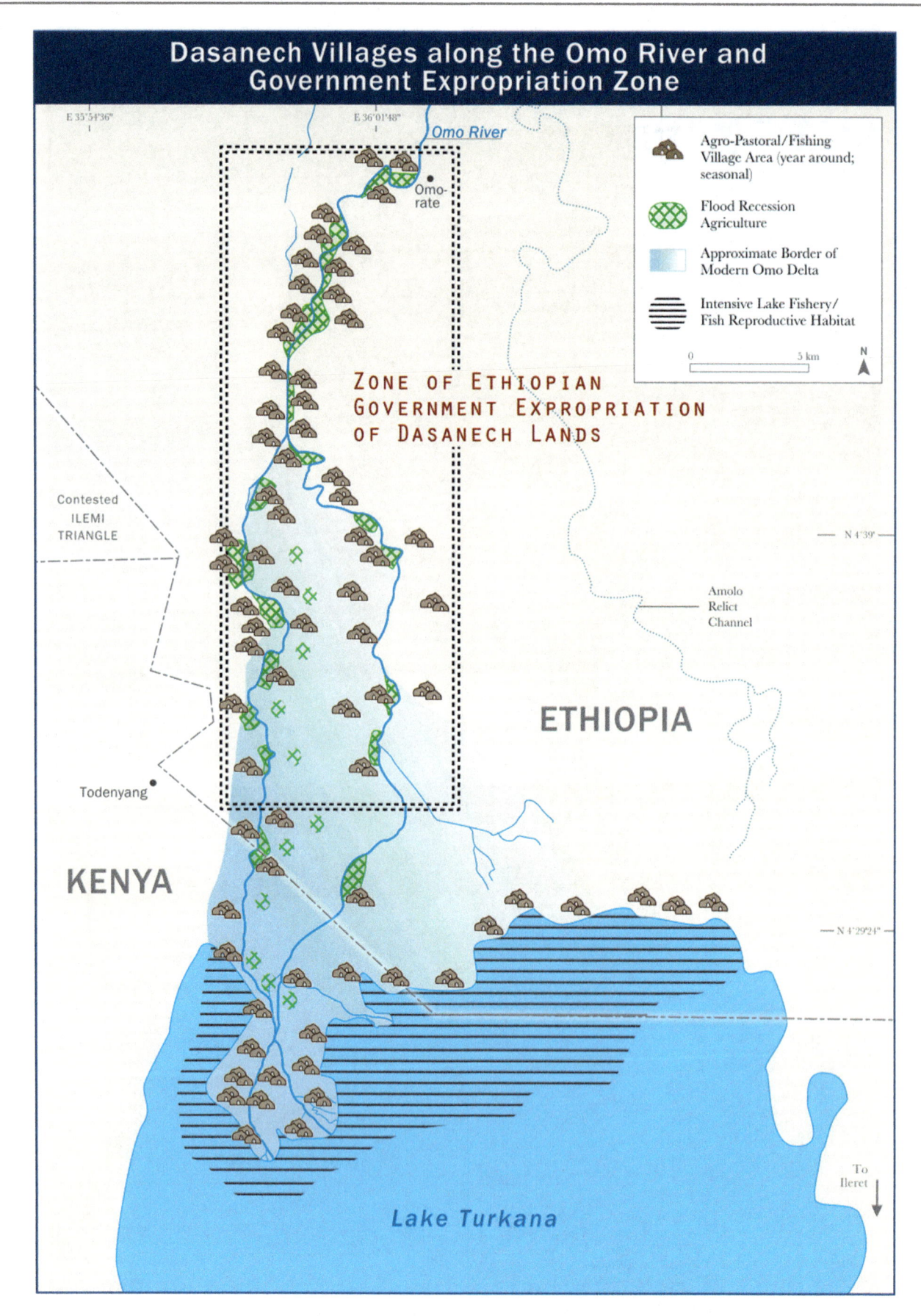

Fig. 7.9 Zone of Ethiopian Government Expropriation of Dasanech villages and livelihood areas

Table 7.1 Dasanech village complexes along the Omo River: 2009–2010

West bank	East bank	Modern delta
Goto	Afuor	Lomosia
Damish	Kapuse	Ngymoru—Lulung
Akudingole	Tieli rieli	Ediporo
Nyemomeri	Lobele	Kipur-cheria
Atalago	Edete	Andora/Ililokelete
Salany (Salin)	Lobaoi	Bokom (both banks of channel)
Lochuch	Apaluka—(largely fishing)	Lonyangereng
Koro	Derish	Nakabila
Bokom	Aluuli	Chongochongo
Olmin	(Loyere, Nyikiki—east of Omo River zone)	Aachuun
Gabite		Jiete-Konya
Toltale		Budori
Malsipi		Nakoida
Terishichess		Araloput
Tuushe		Lokielinya
Turite		Kaakulu
Lopelebin		Koranyilutu (Koro Nyingabite) Naichari
Naakale		

Three areas of villages were identified: west bank, east bank and active (modern) delta

> While these villages (many of which are included in the map of Fig. 7.9) were recorded during the dry season and some of them are seasonally mobile or have shifted altogether, their presence during all or part of the year indicates the extreme dependency of the Dasanech on riverine resources.

- **The economic decline of the Dasanech, the dominant sequence of which is summarized Fig. 7.7, is evident from household data collected by this writer, in 1970/1972 and 2009/2010**. Of 75 households surveyed in the 1970s, information was updated for 35 of them in the latter period (Table 7.2). Households from the original survey were randomly selected from four of the six major pastoral settlement areas indicated in Fig. 7.5.

A number of patterns are evident from these household timeline data:
- All of the 35 households recorded had relocated—mostly to the riverine/delta zone—due to major changes in their key production activities.[5] Most had diversified their production from herding to include flood recession agriculture or fishing, or both. Nearly all households moved part or all of their members to the riverine/delta region, seasonally or year round. Significant shifts in authority and other social relations have accompanied these changes (Fig. 7.10).
- Precipitous herd decline has overwhelmingly driven this economic transformation (with minor exception). Some households lost all livestock; most lost a quarter to two-thirds of their cattle, while small stock losses varied considerably.
- As cattle herds have declined, Dasanech herders have relied more on small stock since they are far more adapted to degraded pastures and to days without watering. Dasanech long-term cultural preference for cattle over small stock has been dominated by this necessary shift.

Dasanech herd owners on the west bank and in the delta were interviewed concerning their loss of livestock to disease. A typical response was:

Only a few people near Omorate got help for livestock diseases. Those living in the delta and the whole west bank received no help from the government.

- Some herd owners diversified their economic production and managed to partially rebuild small stock herds though barter or sale of grain or fish.

---

[5] The relative similarity or difference between the specific household in question and neighboring ones from the original sample was also recorded. Names of household heads are excluded for the protection of individuals.

Table 7.2 Dasanech household wealth status and livelihood change. Households from west bank of the Omo River: 1972 versus 2009

	1972					2009				
Household Number	Village Location P-Plains R-River D-Delta	# Cattle	# Goats Sheep	Farm	Fish	Village Location	# Cattle	# Goats Sheep	Farm	Fish
1	P	140	180	No	No	P	38	30	Yes	No
2	P	150	45	No	No	P/R	40	48	Yes	No
3	P	60	0*	No	No	R/D	18	30	Some	Some
4	P	28	65	No	NO	R/D	5	22	Yes	No
5	P	47	5*	No	No	D	0	0	No	Yes
6	P	280	350	No	No	R/D	15	0*	Yes	Yes
7	P	44	60	Yes	No	R/D	18	24	Yes	No
8	P	120	210	No	No	R	35	68	Yes	No
9	P/R	32	44	Yes	No	D	0	5	Some	Yes
10	P/R	30	12	Yes	No	D	2	0	No	Yes
11	P	310	155	No	No	R	34	50	No	No
12	P	41	15	No	No	R/D	15	33	Yes	Some
13	P	58	40	Some	No	P/R	22	6*	Yes	No
14	P	155	85	No	No	P/R	10	34	Yes	No
15	P	550	200	No	No	P/R	18	400	No	No
16	P/R	50	30	Yes	No	D/R	12	22	Yes	(Yes)
17	P	210	60	Yes	No	R	32	85	Yes	No
18	P	800	110	No	No	R	160	60	No	No
19	P	540	85	No	No	P	105	73	(Yes)	No
20	P	65	38	Yes	No	R/D	40	55	Yes	(Yes)
21	P	75	22	No	No	R/D	14	5	Yes	No
22	P	0	8	Yes	No	D	0	0	No	Yes
23	P	80	130	(Yes)	No	P/R	65	60	Yes	No
24	P	90	45	(Yes)	No	R	20	48	Yes	No
25	P	82	70	No	No	R/D	7	36	Yes	(Yes)
26	P	125	150	No	No	R/D	26	38	Yes	No
27	R	4	23	Yes	Yes	D	0	5	Yes	Yes
28	P	87	110	No	No	R	26	44	Yes	No
29	P	12	15	Yes	No	D	0	3	No	Yes
30	P/R	4	15	Yes	No	D	0	3	No	Yes
31	P	65	50	(Yes)	No	R	16	24	Yes	No
32	P	90	55	No	No	P/R	80	110	(Yes)	No
33	R	12	20	Yes	No	R/D	10	48	Yes	(Yes)
34	P	65	30	No	No	P/R	20	35	Yes	(Yes)
35	R	0	4	Yes	Yes	D	0	0	No	Yes

[a]Lost in raids by neighboring Nyangatom or Turkana

Fig. 7.10 Dasanech male elders in the Omo riverine zone

- Households with remaining livestock now depend on grazing locales within the Omo riverine and active Omo delta lands during much or all of the year. Although most Dasanech are now agropastoral, many of the households with remaining livestock depend on last resort grazing lands such as *dipa*—a locality of the eastern delta with relatively rich pastures. According to villagers, waters from both the Omo River and Lake Turkana sustain these grasslands.

Last Resort Survival: Desperate Dependence on Omo River Annual Flood

➢ **Most Dasanech now have major dependence on the Omo riverine zone—for flood recession agriculture, dry season and 'last resort' livestock grazing, fishing and a host of secondary production activities. All of these depend on the sustainment of the Omo River's annual flood.**

Settlements and major livelihood activities in the delta vary from seasonal to semi-permanent. Most of the latter are located along the river above the maximum flood level. Including in the uppermost portion of the delta. Livestock herding in the delta fluctuates widely with seasonal changes—from short visits, especially during drought months, to year-round presence, depending on environmental and social conditions.

Dasanech households and communities practice flood recession agriculture on annually flooded waterside flats upstream from the delta, on some low riverbanks along the water's edge, and within the active delta (Fig. 7.11).[6] Contrary to the

---

[6]Other opportunities exist for flood recession agriculture, including a few locales where backup of Omo river waters into the terminus of gathering streams (for example, *Kolon*—see Fig. 4.3) or incomplete channels. Some oxbow meanders well north of the Dasanech—for example, in Kara lands (Fig. 1.3) also provide conditions for flood recession agriculture when they are inundated with Omo River waters during high flow periods.

Fig. 7.11 Flood recession agriculture and Dasanech planters. *Top left* Dasanech girl in flood recession farm in delta. *Top right* agricultural plot burned prior to flood. *Center* Omo inside bend with sandy/silt spit—annually flooded with agricultural plot; adjacent forest does not flood. *Bottom left* Dasanech riverside village. *Bottom right* Dasanech tending farm at the river

GOE's assertions, they do not plant in the relict floodplains because flooding does not occur in these vast flats—nor has it for thousands of years. As described in earlier chapters, overbank flooding does not occur upstream from the modern delta, except in a few small localities where there is a break in the natural levee.

Also contrary to GOE and development bank reports, *rainfall in the lower Omo basin is insufficient for farming*, as a brief conversation with any local resident substantiates.[7]

While a relatively small number of Dasanech were practicing flood recession agriculture along the Omo River well back in the twentieth century, flood dependent planting has now become the dominant means of survival for tens of thousands of villagers. In the early years, planting on seasonally flooded riverside flats was a means of minimizing risk through production diversification, rebuilding herds through bartering grain for livestock and fulfilling social obligations (see Chap. 4). Particularly after livestock losses following territorial constriction caused by government actions, extended drought, disease, and raiding by neighboring groups, Dasanech households exchanged grain from their farms along the Omo for small stock and other items. Small grain reserves were widely evident and recorded by this writer in the early 1970s.

➢ **The Dasanech consistently describe their present crisis as one of too little flooding, not excessive flooding.** This is true for both the active delta and the upstream waterside localities—together accounting for the majority of Dasanech settlement in recent years. Chapter 6 details the GOE's false claim of frequent major floods that are destructive of human life and property.

SONT researchers recorded these statements by Dasanech agropastoral and fishing elders.

> • *We have to move with our households and animals to stay close to the river channels and the delta where there is water and grass for us and our animals and where we can farm. Lands east of the river are bare and dry, except when there is good rain. We only find food and water here.* [Female agropastoralist, western edge of Omo delta]

> • *We never had such hunger in my father's and my own time—until recently, when we became old men. Only this hunger can force us to farm and even eat fish! Herders do not eat fish—the fish eaters are the 'dies.' Our times were better. Our land was good for all Dasanech and even a man with fewer animals could eat well from his animals. The animals were healthy, and they gave much milk. We didn't have all this bush—it has come to now when we have lost our land of good grass. I used to stay with my animals in the grasslands. Now I must be at the Omo River where I have learned to farm. I must farm because my family will not eat from our few animals. Even people with many animals don't get enough milk. And when the flood doesn't come to our land, and we cannot farm, we eat fish. I don't want any more of our children to die, so we eat fish.* [Riverside male elder]

➢ **Most Dasanech communities are now dependent on the Omo delta and its immediate environs for their survival—whether for their direct use of delta resources for livestock grazing, flood recession agriculture or fishing, or their indirect but vital use of exchange relations with delta villagers.**

- The southward expansion of the Omo delta to its present 500 km^2 area has coincided with the Dasanech's diversification to agriculture, fishing and wild food gathering (Fig. 7.12). Large numbers of impoverished Dasanech would otherwise have faced catastrophic level conditions of hunger. As noted in previous chapters, the Omo delta was previously a limited land area with 'birdfoot' morphology (Fig. 1.2) and had few localities suitable for planting, livestock grazing and settlement.
- The total area utilized for flood recession agriculture varies widely, depending on flood conditions and multiple other factors. Based on SONT field observations and satellite photos, a medium to high range of 4000–6000 ha is a reasonable estimate for such planting—from Omorate southward to Lake Turkana, including the modern Omo delta.

---

[7]When the ephemeral Kibish River (which dissipates in the dry Sanderson's Gulf, just south of Koras Mt.) has sufficient flow, flood recession agriculture is done along the riverbanks. Presently, large numbers of Nyangatom are settled around Kibish.

Fig. 7.12 Dasanech Life along the Lower Omo River. *Top left* Girls in flood recession farm near the delta; wooded natural levee in rear. *Top right* Goats watering at Omo River. *Center left* Extremely malnourished cattle at watering. *Center right* East bank Dasanech fishers with catch. *Bottom* Agropastoral village near the west bank, in the delta region

Although no precise estimation of the number of hectares utilized for recession agriculture or the population dependent is feasible without ground-based survey and direct interviewing of local residents, an estimate of tens of thousands of Dasanech facing such destruction is entirely reasonable.

Under the present conditions of region-wide political repression and fear of Ethiopian security forces in virtually all villages, combined with the GOE's effective ban on independent research in the area, establishing reliable baseline data for the precise population under threat from loss of flood in recession agricultural lands—like the estimation of the population in general—is not possible. These data are critical for the precise assessment of the crisis unfolding,

- Environmental conditions in the delta are diverse—providing opportunities for multiple production activities that combine to provide the survival of Dasanech communities. This comment by a villager to this writer is typical in its reference to the importance of this aspect of the delta.

  Many people here have some small stock, but they don't give enough milk. So a lot of people have to rely on what they can harvest from their farms. In some places, the crops do well if they get flood from the river, but in some places there is no flood and crops fail, so people have to find other ways to survive, like eating fish or buying grain from those who have it. [Dasanech female elder at village on west bank in delta area]

- Villagers throughout the delta typically are flexible in their production activities in response to their changing conditions. Both settlement locations and resource use patterns within the active delta naturally shift with changes in Omo channel morphology and the river's annual flood as well as other environmental and social factors. A majority of villages in the western portion of the delta have been relocating to its *central and eastern* portions due to the threat of Turkana attack, decreased annually flooded lands and reduced planting locales on the western margin, drainage conditions that favor woody vegetation (shrubs), and tsetse infestation. Villagers with calves often have no option but to have them graze crop stubble (Fig. 7.13) when floods are insufficient to replenish vegetation in their locales.

Dasanech village settlement areas and flood recession agriculture locales active at the time of SONT research are indicated in Figs. 7.9 and 7.14. Many of these locales are now expropriated (see below). The map also indicates desiccation of the region that would result from the radical reduction of the river's flow volume and inflow to Lake Turkana, brought about by the Gibe III dam and large-scale irrigation commercial agricultural along the river.

> **Within the modern delta, a mosaic-like pattern of different vegetation types and water conditions has developed with the recent expansion of the delta—providing habitat for livestock grazing, fishing and food gathering.** The fluctuating and critically important wetland habitat and sharp transition between riverine and upland environments (on the west bank) are shown in Figs. 7.15 and 7.16. The GOE's ESIA (GOE 2009) falsely describes a far more mesic (relatively wet) environment than is in fact the case—even presenting a highly detailed vegetation map indicating an active river channel departing the river just above Omorate to Lake Turkana. This channel (locally termed *Amolo*, is actually a relict one.[8]

- Dasanech planting on low riverside flats (including point bars and low sand/silt spits) and within the modern delta plant a variety of crops, including these traditional ones:

sorghum/millet—Dasanech name—'*ruba*'	Pigeon peas—*gadda*
maize—*nakapono*	other vegetables—*eri*
squash—*bote;* gourds—*turum*	tobacco—*tampo*
sweet potato—*lokoto*	beans—*am haamo*

---

[8]The significance of this misrepresentation is that it suggests the existence of a relatively "favorable" environment for 'alternative' resource use by the Dasanech. In fact, this channel is an ancient one and the channel hasn't flowed for many years. Formerly an active part of the Omo River system, *only* pools of water form during rainy periods. The relict floodplains around the *Amolo* channel are generally poor soils—cracking clays and silts.

Fig. 7.13 Calves in starvation condition grazing in stubble of riverside farm plot

Except for sorghum, which is critical for meeting both subsistence and exchange needs, these crops have long been grown primarily for household consumption. Sorghum has lower water requirements than maize and it is well suited to shifting river conditions—certainly by comparison with maize. Moreover, in a good flood year, planters can produce two sorghum crops and sorghum seeds are available le from previous harvests. Beyond its importance as a household food staple, sorghum is widely traded for small stock (sheep and goats), even cattle—both within Dasanech economy and in the broad arena of interethnic exchange (Fig. 1.6). Sorghum yields are determined by multiple factors including type of sorghum planted, extent and duration of Omo River flood, soil type and land use practices. Experienced Dasanech planters up to fifteen or twenty distinct types of sorghum. "*It all depends on the floodwaters*," according to most respondents.

Several major soil types prevail in the delta zone where flood recession agriculture is practiced. These vary among sand, silt and clay-like textures. Although two of these (locally termed *maal* and *digirte*) are considered by most to be superior, planting is done in a wide range of soils, with different degrees of flooding. Tools are simple—primarily axes (*hoolte*), sticks (*yugeny*) and 'pangas' (*nyewolo*). Labor patterns are flexible, with men generally doing more of the clearing and harvesting while women perform much of the planting and weeding as well as assist in harvesting.

Labor for flood recession agriculture is highly variable among communities. Some have pervasive cooperation in most phases of farming while others have relatively sharp household delimitation of plots accompanied by limited cooperation—primarily for land preparation and harvesting. Problems of crop diseases are sometimes severe, especially from insects and rust. Dasanech villagers consistently state that the Ethiopian government has not helped them deal with these diseases; in fact, most complained of no agricultural services at all from the government. This major hardship for the river dependent Dasanech parallels that of the pastoralists. Contrary to GOE reports, village elders flatly state that they have received no

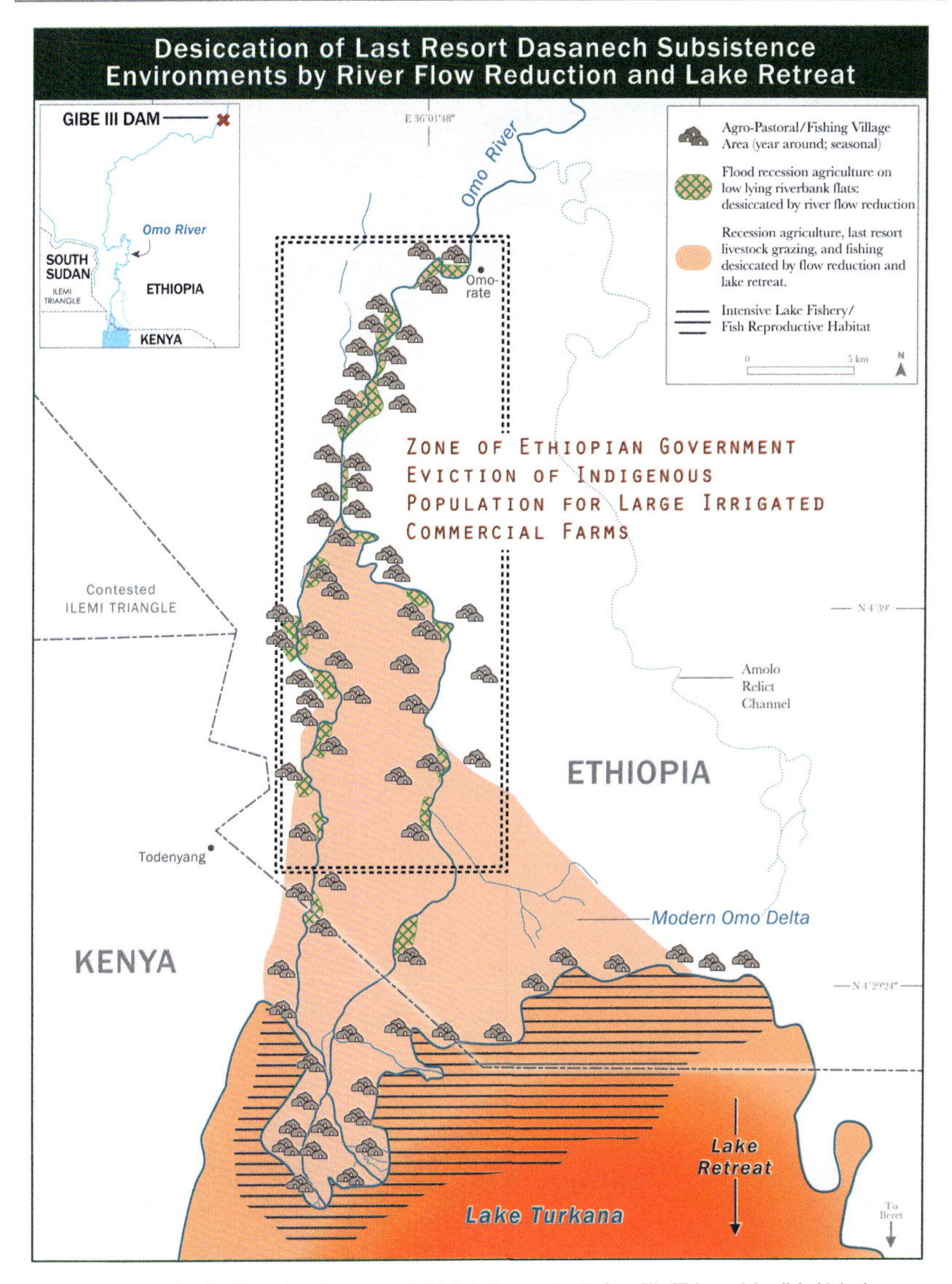

Last Resort Survival: Desperate Dependence on Omo River Annual Flood 131

Fig. 7.14 Desiccation of modern Omo delta and northern end of Lake Turkana predictable from Gibe III dam and dam-linked irrigation systems. Destruction would include flood recession agriculture, last option grazing and fishing habitats

Fig. 7.15 Wetlands at the Omo delta terminus at Lake Turkana. Dasanech cattle grazing

Fig. 7.16 Dasanech village complex at shoreline near northwestern extreme of Lake Turkana, close to Omo delta wetland

Fig. 7.17 Dasanech crossing Omo River at high flood stage for transactions between west bank residents and Omorate traders. Natural levee supports closed woodland—without overbank flooding

government assistance for the rampant and often devastating livestock diseases affecting their herds. Moreover, very few of them have access to the extremely limited NGO assistance in the area.

Dasanech knowledge and management of these conditions has been key to the sustainability of these systems—as distinct from unsustainable ones introduced and commercial style developments brought to the region by the government and private interests. When crops fail in parts of the modern delta, for example, agropastoral villagers may sell or barter some of their remaining small stock to households with more successful harvests in order to obtain sorghum.

With the retreat of Lake Turkana (Figs. 1.2 and 7.14) and expansion of the modern delta in recent decades, tree and shrub growth has increased alongside decreased flow in the main western river channel, according to local residents. Farming has declined in these areas and villages have moved to central and eastern portions of the delta where annual floods facilitating recession agriculture are more likely to occur. Even where flooding occurs in the western delta, crop yields are diminished.

➢ **Reciprocity relations between Dasanech and pastoral and agropastoral households, as well as among the region's ethnic groups, have long been critical to the survival strategies of all**. These include strong east/west bank exchange relations (Fig. 7.17). Many pastoral Dasanech settled near the river plant in the delta region by negotiating labor-sharing arrangements with households there. For example, livestock owned by the delta residents are sometimes sent to stock camps in Kenya-Ethiopia borderlands northwest of the lake with labor provided by herders from west bank villages. In return, west Dasanech of the east bank and eastern delta commonly trade with the Hamar to the east—especially exchanging their sorghum for Hamar small stock. They also acquire knives, axes, earthen pots and hides from the Hamar, who have better access to these highland products. Dasanech settled on the west bank and in the western delta, on the other hand, are more likely to trade their agricultural product (primarily sorghum) for Turkana small stock—conflict conditions permitting. When relations between the two groups are relatively peaceful, Turkana herds may even be permitted to graze in the delta region, in return for wet season grazing by Dasanech herds in Turkana controlled lands.

Without a successful sorghum harvest, the Dasanech in the riverine and delta region are often forced to sell their remaining livestock in order to survive. This has in fact been the case for countless households in recent years. East bank Dasanech agropastoral households, for example, often have no alternative but to market their remaining cattle or small stock in Omorate or other markets frequented by Hamar, Arbore or Kenyan Somalis.

➢ **Wild food gathering is critical for Dasanech survival, especially during harsh times**. These include periods of prolonged drought when livestock milk and other products are reduced or when for example, the omo annual flood is insufficient for agriculture or for successfull fishing. The poorest Dasanech—those without livestock or farm plots—rely on this food source much of the time, particularly in recent times.

Most wild food gathering is done in the riverine and delta zones—precisely the environments most in line for desiccation from the destroyed from the effects of the Gibe III dam and irrigated commercial agricultural enterprises.

➢ **Fishing is done by thousands of Dasanech households and is key to survival of most of them**. It is no longer limited to the 'poorest of the poor' communities and households. The critical role that fishing, as well as recession agriculture, has Agriculture have come to play, in the face of major herd decline, is stated by this elder in the southeastern Omo delta:

We eat fish every day. If others have sorghum, they will cook it and eat it. Some of us have been fishing for a long time —more than those others. Other Dasanech came to join us after they lost their cattle and small stock from drought and disease. Thousands of us are here! Many of our people died because they had nothing to eat—and before they could get here to try to plant or fish.

- Fishing communities utilize their catch for varying combinations of domestic consumption and exchange—whether barter or cash sale in local markets. Dasanech fishers exploit whatever river and lake locales they can access, depending on Omo River flow patterns and annual flood occurrence, seasonal shifts in fish life cycles, availability of fishing gear, security conditions and other factors. Major fishing areas include the lowermost Omo and fringing Omo delta channels and the nutrient-rich waters along Lake Turkana's northern shoreline—waters nourished by the river's annual freshwater and 'pulse' that sustains fish reproductive locales there (Figs. 5.2 and 7.14).
The most common fish caught by the Dasanech at the mouth of the Omo channels and along Lake Turkana's northern shoreline are tilapia and Nile perch. The most common catch species for the Dasanech are the same as those for Turkana fishers (see Chap. 9). Dasanech male fishers also hunt crocodile and hippo at night—the populations of which have radically dwindled to the point of endangerment[9].
Many fishing households still use the simplest of technologies: metal spears and harpoons with string (from barter with other ethnic groups), locally crafted dug-out canoes fashioned from riverine forest trees (Fig. 7.8) or simple rafts constructed from doum palm trunks lashed together (see photos in Chap. 9). Others have wooden boats and nets, with additional gear. Dasanech fishers originally obtained much of their knowledge of fishing, as well as fishing technology, from the Turkana. They now acquire equipment through barter, purchase and capture. During investigations in Turkana villages along the lake's northwestern shoreline, for example, this writer recorded numerous accounts of violence—killings as well as gear thefts—between Turkana and Dasanech fishers.

➢ **Dasanech fishers report rapidly declining fish catch, especially with the incursion of commercial fishing fleets based in Ethiopia—companies promoted and protected by the GOE**. Commercial catch is primarily destined for urban Ethiopian and export markets. Efforts to develop facilities for fish refrigeration and processing were initiated by the GOE for years, with active solicitation of investment by the SNNPR and federal government by the early years of this writer's investigation, when three major companies were active in the lowermost river and in Kenya's Lake Turkana.

---

[9]Hippos and crocodiles, like other riverine zone wildlife (elephant, buffalo, primates including colobus monkey and baboon, a wide range of reptiles, etc.) lived in abundance along a major proportion of the lower Omo river during this writer's early field work (e.g., in the early 1970s). Hunting was nearly incidental until the 1980 s when the GOE began developing Omorate and enterprises along the river, and when firearms became increasingly dominant in communities throughout the region. Much of the wildlife population would require *restoration* if this natural heritage of Ethiopia (and basis for tourism) is to be valued differently from the wholesale destruction underway. Young males—now often with little accountability to elders—often kill for sport.

Company boats easily overwhelm the poorly equipped indigenous Dasanech and Turkana fishers. Villagers in the delta describe enormous fish discard (waste) by the motorized commercial boats, including large deposits of fish bones and other waste, that create major problems for their small nets and also destroy fish reproductive and feeding habitats. One Dasanech fisher responded this way when asked for his view of the foreign fishing boats:

What is bad about them is the amount of fish they kill. Some of them [fillet] the fish right there and throw the waste in the lake, so this makes the lake water poisoned. We get small nets from some of them, but mostly we are losing our fish, so nets don't help us if the fish are gone.

The Ethiopia-based fishing corporations have steadily increased their catch range into Kenya's Lake Turkana, where they are in clear violation of that nation's sovereignty. These company boats extend their fishing ventures as far southward as North Island (Figs. 1.3 and 5.2), often display the Ethiopian flag and are often accompanied by Ethiopian guard boats. They pay no fees to Kenya, nor do they obtain fishing licenses or make catch reports to Kenya's local Beach Management Units.

Turkana representatives from villages along the northwestern shoreline have appealed to the Kenyan government numerous times to expel the foreign fleets, and Kenyan fisheries officials are fully aware of the situation. As of early 2015, the Kenyan government had taken no effective action, despite innumerable requests by Turkana fishers and their representatives—including in the locally based Beach Management Units. Nor have the development banks, in their assessments of lake conditions, fishery status or socioeconomic conditions (AFDB 2009 and 2010) raised the issue of the incursion and impact of Ethiopian commercial fleets in these Kenyan waters on fish stocks and on the worsening interethnic conflict among fishers.

> **What amounts to state-sponsored piracy into Kenyan waters is a matter of international sovereignty, but also greatly worsens the crisis faced by vast numbers of Kenya's Turkana whose livelihood is dependent upon the lake's fishery.**

➤ **Violent conflicts between Dasanech and Turkana fishers who are increasingly desperate to secure catch from these northern waters constitute a major problem in the region. The northernmost portion of the lake accounts for many of the 'hotspots' of conflict expansion** (Fig.5.3).

The frequent violence over gear theft and sporadic killings noted above frequently spreads to pastoral and agropastoral communities in the region (and *visa versa*) with extensive series of reprisals between the two ethnic groups. The frequent points of conflict among fishers—indicated in the map of Fig. 5.3—are also the likely points of major expansion of local conflicts into regional ones. This trend is already well underway and is greatly amplified by the plummeting of livelihood resources as a result of Gibe III dam and irrigated agricultural development.

Governmental, non-governmental and U.N.-based accounts of conflicts among fishers and in the border region generally exclude consideration of the actual *causes* of this mounting crisis. Instead, the 'solutions' prescribed are most often either interethnic 'mediation'—without account of the real dynamics at play—or additional militarization by the Ethiopian and Kenyan governments, or both.

Ethiopian Expropriation and Political Repression of Riverine Communities

The Ethiopian government is engaged in extensive and systematic human rights violations of tens of thousands of indigenous pastoralists and agropastoralists along the Omo River, downstream from the Gibe III dam site, in order to pave the way for new commercial agricultural enterprises. Evictions of villagers and expropriation of their riverine resources by GOE police and militia frequently involve beatings and arrests—and reportedly, torture. Dasanech resistance to their dispossession nevertheless occurs, when possible.

➢ **GOE denials of its expropriation actions are contradicted by available planning documents over many years**. As detailed in Chap. 6, the Ethiopian government excluded mention of its obvious plans for such large-scale commercial farms in its environmental and social impact assessments. Gibe III feasibility and planning documents over the years—including the AFDB-funded Master Plan for the Omo River basin (Woodroofe & Associates 1996), for example—assessed hydroelectricity and irrigation potential. The EIB 2010 assessment included a map of the GOE's projected commercial agricultural development along the river (Fig. 7.18). In 2009, this writer was directly informed of the government's ambitious plan for agricultural development and major irrigation in the lowermost basin during a discussion with a senior agricultural ministry official in Addis Ababa.

In the 1980s, the Ethiopian Derg's commitment to large-scale irrigated agricultural development on indigenous lands was evident from the Ethio-Korean project at Omorate (Fig. 7.9). The GOE and major international aid organizations have prioritized hydrodam development with agribusiness and power production over indigenous land rights since the Koka dam and Awash Valley developments during the post-war Haile Selassie years, despite well-documented disastrous impacts on the Oromo and Afar peoples in the Awash Valley (Bondestam 1974; Carr 1978; Kloos 1982), and elsewhere. (Chapter 2 outlined the early decades of comprehensive planning for dam and dam-linked irrigated agriculture in Ethopia's river basins, including this prioritization.).

The specifics of Ethiopian and foreign ownership of the commercial farm development are relatively well documented along the Omo in the traditional territory of the Mursi and neighboring groups (Fig. 1.3). Hundreds of thousands of hectares are leased or planned for Ethiopian, Indian, and European and other corporate and private investors (Human Rights Watch 2012; Oakland Institute 2011). Information presented by Turton and colleagues at the University of Oxford (available at www.mursi.org) reports that 30 of the 52 Mursi and Kwegu villages (at least 58 % of them) are in areas that are either already delineated for sugar plantation development or are being offered by the government of Ethiopia for private agricultural development. According to this field-based data, the expropriation process involves 73 % (114 of 157) of Mursi and Kwegu agricultural sites. Even two individual sugar block plantations total more than 162,000 ha, with a planned total of 245,000 ha for sugar plantations.

Table 7.3 indicates key concessions documented by the above-mentioned NGOs and researchers during investigations between 2010 and 2013.

> The above figures exclude enterprises underway or planned further downstream, especially the transboundary section of the river. Government prohibition of visitors and investigators has been extreme in the lowermost basin where the Dasanech and Nyangatom also face eviction, particularly since 2009. 'Observations' by international aid groups allegedly investigating possible abuses in the area (including USAID and DFID), moreover, are generally 'facilitated,' or 'informed' by the GOE itself.

The major crops planned for these commercial farms are cotton, sugar and oil palm, with agricultural product directed to international export and Ethiopian urban markets. These crops are high water consumption/chemical-requiring crops. These agribusiness industrial enterprises require major water diversion through canals and irrigation channels, as well as facilitation of chemical/waste discharge into the Omo River—both having disastrous effects on both downstream Omo and Lake Turkana peoples and environments. The extensive irrigation systems are already partially constructed and undergoing expansion. These include a large number of diesel pumps for water diversion and large canals.

➢ **Like the Mursi/Bodi ethnic residing upstream, the Dasanech and Nyangatom face major evictions by the GOE for the establishment of large-scale, irrigated commercial agriculture and supporting infrastructure**. While it is known that private investors along the lower Omo River are from Ethiopia, India, China, Europe and the U.S., but information regarding specific individual, corporate and government owners of these irrigated commercial farms are difficult to obtain due to the GOE's refusal of access to the area by investigators and to the government's policy of extreme surveillance. All of these effectively prohibit local residents from providing information and places them under threat of severe reprisal.

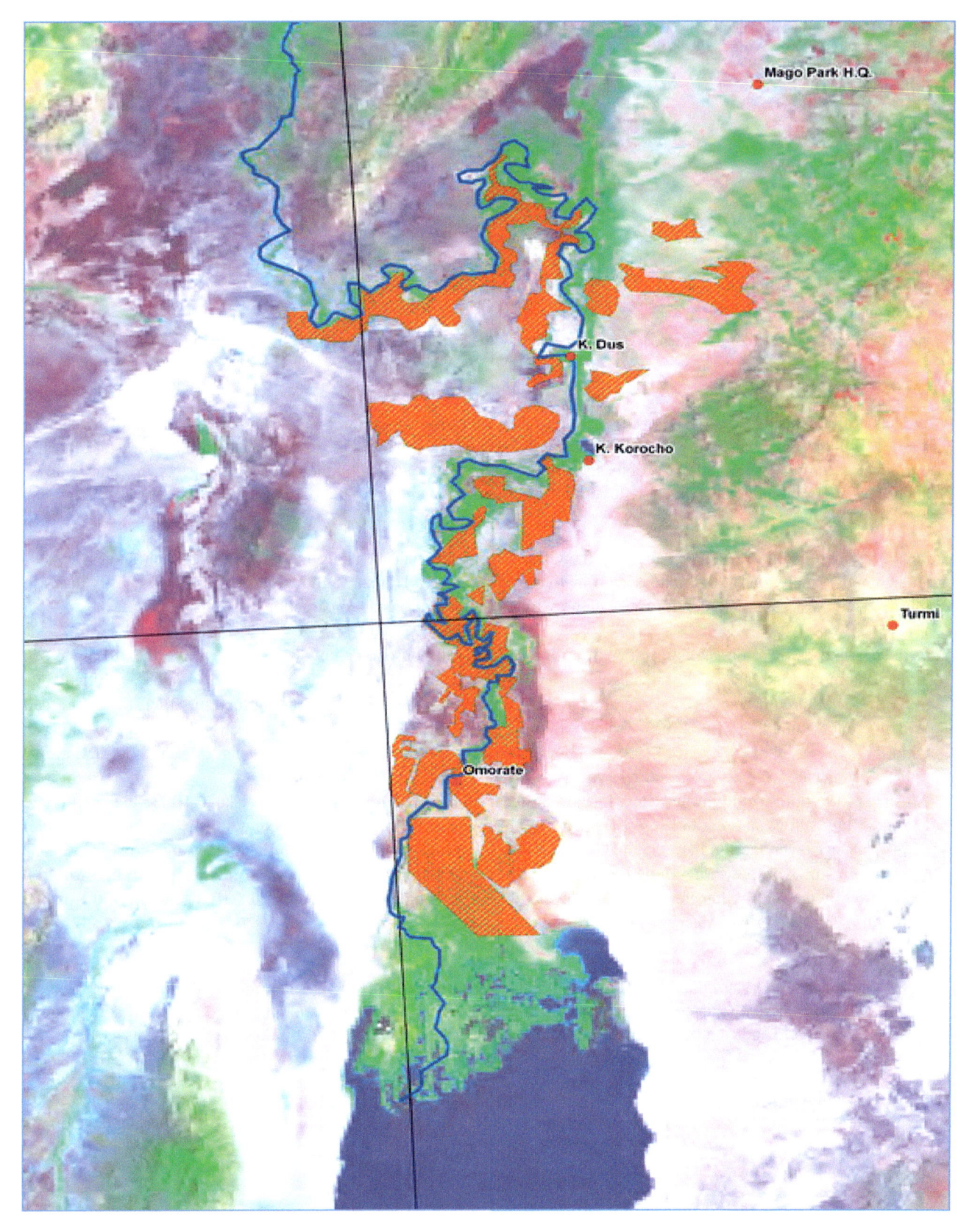

Fig. 7.18 Planned and potential irrigated agriculture in the lower Omo River basin. Some lowermost riverine zone planned farms (expropriation areas) are omitted. *Source* GOE map in EIB 2010

Table 7.3 Selected irrigated agricultural enterprises in the lower Omo basin

Plantation type and size	Owner
Kuraz Sugar Plantation 245,000 ha	FDRE Sugar Corporation (formerly Ethiopian Sugar Corporation)
Koka Oil Palm Plantation 31,000 ha palm oil, sesame, rubber trees	Lim Siow Jin Estate (Malaysian company)
15 smaller land concessions 111,000 ha majority for cotton plantations	Various private companies
Oil Palm Plantation 60,000 ha	Fri El Green Power (Italian company)
Total: Minimum of 445,000 ha	

Company, farm size and crop formation from Human Rights Watch (2012), Oakland Institute (2011)

> Evictions of Dasanech villagers and expropriation of their recession agricultural and last resort grazing lands for these enterprises have been underway for years, along with the construction of irrigation and canal works. Political repression accompanies all of these actions since opposition is not tolerated by the GOE. The impacts are cataclysmic when they are combined with radical reduction of river and lake waters by the dam and irrigation systems.

The underreported estimate of lands seized by the GOE for commercial use along the river in Dasanech and Nyangatom traditional lands—more than 120,000 ha, according to the now outdated Human Rights Watch estimate —is vastly increased if both east and west bank commercial farms along with irrigation and canal construction are considered.[10]

- **Evicted villagers have no realistic survival options**. Nearly all households now have too few livestock to move back into a pastoral life and in any case the rangelands are severely deteriorated. Moving into the modern delta is difficult even in the 'best' of circumstances. There are already tens of thousands of Dasanech residing there or at least claiming delta lands there are suitable for flood recession agriculture.[11] Residents in nearly every Dasanech village complex along the Omo River have either experienced GOE eviction or expropriation themselves or have been impacted by the influx of villagers subjected to these processes elsewhere.
- **Closure of the dam would immediately initiate desiccation of the Omo delta**—the very lands where the expropriated Dasanech take refuge in order to survive (Fig. 7.14). For many of them, fishing has become their *last livelihood option*, along with whatever success they may have in accessing flood recession planting in the already overcrowded delta.
- **Last option survival by fishing along the lowermost Omo and along Lake Turkana's northern shoreline would be eliminated** by the multi-kilometer retreat of that shoreline that would follow closure and early operation of the planned dam with no real reservoir water release.

> Tens of thousands of Dasanech relying on annual flood in the Omo delta or who evicted from lands upstream along the river would face massive scale conditions from the impacts of the Gibe III dam and dam-linked irrigated agriculture. Figures 5.2 and 7.14 indicate the desiccation of delta and lakeshore areas caused by these developments.

The GOE's eviction of Dasanech (and Nyangatom) from their flood recession farms and settlement areas is evidence enough of the government's true development intentions in the South Omo. The government's priority of commercial development

---

[10]SONT researchers recorded specific areas of GOE eviction or expropriation of Dasanech villagers, but for political security reasons, did not attempt to quantify the size of farms being established or planned.
[11]Finding lands in the Omo delta for their livelihood is extremely difficult for these evicted communities, even when they have strong social ties (defined social 'segment,' or clan terms), since the Omo delta lands are already overcrowded. Some of them attempt fishing and move even closer to Lake Turkana's northern shoreline.

over the survival of its indigenous population is starkly evident from its public statements and investor solicitations. The administrator of Debub Omo Zone, in an interview for *Fortune* magazine, stated:

We granted the land to the company along the Omo valley, which is the most suitable area for the plantation of palm oil, to encourage investors to come to the region with the prospect of exploiting this huge potential.

Dasanech village settlements and lands listed in Table 7.4 were among those expropriated by the GOE, according to local resident reporting to SONT researchers, mostly during late 2012 and early 2013.

➢ **The GOE's planning and impact assessment documents describe agricultural development plans as part of 'community development and 'social services' provision—a clear misrepresentation of the reality of the lowermost Omo region as well as GOE policies.** The following false statements in the GOE's downstream impact assessment (GOE 2009) reflect this misrepresentation.

The following examples of such statements by EEPCO officials illustrates GOE fabrication of its development plans for the "benefit" of the Dasanech.

• The South Omo project area is *"hardly inhabited at all except at a widely scattered pattern,"* and that the population density *"at the South Omo project site is below five persons per square kilometer."*[12]

To the contrary, SONT researchers documented thousands of Dasanech living in major village complexes along the west shore of the Omo River riverine zone in proximity to one another, as indicated in Fig. 7.14. This condition is manifest, even to the casual observer visiting the riverside zone.

• The government *"will not displace a single person involuntarily in Gambella, or elsewhere within the country."*

Firsthand accounts by Dasanech elders, describing their evictions from settlements and lands (Table 6.4), directly contradict the government's assertions.

• *An irrigated land of 0.75 ha of land each is prepared for 2050 households. There will not be any land scarcity for any family with a capacity to produce more. Training on improved agronomy practices, technology inputs and livestock management including rangeland will be provided.*

This statement is also false. The communities listed in Table 7.4 are among those whose riverside farming areas or settlement areas have been expropriated. While thousands of villagers are expropriated with nowhere to go, the GOE frequently speaks of 'providing employment' for the local population. A very few Dasanech are incorporated into selected commercial farms, they are in essence wage labor, not 'participants' in local cooperative or community based development. SONT investigation of two major expropriation/farm establishments on the west bank, for example, revealed that only 20 to 30 young men were hired while hundreds of villagers were sent away.

➢ **The Ethiopian government consistently describes its 'consultation' with local communities participatory. Dasanech elders, however, consistently report that, the Ethiopian government has forcefully ordered them to vacate many of their different their village locales, as well as major grazing and recession agriculture lands.**

• Development bank documents refer to the consultations carried out by the GOE, but these rely entirely on the GOE's description of its actions, not that of local residents.
• A *culture of fear* prevails among Dasanech villages. Omorate based GOE militia and security personnel take repressive, even violent measures against individuals and groups protesting their eviction and expropriation.
• Villagers' acts of resistance—even questioning, have been met with swift and sometimes violent action by police or militia.

---

[12] These (italicized) statements were made in a letter from the Minister of Federal Affairs in the GOE, in response to a letter from Human Rights Watch in November of 2011. They are representative of numerous statements issued by the GOE's EEPCO and other key government officials.

Table 7.4 Partial list of Ethiopian government evictions and expropriations of Dasanech villagers

EVICTION OF VILLAGERS AND FLOOD RECESSION CULTIVATORS: WEST BANK OF THE OMO RIVER (2009 – 2012)

Goto: Eviction from extensive flood recession agricultural lands.
Thousands forced to evacuate to find new areas for planting and livestock grazing.
Large commercial farm established.

Damish: Eviction from extensive flood recession agricultural lands.
Most households forced to find new resource areas for planting, livestock grazing.
Large-scale irrigated commercial farm established.

Nyemomeri: Eviction from extensive flood recession agricultural lands.
Eviction of village complex (evangelical missionary operations remain).
Major canal construction westward, creating barrier to livestock movement.
Large commercial farm established; water works. (A few Dasanech - as wage labor).
Population forced to find new areas for subsistence
Villagers remaining in the region highly subject to Turkana attack.

Akudingole: Eviction from extensive flood recession agricultural lands.
Villagers forced to find new areas for livestock, flood recession agriculture, fishing

Salany (Salin/Selegn): Eviction from flood recession agricultural lands.
Highly vulnerable to attacks by Turkana; plains severely degraded by livestock overgrazing due to exclusion from riverside lands.
Many villagers forced to move, primarily into modern Omo Delta.

Kolon Lochuch: Most villagers evicted from flood recession plots.
Village remaining, but a substantial percentage of the population forced to leave in search of new planting locales.

EVICTION OF VILLAGERS AND FLOOD RECESSION CULTIVATORS: EAST BANK OF THE OMO RIVER & MODERN DELTA INTERIOR

Afewerk/Afor: Government-run large farm – established early 2006.
Kapusie: Government-run large farm — established in 2006/2007.
Ediporon: In the modern Omo Delta; no agricultural land, but impacted by the influx of evicted villagers and their livestock into the delta.
Bokom (both west shore & delta interior villages): impacted by new settlers and their livestock. Flood recession agriculture land heavily impacted by influx of evicted Dasanech from villages upstream.

Source SONT interviews with west bank Dasanech elder residents, 2010 June–2013 January

By local accounts, such government 'consultations' have consisted of meetings where a few government representatives and a number of 'trusted' local residents, with officials explaining the 'major benefits' that will come to the villagers as the result of the Gibe III dam and large-scale irrigation systems. One Omorate official stated to this writer that he had actually filled out the forms of "community approval."

> The Dasanech (and Nyangatom) have had no real experience with small or medium scale dams, let alone knowledge of a megadam and how it would affect their lives.[13]

➢ **There is historical precedent for the expropriation of Dasanech lands and eviction of villagers along the Omo River.** Following the overthrow of Haile Selassie in the mid-1970s, the new military regime greatly increased the government's presence in the lower Omo region. Until that time, direct government presence had little impact except when pastoral groups had been forced out of certain territories by Ethiopian and Kenyan government forces.

- The government's establishment of the 'frontier-style' town of Omorate along the east bank of the Omo (Fig. 1.3) furthered the displacement process of prior years and led to the rapid incursion of traders and other outsiders. This caused serious incidences of HIV infection, alcoholism, prostitution, and other social problems previously unknown to the indigenous communities. Within a short time, Omorate became the hub of large-scale agricultural development. The program used the rationale of settling the pastoralists. Highland agricultural officers in Addis Ababa viewed this development as "for their benefit"[14]

  By the 1980s, Omorate emerged as the Ethiopian government's police and administrative center, as well as a town of trade. This has produced unprecedented mixing of Dasanech, Nyangatom, and Kara people and drawn substantial numbers of individuals from other ethnic areas of Ethiopia.

  The 1980s incursion by the national government for agricultural 'development' and new police/administrative presence came on the heels of many years of Christian evangelical missionary presence along the lowermost Omo—a presence the government facilitated by alienating riverside common property of the Dasanech. Small numbers of local residents were included in the missionary program, with certain conditions (villagers report have to become Christianized.) Non-indigenous crops (bananas, tomatoes, mangos, cassava, etc.) have been grown rather than the traditional sorghum, beans, sweet potato, pumpkin, maize and other crops consumed by local villagers. During SONT's work, villagers had to pay fees for use of missionary provided windmills.[15] The impact on local systems of traditional land tenure, food, and other product exchange, and social relations have been significant.

- The large Italian palm oil plantation being installed just north of Omorate, on the east bank of the river, is in the same locale as the Ethio-Korea Joint Agricultural Development Project—an irrigated commercial cotton plantation established in the 1980s under the Derg. Like its predecessor, the palm oil plantation is situated in silty clay soils of the ancient (relict) floodplain. These soils form numerous sinkholes and cracking networks, some of which extent to depths of more than four meters and from major landscape features easily visible from satellite photos and aerial views. The enormity of these cracks in the silty clay soils of the relict floodplains soils is evident in Fig. 7.19. Combined with the specific soil texture and high evaporation rates, these cracking features favor major salt accumulation.

  The Derg's Korean-sponsored irrigated cotton venture was a failure—one repeated under subsequent non-governmental organization management with major salt concentration, radically decreased soil fertility and invasion by intractable non-indigenous plants unpalatable to livestock.

---

[13]Some Dasanech (and Nyangatom) herders are even misled by having seen very small dams (for example, 30–50 ft. ones) built across streams in nearby northwest Kenya (one is just outside of Lokitaung (Fig. 1.3), where many have traveled for trading), or in the Ilemi. Herders view all of these dams negatively, since they block stream flow and vegetation downstream that are vital for their livestock. These experiences obviously offer no basis for comprehending the enormity of a structure like the Gibe III dam.

[14]This writer personally discussed plans for such development with officials from the Ministry of Agriculture. They outlined their objectives of settling the Dasanech (they persisted referring to them with the longstanding Amhara term, '*Geleb*',) teaching them to "grow tomatoes" and "drop their primitive ways".

[15]Personal communication by this writer with missionary representative on the east and west banks of the Omo River.

> The planned large-scale agriculture in this locale is likely to suffer the same fate as the previous cotton plantation. These soils are essentially the same in locales along the river where the GOE plans other irrigated commercial farms—as distinct from the annually flooded flats where the Nyangatom and Dasanech have long carried out recession agriculture.

➢ The GOE carried out another large expropriation of Dasanech communal grazing land on the east bank—a roughly 10,000 ha unit south of Omorate appropriated by an Ethiopian highlander favored by the Derg. This enterprise was abandoned in 1991, following the overthrow of the military government.

This land was later partly utilized by the present government to settle allegedly "displaced people" by the large flood of 2006 and for commercial production. Individuals at this plantation reported to SONT researchers that they were ordered into the new settlement by the government, despite their wish to return to their traditional village areas after the 2006 flood subsided. Their village areas had remained intact and many were not even flooded, according to residents, yet they were ordered to move. By the new government plan, longstanding traditional tenural relations with common property and traditional exchange systems were ended. By early 2010, Dasanech farmers here reported that they were required to grow specified crops for the government's storage facility or for government controlled marketing in Omorate. The GOE states that its grain storage near Omorate, on the east bank, is "for Dasanech use" during times of hunger. Local elders insist that hunger already prevails and that there is little or no assistance from the government, even in the worst of hunger periods. The Dasanech in these government schemes view themselves as largely forced labor for the government. Since researchers other than those under strict control and independent observers are prohibited from working in the area, these descriptions from Dasanech villagers lack further detail.

> GOE expropriation or eviction actions are absent from all GOE planning and impact assessment documents for the Gibe III project, as is information regarding the GOE's active solicitation of private investors for commercial Agriculture and other industrial development in the South Omo.

➢ **Increasing hunger and lack of recovery options for thousands of Dasanech eicted from their riverine locales have intensified hostilities between the Dasanech and their northern Turkana and Nyangatom neighbors, who face similar conditions**. Livestock raising, thefts of fishing gear and boats, armed conflict and killings are widespread.

Locales in the transboundary region already with frequent conflict are shown in Fig.5.3. They include:

- Northernmost waters of Lake Turkana and the Omo delta region.
- Villages near the northwest lake edge, including Turkana settlement at Kenya's Todenyang and in adjacent Ethiopian lands.
- Ethiopia-Kenya borderland grazing (stock camp) areas northwest of the lake.
- Eastern Ilemi Triangle stock camp/herding locales.

Dasanech conflicts with the Nyangatom are most frequent in lands around:

- Koras Mt./ Kibish River grazing, watering and settlement areas.
- Eastern Ilemi Triangle herding and stock camp locales.
- Contested locales near the Omo River.

Violence among the region's ethnic groups has frequently provoked intervention by local government security forces based in the border areas of Ethiopia and Kenya. In the case of Ethiopia, these events frequently lead to increasing repression and harsh measures by the Ethiopian police or militia—actions that are viewed by the Dasanech as part of the GOE's evictions of their communities and continual expropriation of water and land resources vital to their survival.

Although government, international aid agency, United Nations and ecumenical statements concerning conflict in the Ethiopia-Kenya-Ilemi Triangle/South Sudan tri border region consistently cite or clearly imply 'indigenous' inter-ethnic conflict 'natural' to the area. This is a major distortion of the historical reality of relations among these groups.

Fig. 7.19 Cracking silty clays in relict floodplains, near planned irrigated commercial farm

While it is true that conflicts have long been common among the Nyangatom, northern Turkana, Dasanech and a number of adjacent groups, two overarching realities prevail:

- Relations among ethnic groups have long involved sporadic conflicts, but within a framework of broad, regional social and material exchange systems, as well as sharing of grazing and water resources, among others.
- Major arms trafficking in the region, especially that involving the conflict in South Sudan, has radically increased the frequency and intensity of violence.

Literature Cited

Almagor, U. 1978. *Pastoral partners: Affinity and bond partnership among the Dassanetch of South West Ethiopia.* Manchester: Manchester University Press, 258 pp.

Bassi, M. 2011. Primary identities in the Lower Omo Valley: Migration, cataclysm, conflict and amalgamation, 1750–1910. *Journal of Eastern African Studies* 5(1): 129–157.

Bondestam, L. 1974. Peoples and capitalism in the Northeast Lowlands of Ethiopia. *Journal of Modern African Studies* 12: 428–439.

Carr, C.J. 1977. *Pastoralism in crisis: The Dassanetch of Southwest Ethiopia.* Chicago: University of Chicago, Department of Geography Papers, 319 pp.

Carr, C.J. 1978. *The Koka Dam, agribusiness and marginalization of the Ittu Oromo pastoralists in the Awash Valley of Ethiopia.* Report to National Science Foundation, 170 pp.

Carr, C.J. 2012 Dec. *Humanitarian catastrophe and regional armed conflict brewing in the border region of Ethiopia, Kenya and South Sudan: The proposed Gibe III Dam in Ethiopia.* Africa Resources Working Group (ARWG), 250 pp. https://www.academia.edu/8385749/Carr_ARWG_Gibe_III_Dam_Report.

European Investment Bank (EIB). 2010 Mar. *Sogreah consultants, independent review and studies regarding the environmental & social impact assessments for the Gibe III Hydropower Project.* final Report.

Ethiopia, Government of (GOE), Ethiopian Electric Power Corporation (EEPCO). 2009. *Agriconsulting S.P.A., mid-day international consulting, level 1 design, environmental and social impact assessment, additional study of downstream impacts.* Report No. 300 ENV RAG 003B.

Human Rights Watch. 2012. *What will happen if hunger comes? Abuses against the Indigenous Peoples of Ethiopia's Lower Omo Valley.* http://www.hrw.org/sites/default/files/reports/ethiopia0612webwcover.pdf.

Kloos, H. 1982. Development, drought, and famine in the Awash Valley of Ethiopia. *African Studies Review* 25(4): 21–48.

Oakland Institute. 2011. *Understanding land investment deals in Africa.* Country Report: Ethiopia.

Open Access This chapter is distributed under the terms of the Creative Commons Attribution-NonCommercial 2.5 International License (http://creativecommons.org/licenses/by-nc/2.5/), which permits any noncommercial use, duplication, adaptation, distribution and reproduction in any medium or format, as long as you give appropriate credit to the original author(s) and the source, provide a link to the Creative Commons license and indicate if changes were made.

The images or other third party material in this chapter are included in the work's Creative Commons license, unless indicated otherwise in the credit line; if such material is not included in the work's Creative Commons license and the respective action is not permitted by statutory regulation, users will need to obtain permission from the license holder to duplicate, adapt or reproduce the material.

Nyangatom Livelihood and the Omo Riverine Forest

Abstract
Nyangatom agropastoralists settled along the Omo River to the north of the Dasanech rely primarily on flood recession agriculture on riverside flats, with subsidiary fishing and exploitation of forest resources. These Omo River dependent communities maintain complex social and material exchange with other Nyangatom settlements in both the Kibish River-Koras Mountain area at Ethiopia's western border and in the Ilemi Triangle-South Sudan, where they share lands with the Toposa ethnic group. Nyangatom pastoralists and agropastoralists frequently clash with Dasanech and Turkana herders over grazing lands and water resources. The extensive Omo riverine forest—the last such pristine forest within semi-arid Sub-Saharan Africa—requires substantial soil moisture retention from the Omo River's annual flood. Cessation of the flood would quickly promote the death of the forest and destruction of its abundant wildlife and resources essential to Nyangatom survival. Thousands of Nyangatom living along the river would suffer immediate disaster from the effects of Gibe dam closure and dam enabled irrigated agricultural enterprises. Like the Dasanech, the Nyangatom also are subjected to major expropriation and repression by the Ethiopian government, as well as major cutting of their forest by the Ethiopian government and its allied development interests.

Nyangatom Omo Settlements and Dependence on Riverine Resources

➢ **The Nyangatom, like the Dasanech and other neighboring pastoral groups, have a complex and highly adaptive survival strategy system.**[1] Their economy, like that of the Dasanech, depends on both Omo riverine and upland plains environments. Nyangatom settlements extend over a broad area—from the riverine forest along the Omo River westward to the Kibish River near Koras Mt. and well into the Ilemi Triangle where they coexist with the closely related Toposa.[2]

The exclusion of all indigenous groups from the contested Ilemi Triangle for several decades effectively created a buffer zone among ethnic groups, as viewed by the Ethiopian and Kenyan administrations. Long shared by the region's pastoralists and only minimally impacted by them because they had wide-ranging options for seasonal herding, the Ilemi supported rich wildlife populations and relatively pristine grassland environments (Carr 1977). Expulsion of the Nyangatom from their lands in the eastern Ilemi Triangle, in essence, split them into two segments: a largely pastoral one based in the Ilemi and an agropastoral or agricultural one settled along the Kibish and Omo rivers.

---

[1] The Nyangatom were long referred to as the "Donyiro" and "Bume" by outsiders. A collection of detailed description and analysis can be found in Tornay (1979, 1980, 1981), Tornay et. al. (1997) and Tornay and Tvedt (1993).

[2] Although this writer once had easy access to Nyangatom villagers, it is now extremely difficult to interact with villagers, due to intense surveillance by the GOE and pervasive fear of reprisal by local police. Villagers, pre no recent access has been possible, due to GOE restrictions. Consequently, information presented in this chapter is based primarily on interviews with Nyangatom at the Kibish River and in Omorate.

Movement between the two settlement poles and labor/product exchange relations between the Nyangatom's contrasting food production systems was disrupted although the group has maintained as much cooperation and interaction as their situation permits. The radical reduction of pasture and the split between settlement areas that has resulted from the governments' exclusionary policies have increased conflict between the Nyangatom along the Omo and Kibish Rivers (Figs. 1.3 and 8.1) and their Dasanech and Turkana neighbors. By contrast, the Kenyan authorities administrating the eastern Ilemi Triangle for decades have only moderately enforced the agreed upon exclusionary policy as it applies to the Turkana. Many in the region report that the GOK even encourages Turkana movement into the Ilemi lands that Kenya has long claimed.

The high concentration of livestock caused by the governments' policies in the Ilemi forced the Nyangatom (and Dasanech) to crowd their herds into lands between the Kibish River and the Omo River. As a consequence, these pastures became severely overgrazed and ecologically degraded, causing major new livestock mortality and herd decline. Nyangatom pastoralists have faced continued deterioration of grazing areas—a trend worsened by prolonged droughts affecting all transboundary groups. Nyangatom pastoralists interviewed by SONT describe major livestock losses—similar to those reported by the Dasanech and the northern Turkana (see Chaps. 7 and 9).[3] Livestock herded around Koras Mountain. and along the Kibish River are watered at the seasonally flowing Kibish River. During drought seasons, villagers dig water holes in the dry riverbed of the Kibish (Fig. 8.2). These watering holes, frequently extending to 7 or 8 m depths.

Nyangatom villagers along Omo River have resided in well-established settlement areas—generally near the water—where they rely on flood recession agriculture, with a variety of subsidiary production activities. Huts are semi-permanent and are constructed with grass over a frame of branches (Figs. 8.3 and 8.4). Households owning livestock typically send them to stock camps in the upland plains, especially during the many years of tsetse fly presence in the riverine zone. These Nyangatom pastoralists have long concentrated in the Koras Mt.-Kibish River, where they Dasanech herders, and in the Nyangatom–Toposa grazing lands within the Ilemi. There are constant treks back and forth among the Nyangatom at the Omo River, and those along the Kibish River and in the Ilemi (Fig. 8.1). Nyangatom stock owners, like their Dasanech and Turkana counterparts, maintain highly flexible seasonal movements of their livestock herds in response to environmental, socioeconomic and security conditions.

Flood recession agriculture by the Omo-dwelling Nyangatom is primarily on seasonally flooded point bars and river silt/sand flats along the Omo River. Overbank flooding does not occur upstream from the active delta. Crops grown by these villagers are basically the same as those planted by the Dasanech (see Chap. 7), with sorghum and maize as main staples. Grain product is stored in high overhead granaries (Fig. 8.3) that add protection from wildlife. Numerous Nyangatom settled at the Omo River, particularly the poorer households, have taken up fishing—mostly in river waters upstream from the Dasanech, either in dugout canoes or along the shoreline. Until their recent acquisition of gear from merchants and neighboring groups, Nyangatom fishers have used simple technologies including ropes, harpoons, locally constructed nets and small blades (through exchanges with neighboring groups). The Nyangatom complain of decreased fish catch following the incursion of GOE-supported commercial fishing enterprises in the Omo River and Lake Turkana (see Chaps. 7 and 9). The acceleration of commercial fishing in the region, which villagers describe as destructive of fish reproductive habitat as well as catch levels, has been occurring just as Nyangatom villagers had had to rely more on fishing in order to cope with their economic decline.

Nyangatom residents depend on riverine habitat for a host of production activities subsidiary to recession agriculture. In addition to livestock raising, these include wild food gathering, hunting, fishing, beekeeping and boat-making and household small item manufacture for exchange (see Table 8.1 and Chap. 4). As with sorghum and maize growing, these activities are for both domestic consumption and exchange (Fig. 1.6). These secondary types of production take on major importance during times of high stress, including from failed or insufficient Omo River flooding for recession agriculture, crop losses by pest invasions, prolonged drought periods in the upland plains or loss of livestock from disease, raiding by neighboring groups or loss of access to resources due to government.

➢ **Nyangatom communities along the Omo River have no realistic alternatives for their survival—certainly no options for new settlement or major resource** access since they are bordered on all sides by ethnic groups experiencing similar losses of land and resources and with whom they frequently have hostile relations, including the Suri, Kara, Hamar, Dasanech and Turkana (Fig. 1.3). In the face of recent decline and government expropriations, many young Nyangatom men

---

[3]Research by Bassi (2011) in recent years provides some excellent perspective on Nyangatom cultural history.

Fig. 8.1 Nyangatom trek from Omo River villages to Kibish River and Ilemi Triangle settlements

respond by traveling to South Sudan where they acquire arms from liberation forces (most identify the SPLA)—later returning to their homelands along the Omo and Kibish Rivers. This dynamic has weakened traditional authority systems and intensified conflict among transboundary ethnic groups, according to Nyangatom elders.

Nyangatom relations with the Turkana, with whom they are closely related in linguistic and cultural terms, have fluctuated greatly over the years, with peaceful periods punctuated by raiding and violence. The situation has recently inverted, with frequent hostilities erupting, particularly within the context of the Kenya government's 'encouragement' of Turkana re-entry into the contested Ilemi Triangle. The situation is at least as volatile between the Nyangatom and the Dasanech, with their overlapping traditional resource areas and the GOE's expropriation of both groups along the Omo River. One area where clashes are often extreme is around Koras Mt. and the lowermost Kibish River, where some of the few stock watering and grazing alternatives to the Omo River environments exist.

> Nyangatom dependence on the Omo River for recession agriculture and fishing, along with their reliance on the riverine forest for secondary subsistence activities are ignored in the GOE impact assessment (2009b) and is barely mentioned in the 2010 EIB report.

Construction of the Gibe III dam and large-scale irrigated agricultural enterprises would essentially dismantle Nyangatom livelihood along the Omo River. This destruction would occur in at least three ways:

(i) The predicted 60–70 % reduction in Omo river flow volume during the years of reservoir fill and early operation of the dam promises cessation of annual flood sufficient for recession agriculture in most locales.

Fig. 8.2 Nyangatom man and woman at (5 m deep) watering hole dug in Kibish riverbed during the dry season. Kibish waters are the main alternative source of water by Omo River dwelling Nyangatom and well access is a source of conflict with the Dasanech and Turkana

Fig. 8.3 Nyangatom family in village alongside the Omo River's west bank. Household granary and food storage on elevated platform, with day hut and thorn fence in rear

Fig. 8.4 Nyangatom in agropastoral villages along west bank of the Omo River. *Left* Woman with son at mid-day in village. *Top right* children in central square of village. *Bottom right* aerial view of semi-permanent village (chicken hutch and fence repair visible)

Table 8.1 Nyangatom livelihood activities dependent on riverine habitat

Production activity	Location
Flood recession agriculture	Riverside point bars, waterside flats
Livestock watering, grazing	Woodland/transition zone, waterside
Wild food gathering	
Fishing	Riverine forest/woodland
Beekeeping	Omo River waters
Hunting	Riverine forest, woodland and transition
Dugout boat-making	Riverine forest

> **Radical river reduction would terminate Nyangatom flood recession agriculture along the river—parallel to such destruction in Dasanech lands downstream. Like the Dasanech, Nyangatom villagers insist that they receive too *little* flooding by the Omo River, *not* excessive flooding**.

(ii) The drop in water level and associated major impacts on the river's oxygen, sediment and nutrient replenishment would destroy fish habitat in the river and disrupt the seasonal migration of fish upstream from Lake Turkana and delta waters, destroying Nyangatom fishing livelihood.

(iii) The well-developed Omo riverine forest in Nyangatom territory would be destroyed by the radical drop in the river's flow volume and the effective elimination of the annual flood—in turn destroying the key subsidiary activities of wild food gathering, hunting and beekeeping.

Nyangatom communities along the Omo would face widespread hunger and desperation—conditions exponentially worsened by the GOK's expropriation of their lands and clearing of their forest for new commercial scale irrigated agricultural farms. Unable to be mitigated by long-standing exchange relations between Omo riverine zone dwelling Nyangatom with their counterparts settled around Kibish River (Fig. 7.6) and in the Ilemi Tri-angle. Under such dire conditions, the most likely survival option for the Omo Nyangatom would be to migrate back to the Nyangatom and Toposa settlement area in the Ilemi Triangle. This desperate movement would inevitably intensify the already widespread armed conflict in the region as ethnic groups compete for what resources remain (Figs. 5.3 and 5.4).

Like their Dasanech neighbors downstream, the Nyangatom are already experiencing the dismantling of their survival system due to actions by the Ethiopian government. These actions include eviction of settlements and expropriation of their planting lands in favor of the large commerical irrigation agriculture. Recent reports by the Nyangatom include accounts of major such measures by the GOE.

Fate of the Forest: Nyangatom Survival and Ethiopia's Heritage

➢ **The Omo riverine forest and woodland is the last remaining pristine riverine forest in the drylands of Sub-Saharan Africa** (Fig. 8.5). The GOE assessment misrepresents these low altitude riverine forests as basically equivalent to those upstream at higher elevations, when they are in fact unique and without such substitute plant communities upriver. Comparable riverine forests in semi-arid/arid regions in Sub-Saharan Africa have already been eliminated by river flow reduction from large hydrodam construction—dams much smaller than the Gibe III. In the African Horn and East African region alone, the riverine forests along Ethiopia's Awash River and Kenya's Tana and Turkwel rivers have already been largely destroyed by hydrodam and associated developments.

Death of the forest along the lowermost Omo would result from radical reduction of the river's flow, since the forest's adaptation to the high/low extremes of flow would be destroyed. Highly sensitive root systems of riverine forest have evolved in response to fluctuations of subsurface moisture and nutrient replenishment provided by the annual flood's permeation of the natural levee soils, since overbank flooding does not occur upstream from the modern Omo delta. The root systems of the forest tree and shrub species depend on a necessary period of retention of river waters by soils—that is, the *'residence time'* of subsurface soil moisture.

Fig. 8.5 Omo riverine forest in Nyangatom region. *Top photos* Buttressed large fig and other shallow-rooted forest trees. *Center* Nyangatom man on sandy/silty spit on inside river bend; flood recession agricultural plot and non-flooded riverine forest in background. *Bottom* Straight channel with thin recession agricultural plots along shoreline

Specific soil moisture *residence time* is dependent on the Omo River's annual high water stage, including its substantial duration. The Gibe III dam and dam enabled irrigated agricultural development would eliminate the 'residence time' necessary for the survival of the forest.

The rich wildlife populations in the Omo riverine forest, nearly undisturbed for centuries, would be exterminated by the effects of the Gibe III and dam enabled irrigated agricultural schemes due to the elimination of their forest habitat. Wildlife in the Omo riverine (or 'gallery') forest zone, includes the Nile crocodile, hippopotamus, elephant, buffalo, lion, leopard, kudu, monitor lizard, Colobus monkey, grivet monkey, baboon, bushbuck, and a host of water-loving birds, including the fish eagle. Wildlife experts at the University of Addis Ababa describe the area as the second richest wildlife area of Ethiopia, underscoring its importance to Ethiopia's natural heritage, with major potential for park and tourism development. These populations would be rapidly eliminated by forest destruction.

The riverine forest is a highly delicate biotic community, with emergent trees extending to 30 m, with a secondary level of spreading shrubs. All major taxa are included in the species list of Appendix B. Large portions of the landward side of the forest are dominated by shrub thicket, with abundant vines and succulents. Some inside bends receive sufficient subsurface inundation of Omo River floodwater to create grassland swamps. Ecological studies in the Omo riverine zone (Carr 1998) detailed a variety of vegetation types ranging from forest to closed woodland, open woodland and different types of grassland (e.g., grasses with and without scattered trees and shrubs, and with different amounts of herbaceous ground cover). Shrub thickets are common throughout the 'transition zone' from forest/woodland to the broad relict floodplains (Fig. 8.6).

Fig. 8.6 Transition zone between the Omo riverine forest and adjacent drylands. Scattered trees and shrubs with discontinuous grasslands prevail. Like the forest, this zone is not flooded but is sustained by subsurface moisture from the Omo River's annual flood

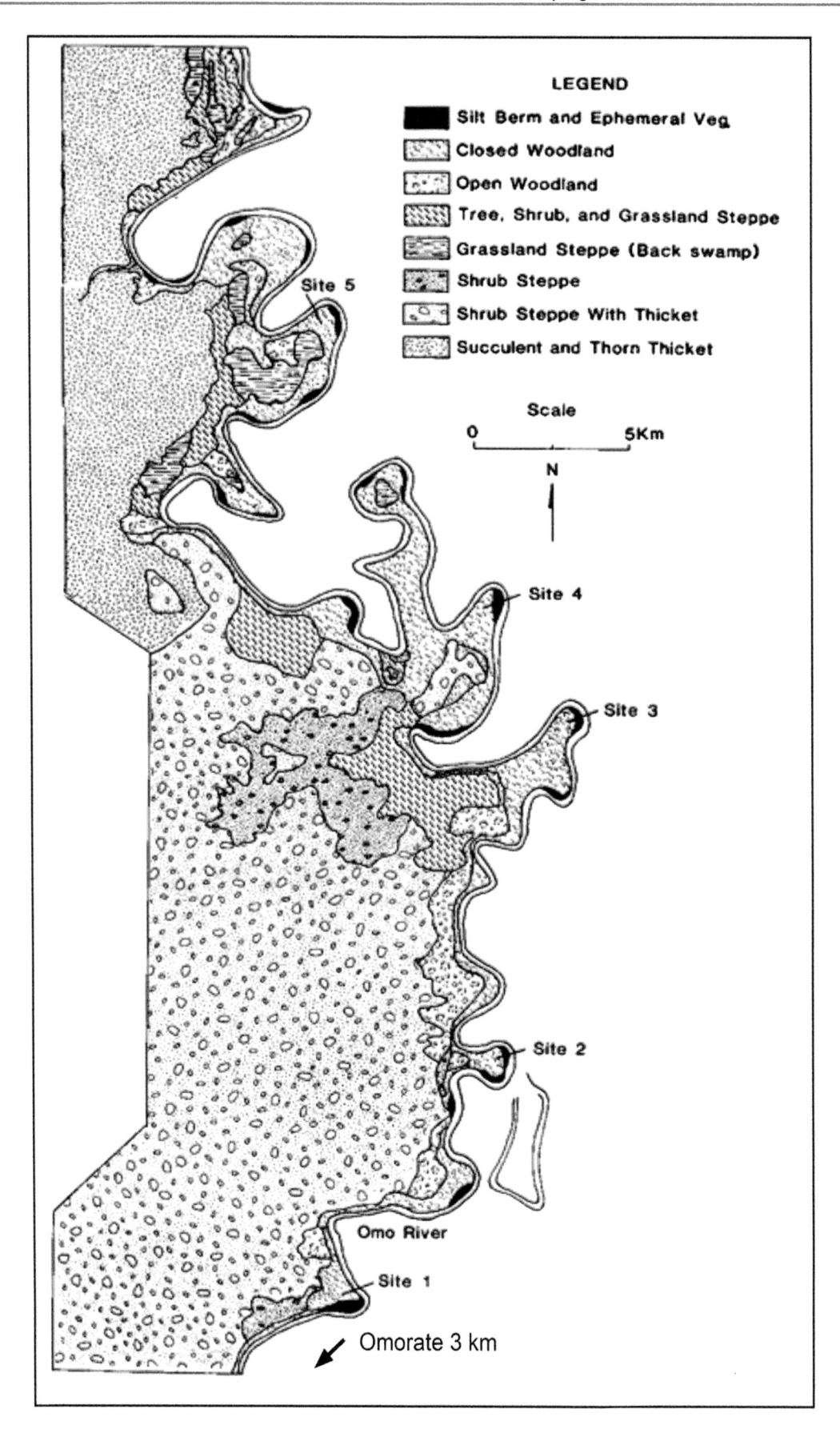

Fig. 8.7 Riverine vegetation types along the lower Omo River. *Source* Carr (1998)

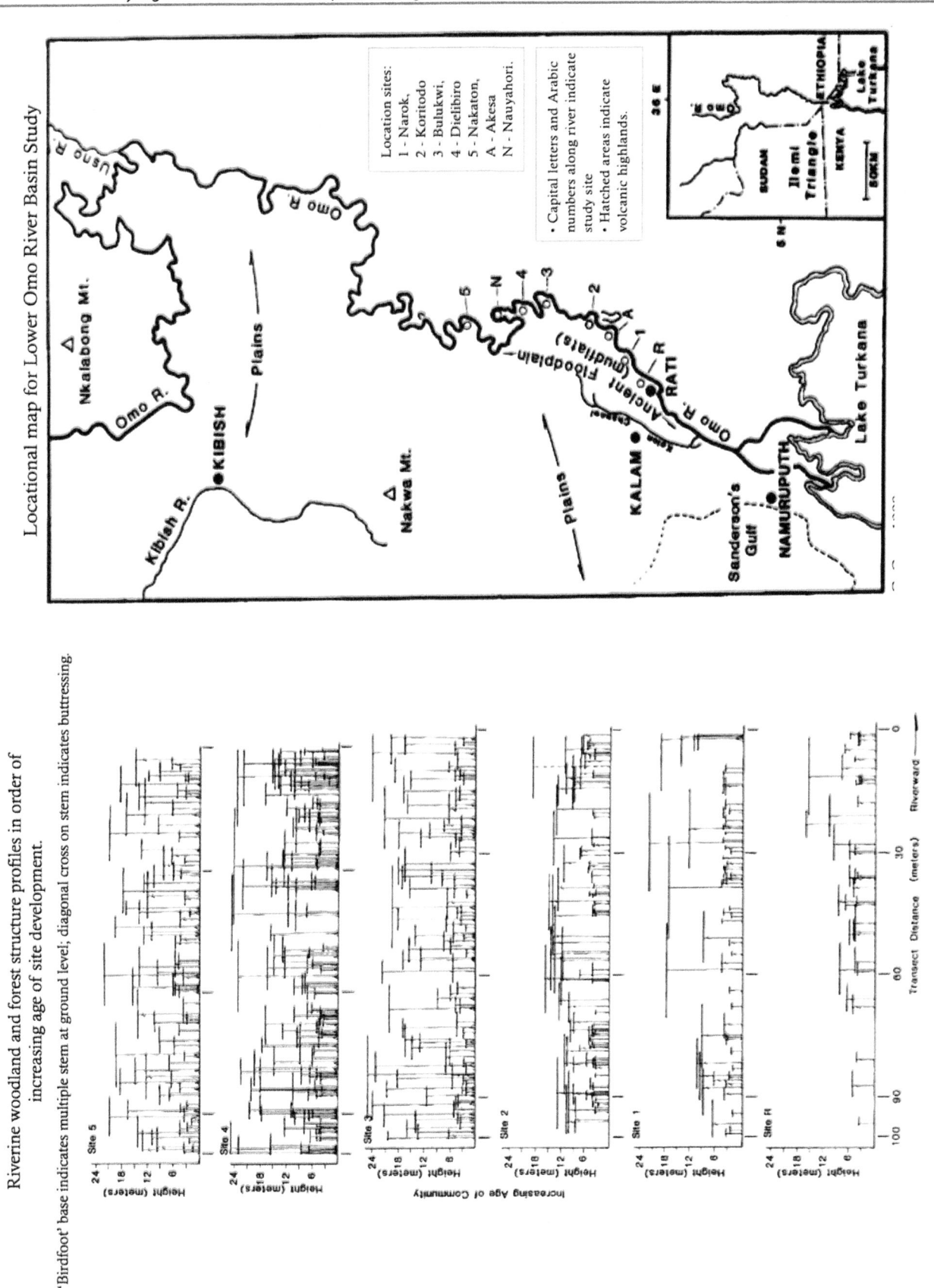

Fate of the Forest: Nyangatom Survival and Ethiopia's Heritage

Fig. 8.8 Riverine forest development along the lowermost Omo River. *Left* Riverine forest and woodland studies with young to mature forest. Vegetation development along a gradient of river natural levee exposure (from south to north) following Lake Turkana retreat. *Right* Location map of study area. *Source* Carr (1998)

Complex depositional patterns and soil/water conditions produce a mosaic-like pattern of vegetation types—increasing the range of grazing potential for livestock types with different nutritional needs and sensitivity to changing water and disease conditions. Figures 8.7 and 8.8, taken from this writer's studies of the Omo riverine forest (Carr 1976, 1977, 1998), summarize the complex development of riverine forest and woodland along the Omo River's as well as its transition to adjacent dryland vegetation communities. A south to north gradient of forest development (from younger to older plant communities) is evident from detailed plant ecological studies at the sites indicated in Fig. 8.8.

The Gibe III dam and dam enabled large-scale irrigation agriculture would cause the cessation of flooding of riverside flats where the Nyangatom carry out their main survival activity—flood recession agriculture. These developments would also destroy the vast majority of riverine zone vegetation types shown in maps.

This destruction of vegetation, in turn, would eliminate Nyangatom (and Dasanech) last remaining areas pasturage for livestock grazing, wild food gathering, hunting and other activities that are their only means of survival during the most severe hardship periods.

Literature Cited

Carr, C.J. 1976. Plant ecological variation and pattern. In *The Lower Omo Basin, Earliest man and environments in the Lake Rudolf Basin*, eds. Y. Coppens, F.C. Howell, L.L. Isaac and R.E.F. Leakey. Chicago: University of Chicago Press.

Carr, C.J. 1977. *Pastoralism in Crisis: The Dassanetch of Southwest Ethiopia*. University of Chicago Department of Geography Papers, 319 pages.

Carr, C.J. 1998. Patterns of vegetation along the Omo River in southwest Ethiopia. *Plant Ecology* 135(2): 135–163.

Tornay, S. 1979. Armed conflicts in the lower Omo Valley, 1970–1976: Nyangatom society. *Senri ethnological studies* 3: 97–117.

Tornay, S. 1980. The hammar of Southern Ethiopia, conversations in Dambaiti-Strecker. *I. Homme* 20(2): 99–108.

Tornay, S. 1981. The Nyangatom. An outline of their ecology and social organization. In *Peoples and cultures of the Ethio-Sudan Borderlands*, ed. M.L. Bender, 137–178. Michigan: African Studies Center, Michigan State University, East Lansing.

Tornay, S., K.W. Butzer, J.A. Kirchner, G.K. Fuller, A. Lemma, T. Haile, T. Bulto, T. Endeshaw, M. Abebe, S. Eshete, G.P. Petros, and D.H. Van. 1997. The growing power of the Nyangatom. University of Chicago.

Tornay, S., and T. Tvedt. 1993. More chances on the fringe of the state? The growing power of Nyangatom; a border people of the lower Omo valley, Ethiopia (1970–1992), Conflicts in the Horn of Africa: human and ecological consequences of warfare, 143–163.

Open Access This chapter is distributed under the terms of the Creative Commons Attribution-NonCommercial 2.5 International License (http://creativecommons.org/licenses/by-nc/2.5/), which permits any noncommercial use, duplication, adaptation, distribution and reproduction in any medium or format, as long as you give appropriate credit to the original author(s) and the source, provide a link to the Creative Commons license and indicate if changes were made.

The images or other third party material in this chapter are included in the work's Creative Commons license, unless indicated otherwise in the credit line; if such material is not included in the work's Creative Commons license and the respective action is not permitted by statutory regulation, users will need to obtain permission from the license holder to duplicate, adapt or reproduce the material.

Turkana Survival Systems at Lake Turkana: Vulnerability to Collapse

> We have lost our livestock and much of our lands. Now we must fish, or we will die.
> [Turkana male elder from lakeside village near Ethiopia-Kenya border]

Abstract

The pastoral economy in transboundary Turkana lands has drastically declined in recent decades, largely due to the effects of colonial and post-colonial policies. Faced with radical herd losses, thousands of Turkana households have moved to Lake Turkana's western shoreline for fishing and/or herding. This population—largely uncounted—is extremely vulnerable to loss of accessible Lake Turkana water, fisheries resources, and lakeside grazing. The Gibe III dam and irrigated agricultural plantations along the Omo would cause major shoreline retreat and eliminate the Omo River's annual flood 'pulse' of fresh-water and nutrients into the lake. Major loss of fish reproductive habitat and fish stocks as well as potable water, along with desiccation of lakeside environments essential to livestock and people would result. As conditions worsen, a general movement of fishing and fishing/pastoral villages southward toward Ferguson's Gulf—itself drying out—and around towns, in search of relief aid or survival opportunities, is likely. With no practical means of continued livelihood, hundreds of thousands of Turkana fishers and pastoralists would face region wide hunger and conditions for disease epidemics. Cross-border conflict between these Turkana and their northern neighbors would sharply escalate, especially in the face of regional arms trafficking. Northern and central Turkana protests and pleas for help have so far been ignored by the Kenyan government which continues to militarize the region.

Northern Turkana Pastoralists: The Long Decline and Migration to the Lake

> **The major human and livestock disease epidemics extending across the Sahel and eastern Africa during the latter years of the nineteenth century produced famines in Ethiopian, Kenyan and Sudanese drylands.** Both written and oral accounts of the period describe frequent raiding of Turkana livestock by their pastoral neighbors, including the Pokot and Dasanech (Fig. 1.3).

Colonial travelers describe major expropriations of Turkana lands as well as livestock during the British colonial administration (Gulliver 1955; Lamphear 1988, 1992; McCabe 1990, 2004; Collins 2006; Hogg 1982).[1] British domination of the Turkana unfolded largely from its colonial base in Uganda.[2] Its interests were largely focused around the Nile River region and as part of its strategy in the region Britain claimed the northern end of Lake Turkana.[3]

---

[1] The bulk of literature regarding early Turkana history pertains to the southern and central regions.
[2] A small British post near the Uganda border (at Lokiriama) constituted an early foothold in the region, both for military and civil presence. Other military outposts followed.
[3] Lake Turkana was named Lake Rudolf by the explorer Count Teleki, after his patron, Prince Rudolf of the Austro-Hungarian empire (von Hohnel 1938).

Dispossession of the northern Turkana during the late nineteenth century and early decades of the 1900s occurred mostly at the hands of the British colonials, but fighting between the Turkana and neighboring ethnic groups in the transboundary region worsened the impacts of this dispossession. Pressures by Ethiopia's Menelik II, who had territorial ambitions extending to the south end of Lake Turkana, provoked a 'protective' response by the British, who furthered their militarization of the region.

The British defeated the Turkana in 1914–1915 and significantly increased their military presence in the northern region. British forces confiscated massive numbers of Turkana livestock, expropriated large portions of Turkana lands and thoroughly disrupted customary seasonal patterns of herding and exchange throughout the region. The colonials also disarmed the Turkana, greatly weakening their fighting capacity and placing them at strong disadvantage relative to their northern neighbors, particularly the Nyangatom and Dasanech, who had superior access to firearms through their connections in Ethiopia. According to northern Turkana elders' oral accounts, the Turkana experienced similar stresses in their relations with the Pokot and Jie peoples to the west and southwest.

Imposing a hut tax on the Turkana, the British administration used confiscation of livestock as the penalty for nonpayment. Unrest in the region—in part, a reaction by the Turkana to these and other aggressive policies by the government—provoked further reprisals and livestock seizures. Food insecurity for the pastoralists was extreme in these early years (Oba 1992). In their weakened state, the central and northern Turkana faced extreme hunger, even famine— especially during severe drought periods.

The colonial government declared a 'closed district' policy in the region that persisted until the 1970s. Meanwhile, the British moved their headquarters from Lorogumu to Lodwar (Fig. 1.1). A small trading center for decades, Lodwar grew to become the administrative capital of Turkana County—now the largest town in northwestern Kenya, with a population of more than 48,000. The British also established a key military post at Lokitaung (Fig. 1.1), which had been a satellite trading center in the far north of Turkana. Military operations from Lokitaung facilitated the colonials' subjugation of new segments of the Turkana. The post there was the launching point for the British routing of the Italians in the Ilemi Triangle and southwestern Ethiopia in World War II conflict and it became Kenya's center for administration of the Ilemi—long a disputed area between Kenya and South Sudan (see Chap. 4) and also the northernmost extent of the Turkana population.

➤ **Turkana territorial losses in the early decades of the twentieth century sparked overcrowding of herds, and therefore overgrazing and deterioration of their remaining pasturelands**. Region wide increases in stock mortality and herd decline followed. In the years surrounding World War II, northern Turkana elders describe their herd losses as particularly devastating, due to raiding by Dasanech and Nyangatom pastoralists who had acquired new arms from both the Ethiopians and the Italians during their respective occupations of the area.[4] Major herd composition changes accompanied the plummeting livestock numbers among the northern Turkana. For most stockowners, goats and sheep became key components of their herds, since small stock can survive conditions of deteriorated grasslands and diminished water sources far better than cattle. Even camel herds declined, according to Turkana accounts, despite the adaptation of camels to long treks for browse and far lengthier periods between watering (see Chap. 4).

There are different interpretations of the causes of Turkana hunger and herd losses during the post-war years. Most written reports pertain to the central and southern Turkana regions, where conditions vary considerably from those in the north. A combination of factors were at play in the herd declines of the northern Turkana, including the following.
- British seizures of livestock, including as punitive measures.
- Continued taxation
- Exclusion from territories by British colonial actions—causing overgrazing and heightened stock mortality.
- Raiding by neighboring ethnic groups
- Extended drought periods, worsening livestock disease/mortality

---

[4] Accounts of these losses by Turkana elders are in agreement with details provided by a former British officer, Mr. Whitehouse, who figured prominently in the Ethiopia-Kenya-Ilemi Triangle border demarcation process. This writer held conversations with Mr. Whitehouse in the early 1970s.

➢ **Northern Turkana elders describe at least six types of responses to radical herd losses.**

(i) **Natural reproduction of livestock with alteration of herd composition and herd mobility.** This recovery strategy includes using exchange relations to obtain new livestock—especially small stock. Rebuilding herds through reproduction, however, necessitates access to sufficient land in order to separate herds and expand their grazing areas, with reliance on customary social exchange and cooperation patterns. Many locales where Turkana had previously sent some or all of their livestock for pasture and water during times of severe hardship became inaccessible to them, either because of government restriction or the threat of livestock seizure by neighboring groups.

(ii) **Increased reliance on production activities once subsidiary to herding.** Unlike the Dasanech and Nyangatom at the Omo River (and Turkana along the Turkwel River to the south), the northernmost Turkana have no practicable flood recession agriculture opportunities, since watercourses in their lands are relatively small and ephemeral. Along the Turkwel River to the south (Fig. 1.1), some Turkana undertook flood recession agriculture. The stresses noted above have forced the northern Turkana to rely far more on secondary production activities, especially wild food gathering, chicken raising (for both consumption and exchange—see Table 4.2), and charcoal production (primarily for marketing).

(iii) **Livestock raiding.** Like all pastoral peoples in the transboundary region and beyond, the northern Turkana have initiated raiding and seizure of livestock as a customary means of economic recovery. In recent decades, this strategy has been less effective for several reasons, including increased Kenyan and Ethiopian military and government security presence in the region, the extreme violence in such raids since firearms have replaced spears and hand to hand combat, and the reality that all pastoral groups have reduced numbers of livestock.

(iv) **Evacuation to towns and internally displaced persons (IDP) camps within the northern region.** Pastoral villagers have variously exercised this option as a temporary or long-term measure. For the northern Turkana, Lodwar and Kakuma (northwest of Lodwar) have been primary destinations. These 'refugees' seek assistance of any type possible in IDP camps and in nearby towns, but assistance is makeshift and temporary, at best. Thousands more Turkana have joined or formed spontaneous or temporary camps along roads near towns (especially Lodwar) and even in the most remote areas such as the Ilemi and border regions. Camps are few in number compared with the needs of tens of thousands of Turkana facing dire circumstances when their attempted recovery strategies have failed. For most of these Turkana families, the distances required to camps are simply too great for travel. Most northern Turkana seek access to international food aid. For the overwhelming majority of them, such aid is sporadic at best—statements by the government and impact assessments notwithstanding.

(v) **Settlement in aid-funded agricultural projects near Lake Turkana and in other scattered locales.** Very few individuals from the northern region have been incorporated into schemes along the Turkwel; most of these have most of the other have failed—most of them having depended on rainfall which is simply too limited and erratic. (All such schemes visited by SONT members in the northernmost plains had failed.) Recently, foreign aid agencies and Kenyan non-profit organizations have introduced settlement and irrigated agricultural development, for a select number of Turkana, along the Turkwel River—reducing access to the river for other Turkana in that region.

(vi) **Migration to Lake Turkana for fishing and last resort livestock raising.** Northern Turkana pastoralists (like the Dasanech) have long regarded fishing as a last resort means of survival. This attitude prevailed throughout Turkana society, despite their familiarity with El Molo fishers along the lake's southern shores (Figs. 1.3 and 4.6). Since the 1920s, the British colonial government, foreign nationals, aid organizations and later, the independent Kenyan government, have all designed and implemented relatively small Turkana fisheries and settlement projects, particularly around Kalokol. Fishing related activities are now the main means of subsistance at the lake.

As early as 1924, the British Colonial administration formed a settlement (commonly termed 'famine camp' in written and oral accounts) at Kalokol, near Ferguson's Gulf. There they taught fishing to 'displaced' Turkana pastoralists. This new livelihood mode absorbed more Turkana over the next few decades through Kenyan government, missionary and aid efforts as well as by the Turkana's own initiatives (Bayley 1982). At Kalokol, nets and other simple technologies were issued, but little follow-through assistance was given so the project failed, leaving a large number of Turkana stranded. Other fisheries development projects sprang up at Ferguson's Gulf and northward from Kalokol along the lake, at Eliye Springs and near the terminus of the Kerio River (Figs. 1.3 and 4.6).

Programs introduced by Kenyans, Norwegians and Italians were ambitious but mostly ill conceived and short-lived. Despite the high failure rate of the projects themselves, many Turkana became skilled fishers. Most of these Turkana began fishing with simple basket nets, though some soon turned to harpoons and began constructing rafts by lashing together the trunks of doum palm trunks—a technology that persists to the present time. Gill nets eventually became dominant among the fishers and wooden boats are now widespread and in strong demand (see below). Kalokol remains the center of fish collection and trading for the region.

The severe droughts of the 1960s disenfranchised huge numbers of Turkana pastoralists. Many northern Turkana from displaced persons camps ('famine camps') were settled at Lowarengak, a lakeside town near the Kenyan border post at Todenyang (Fig. 1.3). Thousands more Turkana households migrated to this area in subsequent years. Many of them took up fishing or fishing related activities. This northernmost population fluctuated greatly, particularly in response to shifting relations with the Dasanech, with whom conflicts intensified. Lowarengak has remained a key fishing center along the northernmost shoreline.

By the 1980s, when major prolonged droughts and widespread hunger conditions occurred, impoverishment among the northern Turkana pastoralists increased markedly. Major numbers of households, even groups of villages, relocated to Lake Turkana—sometimes moving in stages over years. Villages generally moved in a southward direction along the lake (see map in Fig. 4.6), settling anywhere between Lowarengak (near Todenyang; see Fig. 1.1) very close to the Kenya border, and Ferguson's Gulf. The arriving Turkanas' options for settlement were often determined by their social ties with households already established there. According to accounts by Turkana elders in villages along the northwestern shoreline of the lake, the number of households settling there increased sharply during the almost rainless years between 2007 and 2009—a nearly unprecedented drought in the memory of local residents.

The dire economic situation facing northern Turkana pastoralists in the upland plains is evident from SONT interviews with 90 Turkana household heads in the dryland plains west and north of Lokitaung in 2010 and 2011 (Figs. 1.3 and 9.1). A series of common features of life in this northern region emerged from this survey.[5]

- **Nearly all households remaining in pastoral areas owned inadequate numbers of livestock.** Many attempt to remain actively pastoral, despite having only (many with only a few to 10 or 20). What animals they had were often sent to stock camps—many of them at great distances in search of pasture and water.
- **With most young men off herding, households typically consisted of women, children and older men, at least during much of the year** (Fig. 9.1). Many villages had relocated near towns, including for security reasons or for access to periodically delivered food aid. Most households surveyed relied on a few milk animals—mostly goats (some had a few camels; none surveyed had cows present), chicken raising (typically for selling or consuming eggs), other household or village based commodity production. Wild food gathering accounted for a substantial portion for these villagers. Many Turkana men have sought wage labor opportunities, though few have succeeded in this effort.
- **Two major responses of herd owners to serious livestock losses—increased separation and geographic distribution stock animals (often utilizing new labor and other cooperation arrangements) and various exchange strategies for rebuilding livestock numbers** (particularly those involving small stock). Both of these have been entirely inadequate, however, especially in the face of extreme ecological degradation of upland pastures from overgrazing as well as drought, major threat of attack by neighboring groups, and government restrictions—in the Ilemi region and elsewhere.[6] Raiding of Dasanech and Nyangatom livestock by northern Turkana pastoralists—previously an important component of herd

---

[5]A detailed account of these interviews is the subject of a forthcoming report.
[6]Together with elders from the region, SONT researchers constructed a map of the northernmost lands—particularly the narrowly defined border lands and disputed Ilemi area, and villagers stressed that the "best grazing lands" (mostly slightly higher elevation plains with higher rainfall and grass development) are effectively off-limts to them because of government policy enforcement or danger of losing both livestock and lives. The area around the Kibish River, at the Ilemi/Ethiopia border, is another sought after area for water and livestock grazing by all three groups—Dasanech, Turkana and Nyangatom and was heavily settled by Nyangatom during SONT investigations.

Fig. 9.1 Pastoral life in northern Turkana. *Top* Villagers requesting help for repair of broken water well—a well built by the Catholic Church, which also requires villagers in the region to pay for repairs. *Bottom* Hand-dug well serving thousands of livestock in northern Turkana, near the Ilemi Triangle

recovery—is also insufficient. All of the transboundary area's groups now own firearms, so both the incidence and level of violence level of raiding are extreme in certain localities, particularly those that are indicated in Fig. 5.3. Kenyan and Ethiopian security forces in the region impede some of this violence, but the general trend persists—in fact, escalates, since it *stems from the desperate living conditions of all three groups*. The effects of this conflict are devastating for all concerned.

- **Stock mortality with radical herd decline initiated a major exit of northern villagers toward Lake Turkana**. Most household heads in the upland plains region knew of at least one or two nearby villages whose residents had already departed for the lake. A few said that they too were considering moving in that direction, as well. (Some village heads indicated that they might instead move to a town in order to increase their chances of receiving food aid.)
- **Food and other essentials available in local markets are priced beyond the reach of most households**—even those fortunate enough to generate income from household/village based commodity production.
- **Internal social problems accompanying such economic stress are also on the increase**. Many elders flatly state that they no longer have authority over the actions of their young men and that such problems are a radical departure from earlier times. The decline of customary traditional authority relations most likely results from a combination of influences including government administrative systems imposed on the Turkana, economic disenfranchisement and displacement of communities, access to weapons (providing young Turkana men with a new sense of power and independence), and privatization of Turkana lands and resource 'commons'.

Adaptation from Pastoral to Fishing Livelihood

> The Kenyan government (GOK) has failed to officially acknowledge the major migration by pastoralists to Lake Turkana, nor does it openly recognize the vulnerability of this population to lake level drop—a predictable effect of the Gibe III dam and dam enabled irrigated agricultural development.[7] Development banks also have failed to take these major changes into account in their environmental and socioeconomic impact assessments or other publically available reports. This matter is detailed in Chap. 6

➣ **Tens of thousands of Turkana pastoralists from the upland plains now make seasonal treks to Lake Turkana for livestock watering and for whatever shoreline graze and browse is available** (Fig. 4.7). During severe drought periods, the lake provides last option survival for the tens of thousands of livestock. Countless numbers of small stock are brought to the lake's shoreline when distant areas are no longer available. Except for female camels (cows) with young calves, most camels are herded in the western plains and other upland localities with sufficient browse, although they too are often brought to the lake for watering (Fig. 9.2).

The lakeside environment is so severely degraded that large numbers of livestock perish from the long trek to the lake or the trip back to upland grazing areas (Fig. 4.4), especially during severe droughts. The problem is compounded by the recent introduction of one of the most destructive invader plant species in the world—*Prosopis juliflora* (or 'mesquite').[8] The recently unpalatable *Prosopis* accelerates pasture deterioration (Fig. 9.2) and it is spreading rapidly throughout the transboundary region.

➣ **Lakeside Turkana arrive from a wide variety of upland pastoral areas. Nearly all of them now engage in fishing or pastoral/fishing production activities. They reside in settlements ranging from single household villages to large complexes with hundreds of households.** Most household heads describe having moved from an upland region in stages: for example, from near the Ilemi or around Lowarengak near the Kenya-Ethiopia border, southward to as far as Kalokol and Ferguson's Gulf (Fig. 1.3).

---

[7]Some local officials *privately* acknowledge this major crisis brewing, but political conditions prohibit them from making any public statements that are even potentially critical of the government. In fact, key SONT researchers were warned by these officials that their investigative work as well as dialogue with villagers about the matter were dangerous and could bring repercussion. Partly for this reason, the identity of respondents and their villages are protected by SONT.

[8]This rapidly spreading and nearly intractable mesquite shrub consumes major soil water resources, prevents native rangeland species' establishment and is unpalatable for livestock (Manduab et al. 2011).

Fig. 9.2 Northern Turkana livestock herds watering at the lake. *Top* Goats at lakeside for watering and browse (plants are mostly unpalatable invader species, *Prosopis juliflora*). *Center* Major death of baby goats (several hundred) from lack of water and browse on long trek to the lake). *Bottom left* Young camels with female, in *Prosopis* thicket at lake. *Bottom right* Dead wild ass (donkey) during drought period near the lake

- Most Northern Turkana communities from the upland areas now settled along the lake depend on fishing or mixed fishing/pastoral production for their survival. For the most part, these Turkana have lost the majority of their livestock or retain only a small number of animals (particularly small stock). Except for milk animals with very young offspring, even these livestock are often sent to camps well removed from the lake.
- Those households newly residing at the lake usually learn fishing from others already settled there. Many different social types of social arrangements are made for fisheries and related economic activities—arrangements for catch and village based post-catch tasks, including preparation of pallets for merchant pickup. Some villagers engage in highly specialized activities, such as boat preparation and repair or sail manufacture and repair (Fig. 9.3).

  In addition to being involved in cleaning and drying fish catch, women and children frequently do secondary production activities such as poultry raising, charcoal production and other small-scale manufacture. Although fishing communities have made major shifts in production relations since their pastoral existence, some dimensions of pastoral life have remained strong. Graduallyk these too have been impacted by commercial relations in the region.
- **Tens of thousands of pastoral/fishing Turkana reside in lands slightly removed from the lake but are also dependent on the lake resources and activities for their survival**. Many engage in work there—sometimes fishing boat owners, but often in one of many forms of wage labor or bartering and market activities that generally involve livestock products (live animals, skins and milk). The typically small numbers of village-based livestock are taken to the lake for watering, as well as grazing (when it is available.) With the assistance of local elders, SONT researchers identified the village complexes of both fishing and pastoral/fishing villages between Kalokol and the Kenya/Ethiopia border (Fig. 9.4).

➢ **Technology among Turkana fishers is relatively simple and mostly suited to fishing in nearshore areas—particularly Ferguson's Gulf, the Omo delta, Alia Bay (Fig. 9.4) and smaller bays and inlets**. Wooden boats—universally the most desired item by fishers—are constructed, largely by boat-makers from the Kisumu-Lake Victoria region. Boats are pointed at both ends, constructed from timber planks with a v-shaped bottom, and are propelled by paddles and /or sails (Figs. 9.3 and 9.5). Sails are typically fashioned from plastic food aid bags and the Turkana are highly adept at repairing them. Boats are easily adapted for motors, but few Turkana can afford them. Lack of funds for boat purchase and engines is a universal complaint among the fishers along the lake's western shoreline, as pressure on nearshore resources continues to grows with the influx of households needing to take up this new form of livelihood (Fig. 9.6). A few Turkana have developed expertise in boat-making, though the substantial capital required has favored the Luo builders from the Kisumu region who craft them—at Ferguson's Gulf, especially. Meanwhile, traditional rafts constructed of doum palm trunks lashed together (Fig. 9.5) are used locally along portions of the lake's shoreline where waters are sufficiently quiet. Rafts are extremely common in Ferguson's Gulf, for example.

➢ **Fishing in wide areas of the lake is commonplace for both northern and central Turkana. Fishers with sailboats, for example, form seasonal, or temporary camps along the eastern shoreline** (Figs. 9.3 and 9.4) and ventures into the Omo delta are also taken on during multiple months of the year. Expeditions to the eastern shoreline can last up to a month or more and can be extremely risky, depending on security and weather conditions—like Turkana ventures into waters near the Omo delta. Specific direction, distance and duration of trips are determined by numerous factors, including size and condition of boats, lake currents and prevailing winds, Omo River inflow force (including its annual pulse with freshwater and nutrients), fish stock availability (involving reproduction and feeding rhythm, etc.), relations with other fishers and labor concerns.

Fishers commonly sail to eastern parts of the lake when winds subside and currents in the northern portion of the lake are strong: for example, during the Omo River's annual 'pulse' of inflow to the lake. Seasons of the year are often described differently by residents along the lake—even among individuals in one locale, as reports to SONT members revealed. Moreover, seasonal changes of most concern to pastoralists are clearly different from those of fishers—a reality bringing even more complexity to the reckoning of participants in this relatively new type of livelihood among the Turkana. (Strong differences of interpretation emerged in group discussions of the matter in several shoreline villages, for example, and certainly in inland ones). Table 9.1 presents *only one* such description of seasonality.

Fig. 9.3 Turkana fishing villagers along northwestern shores of Lake Turkana. *Top left* preparing nets before sailing expedition party leaves. *Top right* Pallet of dried fish await merchant pickup at roadside. *Center* Turkana wooden boats inshore with villagers bathing and water-getting. *Bottom left* Fishing expedition of sailing boats leaves for eastern Lake Turkana waters. *Bottom right* villagers repairing sails made of plastic international food aid bags

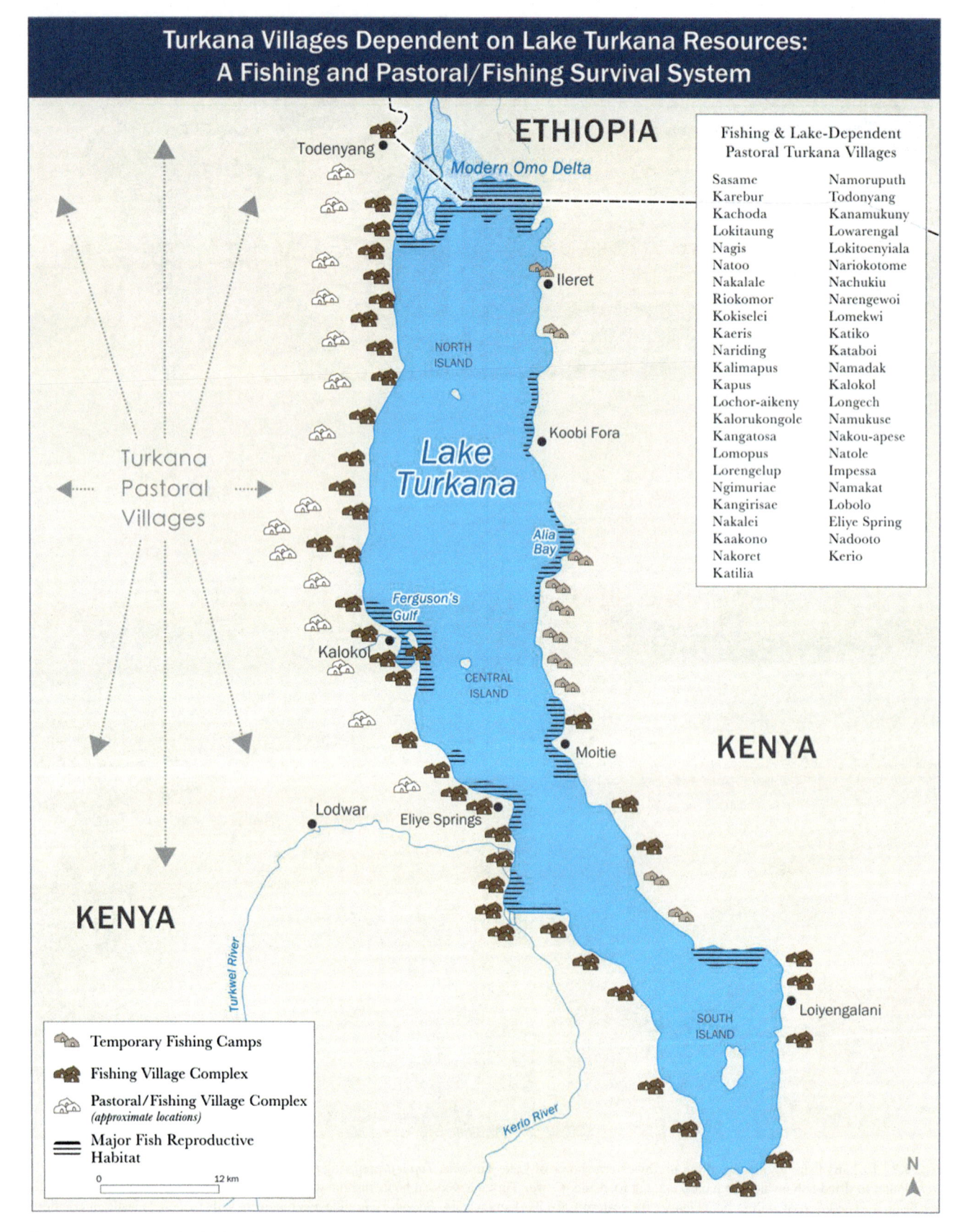

Fig. 9.4 Fishing and mixed fishing/pastoral Turkana village areas at Lake Turkana. Major village complexes are indicated, along with temporary (seasonal) fishing villages along the eastern shore of the lake (another forms at North Island) and key fish reproductive habitats—primarily along shorelines in shallow waters (including near the Omo River inflow, in Ferguson's Gulf and Alia Bay (see Fig. 5.2 for a bathymetric representation of lake level drop)

Fig. 9.5 Turkana fishing village activities at Kalokol and northward along the lake. *Top left* Turkana man with fishing nets and raft of doum palm trunks lashed together—at Ferguson's Gulf. *Top right* Sail boat (with sail constructed of food aid bags). *Bottom left* Clothes washing and interior of non-motorized boat—near Kenya-Ethiopia border. *Bottom right* Local market where households market their prepared charcoal to locals and to travelers

Fig. 9.6 Northern Turkana Fishing Villagers. *Top left* Boys getting water at Ferguson's Gulf. *Top right* Girl with fish caught from doum palm raft in Ferguson's Gulf. *Bottom photos* Family members of fishing village complex near Ethiopia-Kenya border

Table 9.1 Annual seasons described by northern Turkana fishers

---

ANNUAL SEASONS RECOGNIZED BY LAKE TURKANA FISHERS

AUGUST to NOVEMBER: *ANAMAPOLO*
Omo River: Major inflow to Lake Turkana
Currents: Extremely strong southward current from Omo River inflow,
  facilitating movement of west shore Turkana fishers to eastern waters.
Winds: From the north, relatively mild to moderate.
Lake level: Generally at a maximum level in central/ northern portions of the lake,
  Ferguson's Gulf and bays with highest waters and often high fish catch.
    Strong currents around Omo delta are dangerous and avoided. Fish stocks
      in shoreline areas and nearshore are relatively plentiful.
    High level of movement to east shore waters from west shore villagers: up to 1 month trips.
    Quiet backwaters, tiny bays (*'Ahaar'*) forming in modern delta with abundant fishing
      by Dasanech.

DECEMBER to MAY: AKIBONG ANAM
Omo River quiet, much reduced.
Currents: Not strong, slowed considerably.
Winds: Extremely strong, from southeast, in early phase; then quieted.
  Travel on lake possible as winds subside.
Lake level: Swelling gradually dwindling back toward minimum.
Fishing: Most lake travel suspended because of high winds in early phase.
  After winds subside, major travel possible: fishing expeditions to
    lake's eastern shoreline, northern shoreline, etc.

JUNE - JULY: *ALELES NGAITIA*
Omo River: Relatively quiet, low inflow to lake. Currents relatively slow to moderate.
Winds: Moderate, allowing traditional boats throughout the lake.
Lake level: Northern waters only slightly swollen relative to rest of the lake.
  Ferguson's Gulf -- waters reduced, low level.
Fishing: Expeditions throughout eastern lake as well as northern shoreline
    and modern delta (day expeditions, due to Dasanech opposition).
  Ferguson's Gulf – catch levels reduced.
  Destinations dependent on boat type, gear, labor issues.
Note: Many refer to "the time of many birds breeding" for this period.

Fishing seasons contrast with pastoral ones, with fundamentally different factors for livelihood. Monthly periods indicated are approximate and fluctuate with environmental conditions. Much variation in villagers' use of these terms also exists in the northern and central regions

Fishing conditions in the lake would be drastically altered by even short-term cessation of Omo River inflow, which would eliminate the river's annual pulse of freshwater, sediment and nutrients. Major destruction of fish habitat and plummeting fish catch would result—decimating the livelihood of Turkana fishing communities.

Fig. 9.7 Turkana girl and villagers with dried fish pallets set at roadside for transport to market. These pallets here have remained for two weeks, awaiting merchant pickup

As according to fishing elders from villages along the western shoreline, Turkana fishers forming camps along the eastern shore of the lake frequently mingle with El Molo fishers in their targeting of tilapia, Nile perch, and other species. They are at considerable risk of attack, however, by Dasanech from the east shore and by Gabbra herders (Fig. 1.3).

Conflicts between Turkana and Dasanech fishers (described in Chap. 7) are frequent and often involve gear thefts and killings—particularly in the northern lake waters and around the Omo delta where the two groups compete for limited fish stocks during many months of the year. According to all fishers, these stocks are greatly stressed by the large commercial fishing boats based in Ethiopia. Turkana sailboats and gear are generally far superior to the technology available to the Dasanech—another contributing factor to conflict. In a survey of Turkana fishing households in several different village complexes along the northwestern shoreline of the lake, this writer and other SONT researchers recorded numerous accounts of gear theft and killings between Dasanech and Turkana fishers.

As the survival systems of both the northern Turkana and the Dasanech continue to decline and the influx of pastoralists 'refugees' to the lake economy increases, the conditions for violent conflict intensify.

➢ **Fishing for markets is a precarious enterprise for the northern Turkana villagers**. As early as the 1960s, some fishers have sent fish—primarily tilapia and Nile perch—to Kitale, Lake Victoria and other Kenyan markets. Preparation of catch for market has not changed substantially over the years. There are no post-catch facilities for cleaning fish, so fish are commonly cleaned on the sandy shores of the (therefore, deemed of reduced quality in markets), then salted and sun-dried on netting racks strung well above the ground (Figs. 1.3 and 9.9). Dried fish for marketing are stacked and bound onto large pallets and left at the roadside (Figs. 9.3 and 9.7). Pallets are picked up at irregular times by merchant truckers and moved to markets in Kalokol and Lodwar within Turkana, as well as to Kisumu by Lake Victoria and elsewhere in Kenya. The price paid to villagers is entirely set by the merchants and fluctuates widely. These are matters of real distress to northern Turkana fishers, since the number of days between pallet preparation and merchant pickup can be so extended that their financial return for their efforts is miniscule.

While household commodity production—including the common activities of charcoal preparation and chicken/geese raising—is easily incorporated into customary Turkana social relations, commercial relations have generally brought new pressures to the Turkana (Fig. 9.5). Instead of participating in community-based fishing and marketing, for example, a few villagers have now become fish merchants themselves. Local community members view them with some scorn and such privatization style change suggests the potential future impacts of commercial systems on traditional cooperative relationships—should fishing even survive as a means of livelihood in the region.

Fishing Shoreline Communities: Household Practices and Resources

SONT researchers conducted a survey of the practices and outcomes of fishing by households in three distinct settlement areas along the lake, between Ferguson's Gulf and Todenyang (Fig. 9.4). Randomly selected household heads were questioned about their livelihood activities and status. Data for 35 households from the different settlement areas are summarized in Table 9.2.

Major patterns emerging from this survey include the following

- Considerable uniformity exists within and among fishing communities in terms of fishing location and patterns of consumption versus marketing of catch. There is also little variation in types of household commodity production.
- More than half of all households had been settled along the lake for fewer than ten years, 25 % for twenty to forty years and only 2.8 % for more than forty years. In general discussion, many household heads indicated that they remained at the lake following the failure of aid development projects they had been part of, since they had insufficient resources to return to pastoral life.
- Villagers from the northern area (Todenyang, Lowarengak) were from upland pastoral areas such as Lokitaung (Fig. 9.4) and relatively closer to Ilemi pastures. Many from the Kalokol region have arrived from Eliye Springs or lands toward Lodwar and northward while others arrived from villages in the extreme northwestern shoreline area (Fig. 9.8). All households settled along the lake for last resort survival activities after the loss of all or nearly all of their livestock from starvation and disease during drought times or from raids by adjacent ethnic groups.
- All households took up fishing almost immediately after settling by the lake (mostly on others' boats). Of the 35 fishing household heads surveyed, 12 owned (or co-owned) wooden boats. A number of fishers in the Kalokol/Ferguson's Gulf area used doum palm rafts, most of them stating that they could not afford a boat. Nariokotome and Nachukwi (north of Kalokol) were the only village areas surveyed where a substantial number of fishers owned their boats. Very few had been able to purchase an engine—a matter of real frustration among these fishers.
- Food for household consumption consisted primarily of fish, with occasional meat (from purchase with proceeds from fish marketing or from barter with nearby pastoral households). International food aid, primarily in the form of maize meal or powdered milk, has provided occasional temporary relief for households in some locales; others have received no aid at all.
- A minority of fishing households surveyed (approximately 17 %) undertook some type of household-based commodity production, such as preparation and marketing of charcoal, grass mat weaving and chicken raising, in order to purchase additional food (Figs. 9.5 and 9.9).

As described earlier for the region as a whole, most fishing households engage in regular exchange (both barter and sale/purchase) with nearby pastoral or mixed pastoral/fishing villagers. Most needed to market a relatively high proportion of their fish catch, however, and worked cooperatively with surrounding households. The prices paid by merchants for sun-dried and salted fish picked up at the roadside were inconsistent among lakeside villages. Villagers complain that they are powerless in this regard since they have to accept whatever the fish merchant will pay. At the time of the SONT survey, a large pallet of sun-dried mixed species fish brought the low price of about 30 Kenyan shillings/kg in villages near Lowarengak, for example. Nile perch clearly commanded the highest return—about 150 Kenya shillings/kg, except for fish 'maws' (entrails).[9]

---

[9] At the time of survey, USD 1 was equivalent to about 80 Kenyan shillings.

Table 9.2 Household Survey in Lake Turkana Fishing Communities: A Summary

Household number*	1	2	3	4	5	6	7	8	9
Village	Todenyang	Lowarengak	Lowarengak	Lowarengak	Nakitoekonon	Nakitoekonon	Nadoupua	Nadoupua	Lorekawotolem
Number years lived at lake	20	10	4	13	3	12	10	20	8
Prior village areas	Lowarengak	Pringan, Nakitoekonon	Lochwa Arengan Kachoda	Todenyang Lokitaung	Lomekwi Kalokol	Lowarengak Kakuma	Lowarengak Kakuma	Todenyang Kakuma	Lokitaung Kaleng
Reason moved to lake	Livestock loss	Livestock loss	Livestock loss	Livestock loss	Livestock loss	Livestock loss	Livestock loss	Livestock loss	Livestock loss
Major household food **	Fish	Fish	Fish	Fish	Fish	Fish	Fish	Fish	Fish
Own boat	Yes	No	Yes	No	Yes	No	No	Yes	No
Boat type	Wood	Wood	Wood	Wood	Wood	Wood	Wood	Wood	Wood
Fishing: Jan to Apr 2011									
Area fished	Kanamukuny	Ileret	Ileret	Kanamukuny	Ileret Kambi	Lowarengak Ileret	Deep water	Deep water	Deep water
Catch/month: Kg (dried) fish	30	120	200	20	520	40	300	130	230
Fish consumed per month (kg)	30	30	80-100	14	20	60	180	30	50
Number of fish marketed per month (approximate)	70	190	2600	23	300	300	100	80	180
Other areas fished	None	Selicho	Selicho Delta	Ileret, Delta	Ileret	Ileret	Lowarengak	Kanamukuny	Ileret Kanamukuny
Main target fish	Labeo, Tilapia	Labeo, Tilapia, Nile perch	Nile perch	Nile perch	Labelo N. perch	Labelo	Labelo	Tilapia	Tilapia
Fish prep. for market	Sundry (salt)	Sundry (salt)	Sundry (salt)	Sundry (salt)	Sundry (salt)	Sundry (salt)	Sundry (salt)	Sundry (salt)	Sundry (salt)
Market destination	Kalokol Kisumu	Kalokol Kisumu	Kalokol Kisumu	Kalokol Kisumu	Kalokol Kisumu	Kalokol	Lowarengak	Lowarengak	Lowarengak

(continued)

Table 9.2 (continued)

Household water source	Lake	Lake	Well	Lake	Lake	Lake	Lake	Lake	Lake
Own livestock	No	Yes	Yes	Yes	Yes	Yes	Yes	Yes	Yes
Cattle	0	3	8	5	0	0	0	0	0
Small stock	0	10	5	3	27	5	14	33	5
Camels	0	2	4	0	0	0	0	0	0
Other food sources	No	No	No	No	No	No	No	No	No
Household commodity sale	No	No	No	No	Mats	No	Firewood	No	Charcoal
Relief aid	Seldom	Seldom	Seldom	Seldom	Seldom	Seldom	Seldom	Seldom	Seldom
Household number*	**10**	**11**	**12**	**13**	**14**	**15**	**16**	**17**	**18**
Village	Nadoupua	Nadoupua	Nadoupua	Nadoupua	Nadoupua	Nariokotome	Nariokotome	Nariokotome	Nariokotome
Number years lived at lake	30	40	5	3	22	8	13	23	9
Prior village areas	Lowarengak Kakuma	Born	Lowarengak Todenyang	Lowarengak Todenyang	Lowarengak Turkwel	Lowarengak Kakuma	Todenyang	Lowarengak Kakuma	Lowarengak Todenyang
Reason moved to lake	Livestock loss	N/a	Livestock loss	Livestock loss	Livestock loss	Livestock loss	Livestock loss	Livestock loss	Livestock loss
Major household food **	Fish	Fish	Fish	Fish	Fish	Fish	Fish	Fish	Fish
Own boat	No	Yes	No	No	Yes	Yes	Yes	Yes	Yes
Boat type	Wood	Wood	Wood	Wood	Wood	Wood	Wood	Wood	Wood
Fishing: Jan to Apr 2011									
Area fished	Deep water	Deep water	Deep water	Ileret	Ileret	Choro	Deep water	Ileret Choro	Ileret Choro

(continued)

Table 9.2 (continued)

Catch/month: Kg (dried) fish	210	300	25	170	300	100	27	305	160
Fish consumed/ month (kg).	50	140	15	90	140	47	52	30	80
Number fish marketed/ month.	160	220	50	210	430	203	50	490	190
Other areas fished	Ileret	Ileret	Ileret	Kanamukuny	Kanamukuny	Choro	Ileret	Choro, Ileret	Choro, Ileret
Main target fish	Tilapia, Labeo	Nile perch Tilapia, Labeo	Nile perch Tilapia, Labeo	Nile perch Tilapia, Labeo	Nile perch Tilapia, Labeo	Nile perch	Labeo	Nile perch Distichodus	Nile perch Labeo
Fish prep. for market	Sundry (salt)	Sundry (salt)	Sundry (salt)	Sundry (salt)	Sundry (salt)	Sundry (salt)	Sundry (salt)	Sundry (salt)	Sundry (salt)
Market destination	Lowarengak	Lowarengak	Lowarengak	Lowarengak	Lowarengak	Kakuma	-	-	-
Household water source	Lake	Lake	Lake	Lake	Lake	Lake	Lake	Lake	Lake
Own livestock	Yes	Yes	Yes	No	Yes	Yes	Yes	Yes	Yes
Cattle	0	0	0	0	2	0	0	0	0
Small stock	38	8	3	0	43	5	12	26	8
Camels	0	0	0	0	2	0	0	2	0
Other food sources	No	No	No	No	No	Work in mission	Work in mission	Work in mission	No
Household commodity sale	No	No	No	No	No	No	No	No	No
Relief aid	Seldom	Seldom	Seldom	Seldom	Seldom	Seldom	Seldom	Seldom	Seldom
Household number*	**19**	**20**	**21**	**22**	**23**	**24**	**25**	**26**	**27**
Village	Nariokotome	Nariokotome	Nachukwi	Nachukwi	Nachukwi	Nachukwi	Nachukwi	Nachukwi	Kalochoro
Number years lived at lake	5	30	18	10	9	18	10	24	8
Prior village areas	Lowarengak	Lowarengak	Lowarengak	Lowarengak	Lowarengak	Kakuma	Kataboi	Kalokol	Kalokol

(continued)

Table 9.2 (continued)

	Kalokol	Ngomeris	Todenyang	Todenyang	Todenyang		Kalokol	Lobolo	Lodwar
Reason moved to lake	Livestock loss	Livestock loss	Livestock loss	Livestock loss	Livestock loss	Livestock loss	Livestock loss	Livestock loss	Livestock loss
Major household food**	Fish	Fish	Fish	Fish	Fish	Fish	Fish	Fish	Fish
Own boat	No	Yes	Yes	Yes	No	Yes	Yes	Yes	Yes
Boat type	Wood	Wood	Wood	Wood	-	Wood	Wood	Wood	Wood
Fishing: Jan to Apr 2011									
Area fished	Nariokotome	Nariokotome	Choro	Ileret	-	Choro	Ileret	Ileret	Deep water, Kalochoro
Catch/month: Kg (dried) fish	250	430	30	202	-	68	30	42	145
Fish consumed/ month (kg).	60	80	20	30	-	30	60	220	90
Number fish marketed/ month.	370	620	84	120	-	90	310	400	300
Other areas fished	Choro, Ileret	Choro, Ileret	Choro, Ileret	Choro, Ileret	-	Choro	Choro	-	Kerio
Main target fish	Nile perch Labeo	Nile perch	Nile perch Labeo	Nile perch Labeo	-	Nile perch Labeo	Nile perch Labeo	Tilapia, Labeo	Distichodus, Labeo
Fish prep. for market	Sundry (salt)	Sundry (salt)	Sundry (salt)	Sundry (salt)	-	Sundry (salt)	Sundry (salt)	Sundry (salt)	Sundry (salt)
Market destination	-	-	Kalokol Kakuma	Kalokol Kakuma	-	Kalokol Kakuma	Kalokol Kakuma	Kalokol Kakuma	Kalokol
Dinking/cooking water	Lake	Lake	Lake	Lake	Lake	Lake	Lake	Lake	Lake
Own livestock	No	Yes	Yes	Yes	Yes	Yes	Yes	Yes	Yes
Cattle	0	0	0	0	0	0	0	0	0
Small stock	8	8	8	3	10	8	3	9	10
Camels	0	0	3	8	9	3	2	6	0
Other food sources	No	No	No	No	Work in mission	No	No	No	No
Household commodity sale	No	No	No	No	No	No	No	No	Sell mats baskets
Relief aid	Seldom	Seldom	Seldom	Seldom	Seldom	Seldom	Seldom	Seldom	Seldom

(continued)

Table 9.2 (continued)

Household number*	28	29	30	31	32	33	34	35
Village	Kalochoro	Kalochoro	Kura	Kura	Kura	Kura	Namadak	Namadak
Number years lived at lake	13	6	4	7	12	8	20	13
Prior village areas	Kataboi Kalokol	Namadak Kalokol	Lodwar	Lodwar Kataboi	Todenyang	Todenyang	Eliye Springs Kerio	Lobolo, Eliye Springs
Reason moved to lake	Livestock loss	Livestock loss	Livestock loss	Livestock loss	Livestock loss	Livestock loss	Livestock loss	Livestock loss
Major household food **	Fish	Fish	Fish	Fish	Fish	Fish	Fish	Fish
Own boat	No	No	No	No	Yes	No	No	No
Boat type	Wood	Wood	Raft	Raft	Wood	Wood	Raft	Raft
Fishing: Jan to Apr 2011								
Area fished	Deep water, Kalochoro	Deep water, Kalochoro	Kura, Karipun Longech	Kura Longech	Kura Longech	Kura Longech	Namadak Namukuse	Namadak Namukuse
Catch/month: Kg (dried) fish	240	170	400	370	3,000	4'000	2,000	1,500
Fish consumed/ month (kg).	50	78	50	70	40	400	200	50
Number fish marketed/ month.	120	320	2,000 fingerlings	2,400 fingerlings	2,100 fingerlings	16,000 fingerlings	9,000 fingerlings	7,000 fingerlings
Other areas fished	Nariokotome	Kalochoro Nariokotome	Ekwar Adisi (Ferguson's Gulf)	Ekwar Adisi (Ferguson's Gulf)	Ekwar Adisi (Ferguson's Gulf)	Ekwar Adisi (Ferguson's Gulf)	Namukuse Ferguson's Gulf	Lokoro Ferguson's Gulf
Main target fish	Distichodus Labeo	Distichodus Labeo	Tilapia Clarias	Tilapia Clarias	Tilapia Clarias	Tilapia Clarias	Tilapia	Tilapia
Fish prep. for market	Sundry (salt)	Sundry (salt)	Sundry (salt)	Sundry (salt)	Sundry (salt)	Sundry (salt)	Sundry (salt)	Sundry (salt)

(continued)

Table 9.2 (continued)

Market destination	Kalokol	Kalokol	Kalokol	Kalokol	Kalokol	Kalokol	Kalokol	Kalokol
Household water source	Lake	Lake	Lake	Lake	Lake	Lake	Lake	Lake
Own livestock	Yes	Yes	No	No	No	Yes	Yes	Yes
Cattle	0	0	0	0	0	0	0	0
Small stock	8	12	0	0	0	10	13	7
Camels	0	0	0	0	0	5	0	0
Other food sources	No	No	No	No	No	No	No	No
Household commodity sale	Sell mats baskets	Sell mats baskets	No	Seldom	No	No	No	No
Relief aid	Seldom	Seldom	Seldom	Seldom	Seldom	Seldom	Seldom	Seldom

[a]Household head interviewed; [b]Occasional maize meal from food aid organizations

Fig. 9.8 Turkana life at Ferguson's Gulf and northward along shoreline. *Top left* Children on beach, near Kenya-Ethiopia border. *Top right* beach at Ferguson's Gulf. *Bottom left* Goats watering at Ferguson's Gulf. *Bottom right, Lower right* Girl washing clothes from doum palm raft

Fishing Shoreline Communities: Household Practices and Resources 179

Fig. 9.9 Turkana woman drying fish for marketing with secondary geese/chicken—raising

Fish Species and Critical Habitats

Fishing villages all along the west shore depend almost entirely on fishing for their survival, with the number growing all the time. A viable alternative way of survival is rare among them.

[GOK fisheries officer in Kalokol]

➢ Local accounts of important fish and their favored habitats, recorded by SONT researchers summarized in Table 9.3, point to the devastating impacts that the Gibe III dam and dam linked irrigated commercial agricultural development would have shoreline retreat would have on critical fish environments and indigenous fishing.

- Habitats for breeding and early life cycle stages for some of the most important fish species for northern Turkana fishing communities are concentrated along the northern shoreline of the lake and the Omo delta, where the annual flood pulse of the Omo River provides major freshwater and nutrient contributions to the lake—as well as in Ferguson's Gulf, Alia Bay and several other key fishing habitats along the lake's shallow locales.
- **The two most important catch species for the Turkana—Nile perch and Nile tilapia—depend on these habitats**. Tilapia lay their eggs and hatch in grassy or reed areas along the shoreline in Ferguson's Gulf and in other bay waters. Tilapia fingerlings mature along the lake's muddy shores. Nile perch, on the other hand, lay eggs and hatch in deep water, but juveniles feed on tilapia and other species' fingerling populations in the delta and along the shoreline. Their presence in the northern shoreline and delta area corresponds with the Omo's annual flood and annual pulse of lake inflow between early August and December. Nile perch also migrate upstream in the Omo River, where they provide subsistence to the

poorest Dasanech and Nyangatom communities. Three different species of tilapia: *Orochromis niloticus*, *Sarotherodon galilaeus* and *Tilapia zillii*, are caught in the river, dried and sold to the export-oriented fishing enterprises operating from Ethiopia.
- Turkana fishers report a larger number of migrating species than are described in the scientific literature. Of the more than 50 fish species recorded for Lake Turkana, at least 12 are of major significance to the Turkana fishing communities. Hopson (1982) describes four different fish communities in the lake: a littoral assemblage, an inshore assemblage, an offshore demersal assemblage, and a pelagic assemblage. Eleven fish species are endemic to the lake—nearly all of them living in the offshore pelagic or demersal zone (Lowe-McConnell 1987). During the river's annual flood, some of these migrate up the Omo River and breed for various periods (Hopson 1982; Beadle 1981; Lévêque 1997). These species include *Alestes baremoze*, *Hydrocynus forekalii*, *Citharinus citharus*, *Distichodus niloticus* and *Barbus bynni*.
- During the early months of the year, currents come from the north, with relatively light Omo River inflow, so Turkana communities from the western shoreline can access both the Omo delta/northern shoreline areas and the eastern portion of the lake, along with all available areas along the western shoreline. These conditions often have facilitated a relatively high fish catch. Local fishermen report catch levels range from 30 to 4000 kg (kilograms per month) during the comparatively favorable February to April period. This wide range of catch values reflects a multiplicity of factors, similar to those identified earlier for seasonal movements (e.g., access to boats and gear, available labor, number and duration of fishing expeditions, current and wind conditions, and shifts in target fish locales.) Catch values of 1000–4000 kg were recorded only for fishers at Ferguson's Gulf area—primarily those fishers possessing sailboats with engines.

Table 9.3 lists those fish deemed most important by local fishers, along with identification of critical fish reproductive and life cycle habitats, as well as their estimated sensitivities to lake level change. Localities and seasonality of fishing basically conform to the seasonal movements and fish reproductive habitat locations indicated in Fig. 9.4. A view of progressive lake level drop—predictable from Gibe III and commercial scale irrigated agricultural developments—is presented in Fig. 9.10. A closer view of progressive lake level drop—including changes to be expected in the earliest phase of these developments, is shown for the Omo delta and Ferguson's Gulf in Fig. 9.11.

> **Given the density of northern fishing communities, the shoreline fish habitats on which they depend, their already precarious nutritional status and the borderline potability of lake water, Turkana fishing communities are clearly vulnerable to catastrophic level destruction from lake retreat caused by the planned Omo basin developments.**

Counting the Discounted: Northern Turkana Population at the Lake

➢ **The question remains as to just how many Turkana are vulnerable to such livelihood destruction in terms of possible destruction of the lake fishery loss of the lake resources for livestock raising and access to potable water for basic household needs.**

It is apparent from the Kenyan government, development bank and other international agency documents and particularly the 2009 national census by the Kenyan government, that the lakeside population remains vastly underestimated in official records. This fact is basic to the government and banks' failure to account for its vulnerability to demise from the planned developments in impact assessments and other reports. The SONT research project had insufficient resources to be able to accurately assess the numbers of Turkana facing these threats, let alone the population of other vulnerable ethnic groups around the lake—a calculation that must include Dasanech, Gabbra, and El Molo communities (Fig. 1.3). Targeted information gathering from local records (when available), meetings with council of elders members were the only realistic means available for establishing a rough estimate. Moreover, village populations along the lake, as well as pastoral/fishing ones slight removed from the lake, shift rapidly with changing social and environmental conditions both in the lake zone and the upland plains. (Fig. 9.4).[10]

---

[10]Fishing and pastoral/fishing village complexes shown in Fig. 9.4, for example, are mapped in relational terms, since GPS was not available to SONT researchers at the time of survey.

Table 9.3 Lake Turkana fish species and habitats of importance to Turkana fishing communities

Scientific/Common Names	Turkana Name	Area most fished by Turkana and Spawning Habitat	Importance to Turkana Survival System	Sensitivity to Lake Retreat
Tilapia spp, including: **T. nilotica** **T. galilaea** **T. zilii** Oreochromis *niloticus*	Kokine	Delta, near shore, Ferguson's Gulf, bays *Spawn:* Delta, Ferguson's Gulf, shorelines	♦♦♦ Critical Consumption, Marketing	3 Extreme Loss of spawning habitat; desiccation of delta, shoreline, Ferguson's Gulf and bays.
Lates niloticus Nile Perch	Iji	Delta, north shore, North and Central Islands *Spawn:* Pelagic, but juveniles feeding in delta	♦♦♦ Critical Consumption, Marketing	3 Extreme Feeding habitat and catch habitat destruction
Labeo Horrie	Chubule	Delta, nearshore, some throughout lake *Spawn:* Grassy shore areas	♦♦♦ Critical Consumption Marketing	3 Extreme **Loss of spawning habitat and catch habitat destruction**
Distichodus niloticus	Golo	Delta, north shore, shorelines *Spawn:* Delta-Omo R. (grassy shoreline)	♦♦♦ Critical Consumption, Marketing	3 Extreme Spawning and catch habitat destruction
Clarias lazera Catfish	Kopito	Delta, north shore, near shore *Spawn:* Muddy shallow water, grassy reeds	♦♦♦ Critical Consumption, Marketing	3 Extreme Spawning habitat destruction
Synoclontis sp.	Tir	Shoreline *Spawn:* Shoreline	♦♦♦ Critical	3 Extreme Spawning/ juvenile habitat loss
Alestes - including: A. dentex A. baremose A. nurse	Juuze	Delta, north shore, **Ferguson's Gulf, bays,** offshore, flood shallow *Spawn:* Delta, bays, North Island	♦♦♦ Critical Consumption, Marketing	3 Extreme Spawning/ juvenile habitat Destruction (except North Island)
Citharinus citharus	Gesh	Delta, north shore, near shore, general lake *Spawn:* Delta	♦ Significant Consumption (limited)	3 Extreme Spawning and feeding habitat destruction
Hydrocynus forkalii Tigerfish	Lokel	Delta, shoreline, offshore *Spawn:* Delta	♦♦ Major Consumption	3 Extreme Spawning habitat destruction
Barbus turkanae B. bynni	Momwara	Delta, near shore, offshore (schools) *Spawn:* Delta	♦♦ Major Consumption (limited) Marketing	3 Extreme Spawning habitat
Bagrus spp. Balck Nile Catfish	Loruk	Offshore/ demersal. *Spawn:* General lake	♦♦♦ Critical Consumption, Marketing	1/2 Moderate/high
Schilbe uranoscopus	Naili	Delta, north shore *Spawn:* Delta	♦♦ Major Consumption (*northern* region)	3 Extreme Spawning habitat destruction
Cichlidae	Loroto	Deltas, shallow water	♦ Significant Consumption Marketing	3 Extreme
Bagridae- (giraffe catfish)	bulubuluch	Delta only	♦ Significant Consumption	3 Extreme

Key:
-- Importance to Turkana Survival System: ♦ = significant. ♦♦ = major, ♦♦♦ = critical
-- Sensitivity to Lake Retreat: 1 = moderate, 2 = high, 3 = extreme

Identification of taxa and assessment of habitat sensitivity described by local fishermen from villages along Lake Turkana's northwestern shoreline

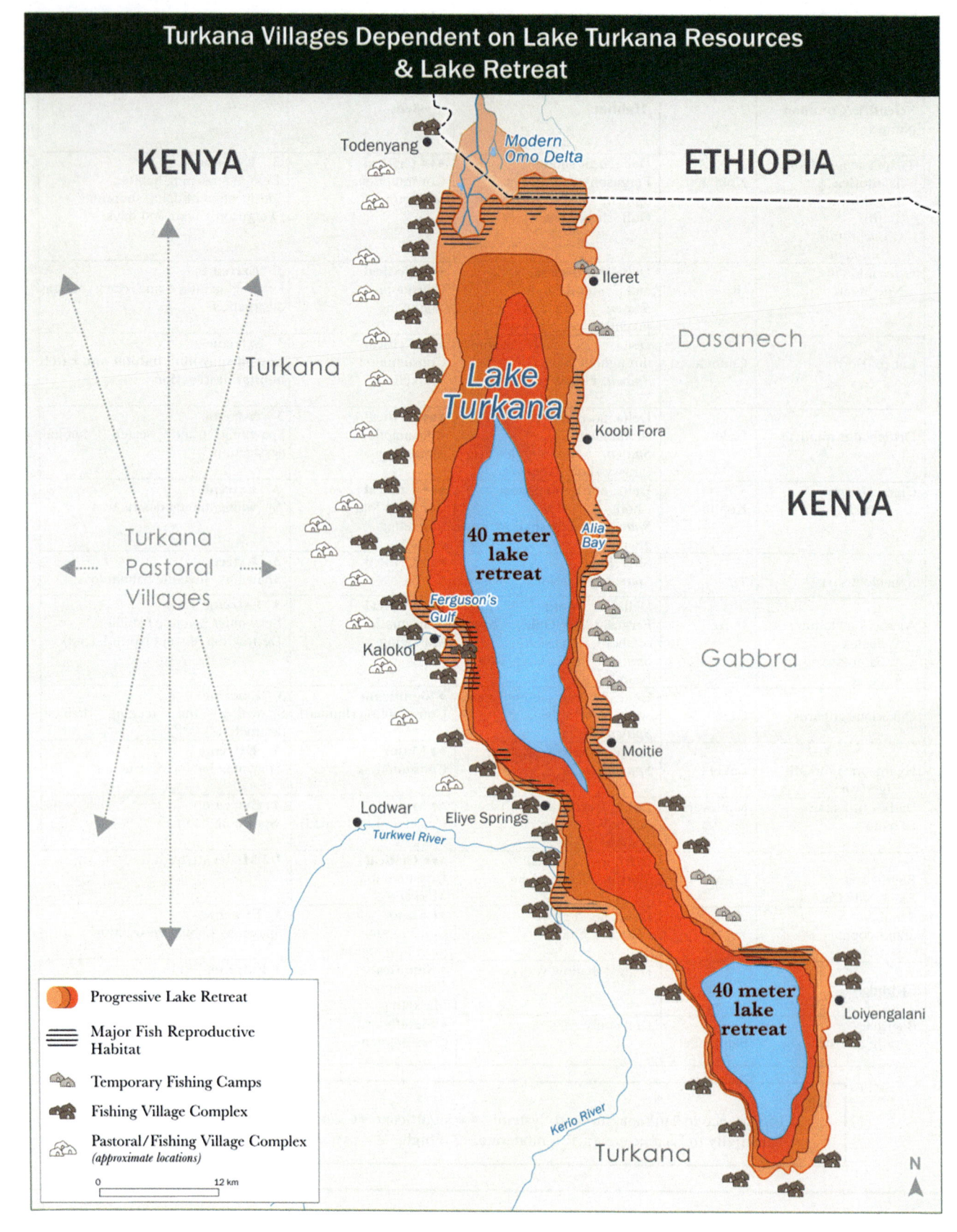

Fig. 9.10 Bathymetric representation of Lake Turkana retreat from Gibe III Dam and linked irrigation agriculture. *Source* ARWG; bathymetric base map from Hopson (1982)

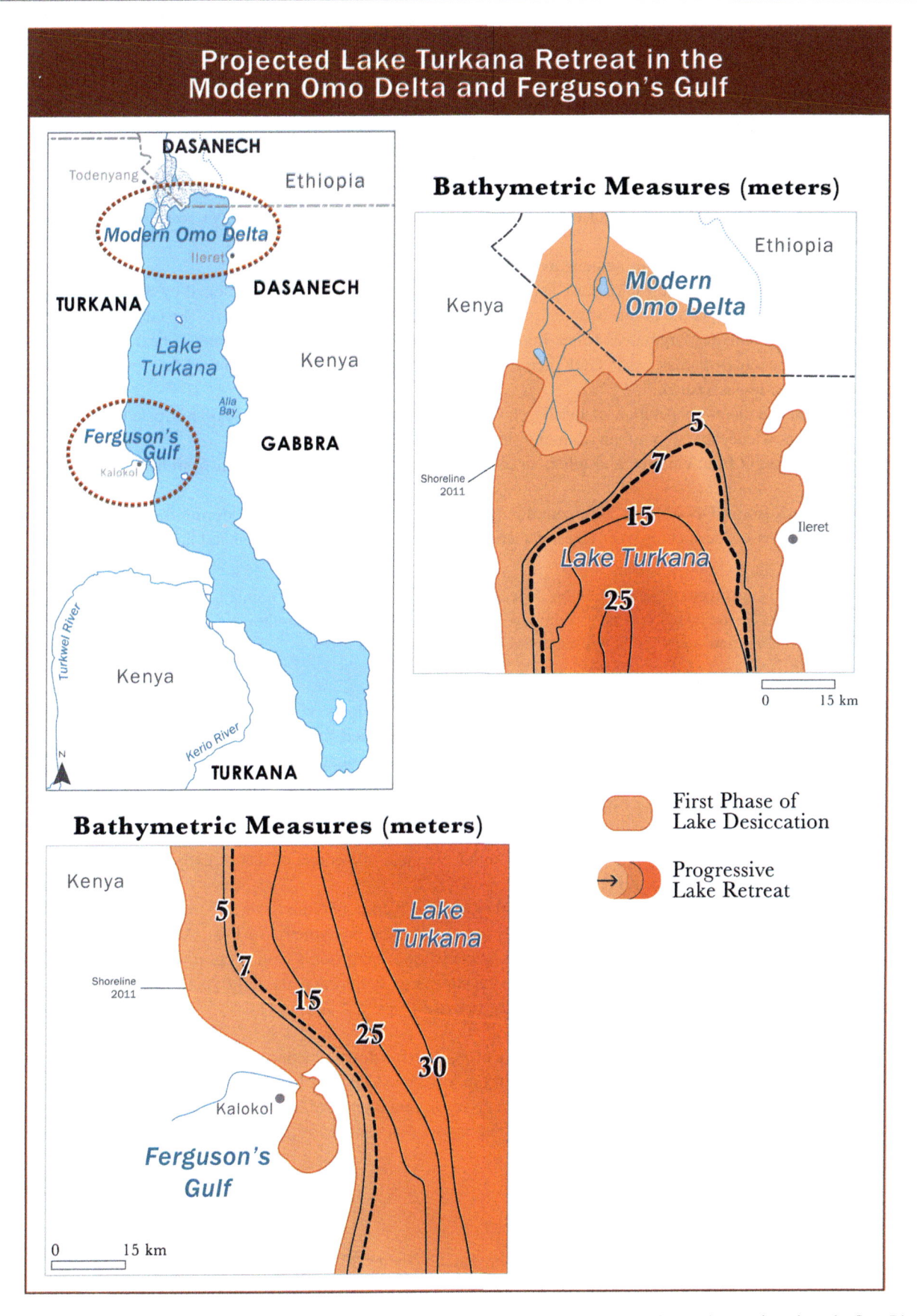

Fig. 9.11 Desiccation of Ferguson's Gulf and the modern Omo delta: projected from Gibe III dam and irrigated enterprises along the Omo River. *Source* ARWG; bathymetric values from Hopson (1982)

Some of the figures obtained were markedly below certain non-governmental organization estimates—even radically so, as in the case of Kalokol town, which was recorded by several nonprofit groups as 55,000 whereas the SONT estimate (from consultations with local elders and aid agency figures) was closer to 30,000. Since the population of Kalokol fluctuates greatly with economic and environmental conditions, this is not surprising. Other estimates, such as those by Oxfam for a number of village/town locales along the lake—recorded in 2007 (*prior* to the 2007–2010 drought), were lower. *The highest population figures, however, provide the best indication of the central and northern Turkana's vulnerability to the decline or disappearance of Lake Turkana's waters and living resources.*

Local Turkana administrators and council of elders members consistently described to SONT researchers that the GOK's 2009 census takers:

- Recorded populations only in major centers, avoiding the more populous outskirts.
- Avoided rural areas near the lake where most Turkana live (all local administrators questioned by SONT attest to the fact that there are often more people living between main village complexes than within them). The GOK Census states that census takers recorded very large areas (often hundreds of square kilometers around towns). Local officials questioned gave contrary accounts—namely, that GOK census takers did not record the large and diffuse populations in lands surrounding those towns visited.
- Did not request the direct cooperation of locally chosen administrators—individuals who are trusted by local residents and far more knowledgeable about their communities and population sizes.
- Recorded information from children and others unlikely to report accurately, rather than from heads of household.

Population estimates from SONT efforts with community members were taken primarily during the dry season. As noted earlier, the populations of these complexes can fluctuate widely with changing environmental and social conditions.

> Even a conservative estimate of the population dependent on the lake's resources points to at least **300,000 Turkana who are** dependent on the lake's waters—for household members' daily consumption, for fishing and/or for livestock watering and grazing, and many exchange relations involving lake resources.

- **The Turkana population surrounding Ferguson's Gulf is particularly vulnerable to major hunger and disease conditions brought about by the retreat of Gulf waters, accompanied by fishery collapse.** Together with council of elders members, SONT conducted a preliminary survey of villages around the Gulf, and continued data collection at Kalokol and northward along the western shoreline. *This work consistently produced larger figures than those released by the 2009 GOK census.* Moreover, this population is overwhelmingly likely to swell as environmental and economic 'refugees' move southward along the lake and toward the lake from the upland plains. Ferguson's Gulf, along with the northern reaches of the lake, must be anticipated to suffer the most immediate effects of early lake level drop from the developments underway (see Fig. 9.11).
    - Lake retreat caused by the Gibe III dam's inevitable radical reduction of Omo River inflow, even during the presumed reservoir-fill, would desiccate Ferguson's Gulf as indicated in the bathymetric of Figs. 9.10 and 9.11.[11]
    - The extremely shallow waters and biochemical characteristics of Ferguson's Gulf support major reproductive habitats for fish species critical to the Turkana and intensive fishing activity during parts of the year. These critical fish habitats would be eliminated by even the first phase of lake retreat (Figs. 5.2 and 9.11).

As this book goes into print, there is substantial evidence of major river flow reduction from closure of the Gibe III dam and reservoir filling, and early reports from villagers in both the lowermost Omo at Lake Turkana suggest both that recession agriculture in the delta and riverside environments and the river's annual pulse of freshwater, sediment and nutrients into the lake are fully compromised. All communities contacted report fish catches reduced, although to date, no systematic

---

[11] Gulf locales measured 5–6 m in depth as recently as 2005 were only 2.5–3 m by 2013, according to local fishery officer reports to SONT researchers. The mouth of Ferguson's Gulf, once more than 1800 meters wide (according to figures quoted from the East African Common Services Organization in the early 1960s), has become so nearly closed and shallow that even small wooden vessels typically could not pass through the mouth and had to remain in the main lake waters (Fig. 9.11).

investigation has been possible. The possible role of climate change in worsening the lake level impact from Gibe III dam and irrigated agricultural development is unknown, but it is likely to be substantial in the longer term.

➢ **Turkana fishing villages extend around the (Longech) spit at the Ferguson's Gulf opening to the lake, and continue into lands between the Gulf and the town of Kalokol, where nearly all villagers are engaged in fishing activities in one way or another**. Many of them bring livestock and are part of the pastoral/fishing complexes described earlier. To the extent possible, herd owners keep their small stock locally (Figs. 4.6 and 9.2) and groups of thousands of small stock can be seen trekking to the lake for watering any grazing that is available. Camels are typically sent inland and are only brought to the lake for watering when necessary.

Council of elders members at Ferguson's Gulf drew a map of villages present at the time of SONT's visits in late 2012, as shown in Fig. 9.12. The supplementary Google Earth image of Ferguson's Gulf indicates the high level of accuracy with which these individuals (relationally) represented their villages, despite no previous experience with maps. Based on estimates earlier submitted to local officials, elders reported the population totals listed in Table 9.4. While figures for individual locales sometimes fluctuate greatly with shifting economic, environmental and other conditions, the overall totals are representative of the settlement presence and Turkana dependence on the Gulf's resources.

Most of these villages were not even included in the GOK's census; others were greatly underestimated. For example, the population for Namukuse village area was under-represented by at least 40 % in the census. For an accurate estimation of the

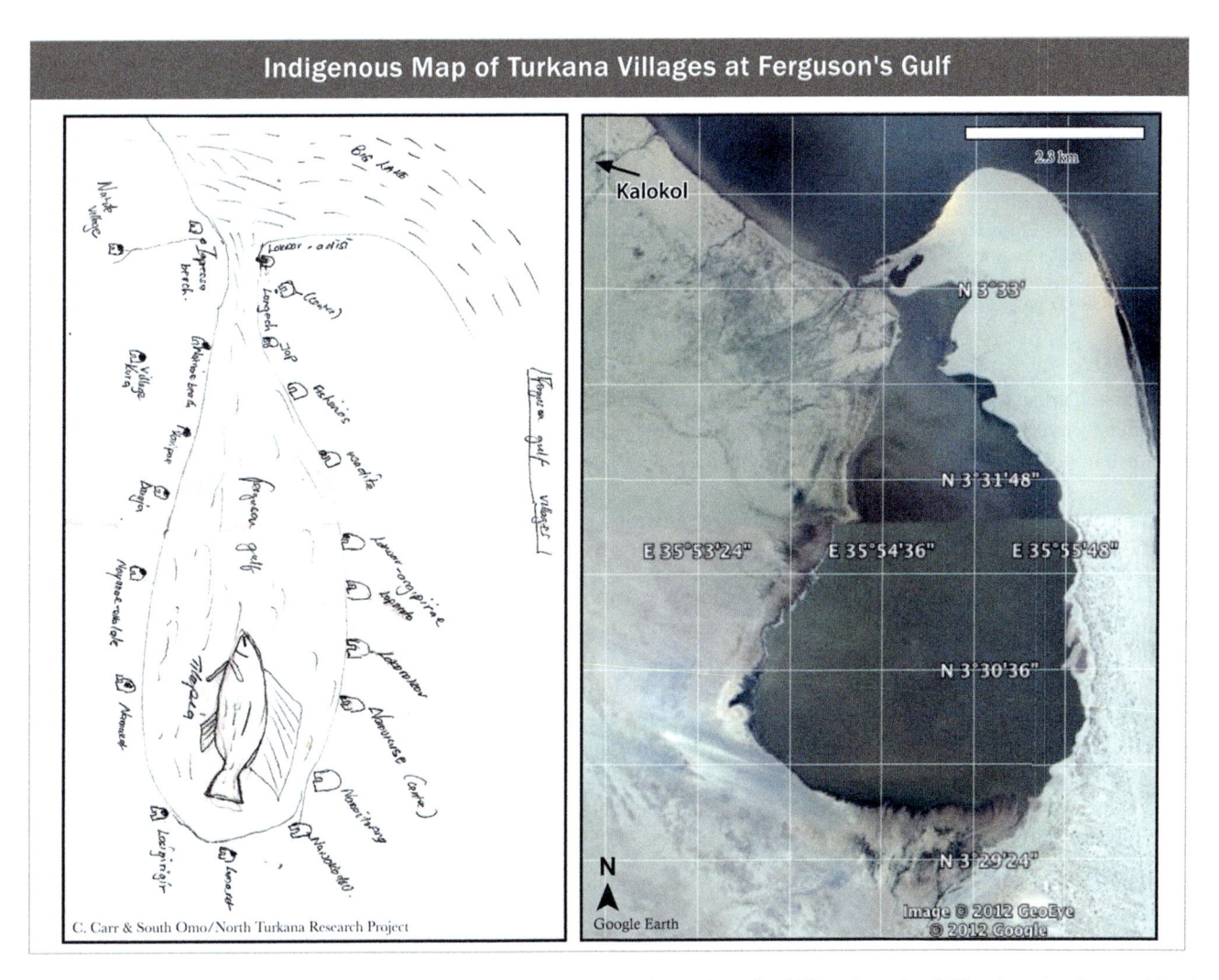

Fig. 9.12 Indigenous map of Turkana villages at Ferguson's Gulf. *Source* Local Council of Elders from the Gulf region—meetings and field reconnaissance with SONT researchers. Satellite map added

Table 9.4 Population estimates from Ferguson's Gulf region

Longech[a]	12,000
Lomaret	500
Jap	2000
Losigirigir	200
Village (South. of Jap)[b]	1800
Namakat	1000
Wadite	3000
Nayanae ekalale	500
Lokwar angipirea	1500
Daraja[c]	650
Loporoto	800
Karepun	700
Lokorokor	3000
Natirae	1500
Namukuse	10,000
Kura	3500
Village-near above	900
Impressa	5000
Nawoitorong	1000
Natole	4000
Nawokodu	500
Total population	54,050

Source Council of Elders (from prior accounts reports to aid officieals) as reported to SONT researchers in field-based meetings
[a]Cholera outbreaks known
[b]Former GOK fisheries camp
[c]Former NORAD project locale

vulnerable Turkana population actually dependent on the Ferguson's Gulf region, the populations of Kalokol town, its immediate environs, and the area between Kalokol town and Ferguson's Gulf must be added. Estimates from local administrators (in private consultation) and council of elders members were:

Kalokol town 11,500 (GOK census figure; larger environs not recorded)
Kalokol (outlying)—4500 (incl. Nakiria—2300)

This estimate excluded additional population segments essential to include, but such a survey was beyond the logistical capabilities of the SONT team. These population components include:

- Thousands of pastoral/fishing Turkana in settlements slightly more removed from the lake but fundamentally dependent on its resources for their survival (Fig. 9.4).
- Villagers diffusely settled between Kalokol town and Ferguson's Gulf.

Based on above estimates for the immediate Ferguson's Gulf, area combined with Kalokol and its outlying communities, the affected resident population in the Gulf area was at least **70,050**.

In all probability this estimate is a conservative one, due to the continued migration to the lake since the time of this SONT survey.

> The drying out of the Gulf would produce crisis level of hunger for both the population residing around the Gulf itself and for multiple thousands of Turkana living in locales slightly removed.

The Turkana population residing along the western shoreline is not only acutely vulnerable to economic collapse: it is also threatened with major disease epidemics, including cholera. According to U.N. data, cholera in the Turkana region is already one of the highest in Kenya, with recorded outbreaks along the western shoreline of Lake Turkana, especially around Kalokol (Africa Health 1998).

➢ **Population estimates for towns along the shores of Lake Turkana between Kalokol and the Ethiopia/Kenya border** (Fig. 9.4) **present a similar picture of major exclusion by the Kenya government's 2009 national census.** Along with nearby large village complexes, these areas were recorded by SONT members with assistance from local government administrators, council of elders members, and Beach Management Unit members (local residents who are government appointed). The results are summarized in Table 9.5; data excludes populations of villages scattered between population centers—areas that must be included for a minimally acceptable population estimate for the northern Turkana region. Local officials describe the density of people and livestock in these areas as considerably swollen following major stress conditions in upland plains to the west and northwest. *Both administrators and elders reported that most of these village areas were not visited by government census takers.*

Table 9.5 Population estimates for towns and village complexes along Lake Turkana's western shore from Kalokol to the Kenya-Ethiopia border

Kangaki	2000
Lokalale	650
Lomekwi	3000
Ngingolekoyo	1200
Nachukwi	5000
Kangatukusio	320
Kataboi	9000
Kaitengiro	500
Katiko	8000
Toperenawi	3000
Nasechabuin	3000
Kalotumukol	1700
Nalukowoi	350
Nariokotome	5000
Kaitio	200
Kokiselei	950
Kalochoro	290
Narengewoi	1300
Nayanae engol	800
Namarotot	500
Nadoupua	800
Lokapetemoi	300
Namadak	4000
Todenyang/Arii	10,300
Lowarengak	7000
Lokitonyalla	2300

Source Local government officers, and Council of Elders (from prior accounts reports to aid officieals) as reported to SONT researchers in field-based meetings

Based on these locally derived estimates, the indigenous population in the shoreline area between Kalokol and Todenyang (Fig. 2) is at least **71,460**

> Combining the above rough estimates, the indigenous population in the shoreline area of Ferguson's Gulf, extending northward Lake Turkana to the Kenya/Ethiopia border (near Todenyang) is at least: **141,000**

The true population of those Turkana who are vulnerable to destruction of their survival means from the effects of the Gibe III dam and irrigated agriculture on Lake Turkana is far greater, however. This population includes those fishing and mixed pastoral/fishing villagers who reside slightly more removed from Lake Turkana, but who nevertheless depend on it for their survival, through:

- Work as fishers—generally working for boat owners, or in post-catch fisheries related work
- Trading for fish, offering livestock products (meat, milk, skins, live animals)
- Livestock watering and lakeside grazing.

While there are no estimates for this pastoral/fishing population, which was apparently largely omitted from the census, many thousands of villagers clearly reside throughout this zone (villages are listed and relationally mapped in Fig. 9.3). A population estimate of at least **200,000** for those Turkana residing either along the lake or slightly removed from it is likely a conservative one.

> The unreliability of the GOK's population census does not alter the reality that the total indigenous fishing and pastoral population depending on Lake Turkana for their survival is far greater, even excluding the tens of thousands of Dasanech residing (within Kenyan borders) in the modern Omo Delta (see Fig. 1.2) and around the lake's northeastern shoreline, as well as El Molo, Rendille, Gabbra and other peoples around the southern and eastern lake.
> **In sum, the total regional population facing a survival crisis from their dependence on Lake Turkana should be presumed to be at least 300,000.**

The extreme vulnerability of this population is compounded by the continuing decline of Turkana the region's pastoral sector. Since at least a significant proportion of this population possibly remains uncounted by the Kenyan government in its 2009 national census as well, the region's looming disaster scale impacts of the developments underway cannot be ignored in national and international policy institutions and civil society.[12]

With comparable crises to that of Kenya's Turkana unfolding among the Dasanech and Nyangatom peoples as well, cross-border armed conflict among these groups and their neighbors (Fig. 5.3) **can be expected to escalate—thus worsening the armed struggle underway in South Sudan.** (Numerous young men from the region—especially Nyangatom and Turkana—have already joined insurgent groups in South Sudan—some of them returning with new arms.)[13]

> Active policy decisions by the Ethiopian government, the Kenyan government and international development organizations—particularly the World Bank and the African Development Bank—raise the specter of violation of internationally recognized human rights. These violations center around U.N. resolutions regarding the human rights to water, to livelihood and to freedom from political repression.

---

[12]The 2009 AFDB socioeconomic report cited a general lake-associated population of 300,000 (AFDB 2009), but embedded in the body of the report text, without notation of any significance within the context of the planned developments and without mention in the Summary and Conclusions sections. The EIB assessment of the Ethiopian segment includes some fragmentary and ambiguous population estimates for Ethiopia's lower Omo region.

[13]Detailed consideration of the decline of authority relations amongst the Turkana and the pastoral Suri is available in Abbink (2007) and Skoggard and Adem (2010).

Literature Cited

Abbink, J. 2007. Culture Slipping Away: Violence, Social Tension, and Personal Drama in Suri Society, Southern Ethiopia. In: Rao, A., M. Bollig and M. Bock (Eds.). The Practice of War: Production, Reproduction, and Communication of Armed Violence. pp. 53–71. Berghahn Books.

Africa Health. 1998. Cholera in Turkana. 21(1):43.

Bayley, P.B. 1982. The commercial fishery of Lake Turkana. In: *Report on the findings of the Lake Turkana project, 1972–1975*, Edited by Hopson, A.J., Chap. 6, Vol. 2. London: Government of Kenya and the Ministry of Overseas Development.

Beadle, L.C. 1932. Scientific results of the Cambridge expedition to the East African lakes, 1930–1931. The waters of some East African lakes in relation to their fauna and flora. *Zoological Journal of the Linnean Society* 38: 157–211.

Beadle, L. C. 1981. The Inland Waters of Tropical Africa. London, Longman.

Carr, C.J. 2012. Humanitarian catastrophe and regional armed conflict brewing in the border region of Ethiopia, Kenya and South Sudan: the proposed Gibe III Dam in Ethiopia, Dec 2012, 250 p. https://www.academia.edu/8385749/.

Collins, R.O. 2006. The Turkana patrol of 1918 reconsidered. *Ethnohistory* 53(1): 95–119.

Government of Kenya, National Bureau of Statistics, 2009. Population and housing census. http://www.knbs.or.ke/index.php?option=com_content&view=article&id=110&Itemid=483.

Gulliver, P. 1955. The family herds: a study of two pastoral tribes in East Africa: The Jie and Turkana.

Gwynne, M.D. 1969. *A bibliography of Turkana*. London: Royal Geographical Society.

Hogg, R. 1982. Destitution and development: the Turkana of North West Kenya. *Disasters* 6(3): 164–168.

Hopson A.J. 1982, Lake Turkana, a report on the findings of the Lake Turkana project, 1972–1975, Vols. 1–6.

Lamphear, J. 1988. The people of the grey bull: the origin and expansion of the Turkana. *Journal of African History* 29(1): 27–39.

Lamphear, J. 1992. The scattering time: Turkana responses to colonial rule. Oxford: University of Oxford Press, 308 p.

Lévêque, C. 1997. Biodiversity Dynamics and Conservation: The Freshwater Fish of Tropical Africa. Cambridge, UK: Cambridge University Press.

Lowe-McConnell, R. H. 1987. Ecological Studies in Tropical Fish Communities. Cambridge University Press.

Maunduab, P., S. Kibetb, Y. Morimotoa, M. Imbumic, and R. Adekab. 2011. Impact of *Prosopis juliflora* on Kenya's semi-arid and arid ecosystems and local livelihoods. http://www.tandfonline.com.

McCabe, J.T. 1990. Success and failure: the breakdown of traditional drought coping institutions among the pastoral Turkana of Kenya. *Journal of Asian and African Studies* 25(3): 146–160.

McCabe, J.T. 2004, Cattle bring us to our enemies: Turkana ecology, politics and raiding in a disequilibrium system. Michigan: University of Michigan Press.

Mkutu, K. 2003. Pastoral conflict and small arms: the Kenya-Uganda border region. http://mercury.ethz.ch/serviceengine/Files/ISN/124873/ipublicationdocument_singledocument/0d961f0f-e565-4b17-99f7-a256a27389ca/en/Pastoral+conflict.pdf.

Oba, G. 1992. Ecological factors in land use conflicts in Turkana. Pastoral Development Network, Paper no. 33a, 23.

Skoggard, I. and T.A. Adem. 2010. From raiders to rustlers: The filial disaffection of a Turkana Age-Set. *Ethnology* 49(4): 249–262.

Von Hohnel, L. 1938. The Lake Rudolf region: Its discovery and subsequent exploration, 1888–1909. *Journal of the Royal African Society*, London 37.

Open Access This chapter is distributed under the terms of the Creative Commons Attribution-NonCommercial 2.5 International License (http://creativecommons.org/licenses/by-nc/2.5/), which permits any noncommercial use, duplication, adaptation, distribution and reproduction in any medium or format, as long as you give appropriate credit to the original author(s) and the source, provide a link to the Creative Commons license and indicate if changes were made.

The images or other third party material in this chapter are included in the work's Creative Commons license, unless indicated otherwise in the credit line; if such material is not included in the work's Creative Commons license and the respective action is not permitted by statutory regulation, users will need to obtain permission from the license holder to duplicate, adapt or reproduce the material.

Human Rights Violations and the Policy Crossroads

> If our floods go away, what will we do? We have nowhere to go—nowhere to take our. children. You will find our bones here.
>
> [Dasanech agropastoral woman in the Omo River delta]

Abstract The pursuit of Omo River basin development is leading to a major human rights crisis in the Ethiopia-Kenya-South Sudan transboundary region of eastern Africa. Among the principal human rights being violated are those recognized by the International Covenant on Economic, Social and Cultural Rights (ICESCR) Treaty adopted by the U.N. General Assembly. Although the Ethiopian government is most immediately responsible for initiating human rights violations in the region, the Kenyan government and international development banks are variously complicit, collaborative and partnered in these transgressions. The World Bank, African Development Bank and major donor countries continue to support—even legitimate—the development despite predictable destruction of hundreds of thousands of indigenous peoples' livelihoods and major political rights violations, particularly within Ethiopia. *Cumulative and synergistic* effects of the Gibe III megadam and its linked irrigated plantations and energy export transmission system must be integrally considered for adequate social and environmental impact assessment, yet both governments and development banks have failed to act on this mandate. A crossroads in public policy has now emerged: either pursue the present pathway toward massive scale hunger, regional economic collapse and major new cross-border armed conflict or suspend the development underway in order to take genuine account of human rights and proceed in a direction that is accountable to citizens and provides for a sustainable future for the three nations involved.

The Crisis Unfolding and the Human Right to Context

Previous chapters of this book have detailed how and why the Gibe III hydrodam—one of the world's tallest, together with its linked irrigated agricultural and electricity export transmission development—is creating a human rights as well as humanitarian crisis in the transboundary region of Ethiopia-Kenya-South Sudan (Fig. 1.1). The indigenous population in these semi-arid lands is already subject to some of the highest levels of malnutrition, epidemic level disease and armed conflict in Sub-Saharan Africa.

Hundreds of thousands of indigenous pastoralists and agropastoralists, of multiple ethnicities in the region (Fig. 1.3) have been so severely disenfranchised over decades that they have had no recourse but to move to lands along the Omo River or Lake Turkana (Fig. 4.6). There they depend on river or lake waters for varying combinations of livestock raising, flood recession agriculture, fishing and other livelihood activities, as well as for potable water for household use. Government census information and international aid documents grossly underestimate the size of the vulnerable population and the real threat that these developments pose to their continued existence.

Radical reduction in Omo River flow volume and the river's inflow to Lake Turkana. brought about by the Gibe III dam and dam-enabled irrigated agricultural enterprises, spells disaster for these river and lake dependent livelihood activities and the peoples depending on them. The reasons for the massive scale vulnerability and the dynamics involved are detailed throughout the chapters of this book. The overall dismantlement of livelihood systems, including complex survival strategies that have for centuries been able to adapt to their changing circumstances, are now pushed beyond their capacity to adjust.

> For much of the indigenous population, the loss of Lake Turkana and Omo River water resources amounts to a virtual death sentence.

Added to this impending destruction is the plausible threat of Gibe III dam failure from earthquake or seismically triggered landslide events, causing unprecedented human and environmental decimation downstream in the Omo basin and around Lake Turkana. The scientific basis for postulating a 20 % chance of a 7 or 8 magnitude earthquake in the Gibe III dam region during the next 50 years is outlined in Chap. 3. The Ethiopian government, Kenyan government and development banks, in their environmental and social impact assessments for the Gibe III dam, have unambiguously failed to take account of the potentially catastrophic destruction that would result from dam collapse or overtopping due to a seismic event.

If the Gibe III dam and dam-enabled irrigated agricultural commercial farms become operational, major escalation of conflict among ethnic groups—each desperate for access to the region's disappearing resources—would inevitably occur (Fig. 5.3), with major spillover into the Ilemi Triangle and South Sudan that are already embroiled in major conflict. In turn, the Kenyan and Ethiopian governments would increase their militarization of the region (a process already underway), in order to fortify their respective territorial claims and protect the activities of foreign corporations contracted by them to exploit the region's agricultural and petroleum resources. This process is already taken to an extreme in the lower Omo basin within Ethiopia.

Major oil and gas deposits exist throughout much of the transboundary region. Oil exploration activities have been underway for years prior to Ethiopian and Kenyan government public announcement of "newly discovered reserves" (see Appendix). Contrary to Tullow Oil and government statements, these activities have been undertaken without account of their impact on the region's hundreds of thousands of indigenous residents and their environment. As in other areas of East Africa and the Horn region, their actions have ignored international (U.N.) and national government mandates for "free and informed consent" by affected communities. Active exploration by the oil industry and related international aid funded infrastructural development within the region, have provoked local resistance—even militant—despite 'assurances' by industry and government representatives of 'major benefits' that will accrue to local communities. Numerous small *legitimation* efforts by the industry have been introduced in central and northern Turkana, including construction of small communal health small clinics and other facilities, school payments for Turkana children (e.g., by Tullow Oil), and small traveling theater groups dramatizing the positive outcomes of oil and gas development in Turkana lands.[1]

> Transgressions of internationally recognized human rights by the Ethiopian and Kenyan governments, in collaboration with the AFDB and World Bank, center around violations of internationally recognized rights—to an adequate standard of living, to food and the means of producing it, to water and to freedom of political expression.

➢ **The United Nations drafted the International Covenant on Economic, Social and Cultural Rights (ICESCR) in the 1950s, in accordance with the** Universal Declaration of Human Rights (UDHR)—passed in 1948 by the General Assembly. The ICESCR, also referred to as 'the Covenant', was passed by the U.N. General Assembly in December of 1966 and brought into force in 1976.[2]

---

[1] These community 'contributions' are widely considered by Turkana elders to be 'bribes' for complicity with oil operations in their region.
[2] A succinct history of the U.N.'s human rights related multilateral treaties and treaty bodies is available (U.N. Office of the High Commissioner for Human Rights (OHCHR 2012).

> **The ICESCR (Covenant) constitutes a multilateral treaty committing State parties to work toward the granting of economic, social and cultural rights to individuals. It formally recognizes the human right to an adequate standard of living, including the right to food, freedom from hunger and health, among other rights. The Covenant also specifies the steps to be undertaken by all State parties in order to realize those rights.**

The Covenant has 164 State parties and is legally binding of all States signing and ratifying it. (The United States, along with Myanmar, for example, has signed but not ratified the treaty.) Both Kenya and Ethiopia have signed and ratified the ICESCR—in 1972 and 1993, respectively. Those Articles of the Covenant most pertinent to the human rights noted above include the following.

—**Article 1(2)**: *In no case may a people be deprived of its own means of subsistence.*

—**Article 2(1)**: *"Each State Party to the present Covenant undertakes to take steps, individually and through international assistance and co-operation, especially economic and technical, to the maximum of its available resources, with a view to achieving **progressively** the full realization of the rights recognized in the present Covenant by all appropriate means, including particularly the adoption of legislative measures."* [Emphasis added. Note: "achieving progressively the full realization of the rights..." clearly implies no backsliding, or regressing.].

—**Article 11(1)**: *"The State parties to the present Covenant recognize the right of everyone to **an adequate standard of living** for himself and his family, including adequate **food**, clothing and housing, and to the continuous improvement of living conditions.* [Emphases added].

—**Article 12 (1, 2)**: *"The States Parties to the present Covenant recognize the right of everyone to the enjoyment of the highest attainable standard of physical and mental health."* and The steps to be taken by the States Parties to the present Covenant to achieve the full realization of this right shall include those necessary for..."*The prevention, treatment and control of epidemic, endemic, occupational and other diseases.*"

➢ **The Committee on Economic, Social and Cultural Rights (CESCR) was established in 1985 to interpret specific Articles of the Covenant and to monitor its implementation by States parties**. Established by U.N. Economic and Social Council (ECOSOC) Resolution 1985/17, the CESCR has 18 independent experts assigned these tasks. All State parties are required to present regular reports to the CESCR regarding their compliance with the rights established by the Covenant, whereupon the Committee carries out a multistage process of evaluation and subsequent dialogue with member States. Critics note the lack of effective U.N. oversight, largely due to insufficient funding and vulnerability to political pressures.

A series of General Comments are issued by the CESCR, two of which have critical implications in a consideration of human rights violations in the transboundary region.

- **The Right to Food: CESCR General Comment 12**. The CESCR's General Comment 12, issued in 1999, was largely in response to member States' requests following the 1996 World Food Conference for clarification of Article 11 of the Covenant concerning food. Follow-through meetings at the OHCHR and the FAO regarding the human right to food also contributed to the need for clarification. Most notable in the General Comment (GC 12) is the statement affirming the normative content of the Covenant's Article 11, paragraphs 1 and 2.

    —*"The right to adequate food is realized when every man, woman and child, alone or in community with others, **have physical and economic access at all times to adequate food or means for its procurement**"* [Emphasis added.]

    Further interpreting Article 11 of the Covenant, General Comment 12 (GC 12) states,

    —*"The right to adequate food, like any other human right, imposes three types or levels of obligations on States parties: the obligations to respect, to protect and to fulfil... The obligation to respect existing access to adequate food requires States parties **not to take any measures that result in preventing such access**. The obligation to protect requires measures by the State to ensure that enterprises or individuals do not deprive individuals of their access to adequate food. The obligation to fulfil means the State must proactively engage in activities intended to **strengthen people's access to and utilization of resources and means to ensure their livelihood, including food security**."* (Section 15; Emphases added.]

- More than two years prior to the GOE and development bank impact assessments, the U.N. High Commissioner for Human Rights (OHCHR) conducted a 2007 study of the specific obligations of States relative to access to safe drinking water and sanitation, with the conclusion that full recognition of such access should be recognized as a human right.

The OHCHR, in its Fact Sheet 35, has stated the following with regard to the human right to water as it pertains to both agriculture and pastoralism.—*"Water is essential for life, but is also key to food security, income generation and environmental protection."*

—*General Comment No. 15 states that priority should be given to "the water resources required to prevent starvation and disease, as well as the water required to meet the core obligations of each of the Covenant rights."* **Bearing in mind the interdependence and indivisibility of all human rights, it can be said that the right to water ensures priority for water use in agriculture and pastoralism when necessary to prevent starvation.** [Emphasis added].

- **The Right to Water: CESCR General Comment 15.** In November of 2002, the CESCR affirmed the right to water in Articles 11 and 12 of the Covenant and stated that this right is "inextricably linked" to other rights specified in the ICESCR—namely, the right to adequate food and health (CESCR 2002).[3] By the General Comment 15, the CESCR considers that this is not limited to domestic uses but rather, extends to multiple uses of water—including for production. The CESCR did recognize, therefore, that water is required for other purposes including food production and other aspects of livelihood, even though it prioritized the right to water for personal and 'domestic' uses (*ibid.*).

—"*The human right to water is indispensable for leading a life in human dignity.* **It is a prerequisite for the realization of other human rights**." (Sect. 1; Emphasis added.)

—"The right to water is defined as the right of everyone to sufficient, safe, acceptable, physically accessible and affordable water for personal and domestic uses". (*Sect. 2*)

Article I.1 of the CESCR's General Comment No. 15 continues with these statements.

"Parties should ensure that there is *adequate access to water for subsistence farming and for securing the livelihoods of indigenous peoples*" (pg. 4)

"*The Covenant specifies a number of rights emanating from, and indispensable for, the realization of the right to an adequate standard of living "including adequate food, clothing and housing." The use of the word "including," indicates that this catalogue of rights was not intended to be exhaustive. The right to water clearly falls within the category of guarantees essential for securing an adequate standard of living, particularly since it is one of the most fundamental conditions for survival.*"

The General Comment describes States' obligations to realize the human right to water:

"*States must, at a minimum, show that they are making every possible effort, within available resources, to better protect and promote this right. Available resources refer to those existing within a State, as well as those available from the international community through international cooperation and assistance, as outlined in Articles 2 (1), 11 and 23 of the Covenant.*"

Among the other pertinent statements of Comment No. 15 are those "*that individuals should be given equal and full access to information concerning water and the environment,*" (Para. 48).

- **General Assembly Resolution 64/292: The Right to Water.** The United Nations took a major step in advancing the formal acknowledgment of human rights in July/August of 2010, when the General Assembly (GA) adopted Resolution 64/292. The Resolution explicitly recognizes the right to water for drinking and sanitation and was adopted within the context of the Covenant's Article 11 (and the UDHR's Article 25.) For the first time, the U.N. formally recognizes:

—"*The right to safe and clean drinking water and sanitation as a human right that is essential for the full enjoyment of life and all human rights.*"[4]

---

[3]Comment No. 15 was published soon after, as E/C.12/2002/11 on January 20, 2003.
[4]General Assembly Resolution 64/292 (para. 1). *The Human Right to Water and Sanitation*, August 3, 2010.

It also calls on States and international organizations to provide financial resources in order to provide safe, clean, accessible and affordable drinking water and sanitation for all.

Even a relatively strict interpretation of the relationship between the right to food and right to water—for example, one considering water as an input to food production—acknowledges that, "just like food and housing, an adequate supply of water is absolutely essential for an adequate standard of living, as it is necessary for the sustenance of life itself and for ensuring a life with dignity" (Winkler 2012)[5]. As such, it has the same status as the rights to food and housing that are also encompassed under the heading of the 'right to an adequate standard of living.'

It should be further noted that if drinking water, sanitation water and water for food production all come from the *same sources* in a given community, then protecting drinking water effectively means protecting *all* water. This is precisely the case in the lowermost Omo River basin and at Lake Turkana, where communities must utilize the same water for drinking and sanitation as for their major livelihood activities. Any application of General Assembly Resolution 64/292 by the Ethiopian and Kenyan States thus requires them to protect water sources used for all their needs.

- The U.N. Human Rights Council responded to the General Assembly's adoption of Resolution 64/292 with its own Resolution 15/9 on September 30, 2010, which:

    —affirms that the rights to water and sanitation are part of existing international law, and
    —confirms that these rights are legally binding upon States.

The Resolution also calls upon States to develop appropriate tools and mechanisms to 'progressively' achieve the full realization of such human rights obligations (in other words, without 'backsliding').

A number of closely related U.N. based declarations, conventions, international treaties and policy instruments have identified human rights violations in ways applicable to the transboundary region. One of these is the **2007 U.N. Declaration on the Rights of Indigenous Peoples (UNDRIP)**, adopted by the General Assembly in September 2007 (U.N., General Assembly 2007)—just as construction of Ethiopia's Gibe III dam was beginning. An outgrowth of the ILO *Convention No. 169, UNDRIP issues the core assertion that indigenous peoples have rights to self-determination, lands, territories, natural resources and "free, prior and informed consent"* (FPIC). Like the U.N. Declaration of Human Rights, UNDRIP is non-binding for states, but it provides a critical measure for interpreting human rights violations as they impact indigenous peoples.

- **The African Charter on Human and Peoples' Rights was adopted by the Organisation of African Unity in 1981— 18 years after the OAU was formed and came into force in 1986, in order to both "promote and protect" human and peoples' rights**. As an instrument of human rights, the African Charter was comparable to such bodies in Europe. In accordance with Article 30 of the Charter, the **African Commission on Human and Peoples' Rights (ACHPR)** was set up in 1987, with the tasks of promoting and protecting human and peoples' rights and interpreting the Articles of the Charter. Responsibility of protecting rights to hear and rule on complaints as well as considering individual complaints of violations of the Charter. Meeting twice a year, the Commission has 11 members who are *nominated by their own states (State parties)* and elected by the AU Assembly. Presently, two of the eleven members are from Ethiopia and Kenya.[6]

Although the ICESCR is a multilateral treaty that is binding on all members since it came into force, a mechanism for individuals to bring a complaint against a member state when they have experienced violation of the Covenant was only established in 2008, with the U.S. General Assembly's adoption of the Optional Protocol to the ICESCR. As of 2013, enough states (although few African states) had ratified the Optional Protocol so that it did go into force. A Working Group on Economic, Social and Cultural Rights within the ACHPR is actively cooperating with an international network to promote ratification by African members states.

- A major complaint has been brought against the Ethiopian government by the Survival International Charitable Trust (SICT), citing the GOE's violation of at least four Articles of the African Charter—Articles recognizing the right to

---

[5] Broader interpretations of GA 64/292 consider the Resolution to go directly together with the rights to food. Both interpretations consider it to address an 'adequate standard of living.' See, for example, R.P. Hall et. al. (2013).
[6] The present Ethiopian ACHPR Commissioner is Chairperson of the ACHPR's Working Group on Extractive Industries, Environment and Human Rights Violations.

Self-Determination, to *Natural Resources*, to *Development* and to *Environment*. SICT has brought these charges to the ACHPR on behalf of multiple indigenous peoples in the South Omo[7]. At the time of this book's submission for printing in 2015, the case is pending with the ACHPR.

- An important milestone in ACHPR recognition of ICESCR's recognition of the human rights to food, livelihood (standard of living) and water transpired in February of 2015, when the ACHPR adopted Resolution 300 on the Right to Water Obligations. In specifying the obligations of African States in implementing the right to water, the Commission has taken a rather broad view. The Resolution was in part the product of cooperation between the Commission and a group of civil society organizations (CSOs).[8] Resolution 300 explicitly refers to the Commission's guidelines on Economic, Social and Cultural Rights (adopted in Tunis in 2011) which require States:

    —*"to protect water resources from pollution, to prioritize the provision of water for personal and domestic use and to protect the right to water and other related rights, the realization of which directly depends on water resources management."* [Emphasis added.]
    —*"to strengthen natural resources governance...using a human rights approach...with the participation and in the interest of the population."*
    —[Observe] *"Resolution 64/292 of the United Nations General Assembly and Resolution 15/9 of the United Nations Human Rights Council."*

The ACHPR expresses concern over "*the negative effects of overuse...**and other development activities** threatening the rights of present and future generations, the realization of which depends on access to water*," and urges African states to "*protect the quality of national and international water resources and the **entire riverine ecosystem***." [Emphases added]. It remains to be seen how the Commission will apply Resolution 300.

The Ethiopian Government's Violations of Human Rights in the Transboundary Region

The Ethiopian government (GOE) has blatantly ignored ICESCR recognized human rights in its single-minded pursuit of the Gibe III and linked Omo River basin developments, despite predictable massive scale livelihood and human destruction throughout the transboundary region.

GOE violations of ICESCR recognized rights stem from several major dimensions of its actions:

(i) The planning, design and construction of the Gibe III dam, dam-dependent irrigated agricultural commercial farms (and infrastructure for it), and an electricity export system (EHPP)—all without regard for the survival needs of affected indigenous peoples in Ethiopia and Kenya.
(ii) The radical reduction of Omo River flow volume and inflow to Lake Turkana, causing widespread desiccation of the lowermost riverine zone in the lower Omo basin and major lake retreat—in turn, destroying pastoral, agropastoral and fishing livelihoods of hundreds of thousands of residents (see Figs. 5.2, 7.14 and 9.10).
(iii) The forceful, even brutal expropriation of indigenous lands and eviction of communities from their village areas along the Omo River—with ongoing denial of their right to political expression or dissent.

The Ethiopian government's actions in the lowermost Omo basin have consistently violated Articles of the Constitution of Ethiopia (for example, Article 40), its own environmental laws, and government agency as well as ministry operational requirements. Even a partial list of Ethiopian standards being violated must include:

—The Environmental Policy of Ethiopia (EPE) addressing environmental and social accounting.
—The National Conservation Strategy.

---

[7]The complaint filed by SICT does not explicitly address the borderlands between Ethiopia and Kenya, however, nor the hundreds of thousands of indigenous peoples residing around Kenya's Lake Turkana.
[8]The CSOs include the Platform for International Water Law (University of Geneva), WaterAid, the Legal Resources Centre, WaterLex and Green Cross.

—Required procedures of the Environmental Protection Authority (EPA)—since 2013, renamed the Ministry of Environmental Protection and Forestry.[9]
—Proclamations including No. 299/2000 requiring environmental impact assessments to address sustainable development and 'EIAs' to be conducted for policies, programs and plans, as well as development projects.[10]

The Ethiopian government's violations of even its own mandated actions include its failures to:

- Conduct a downstream environmental/socioeconomic impact assessment (ESIA) for the Gibe III dam until more than two years *after* construction began—even then, producing only a geogaphically fragmentary assessment rather than addressing the actual impact area (the transboundary zone) with major omissions, misrepresentations and falsifications (see Chap. 6 for a detailed enumeration and discussion of Ethiopia's ESIAs).
- Consider the *cumulative and synergistic effects* of the linked developments in the Omo basin—the Gibe III dam, its linked irrigated agricultural enterprises and electricity export transmission system.
- Observe indigenous land and resource use rights, including those in Ethiopian land administrative law pertaining to pastoralists. Instead, state sponsored/approved forcible encroachment by private interests into communal lands predominates.
- Genuinely inform and consult populations prior to the inception of development—rather than 'manufacture' 'consent' from orchestrated and government (or donor) controlled gatherings.
- Implement a valid assessment of the efficacy of 'compensation' or 'mitigation' for the hundreds of thousands of indigenous residents, even if such were a genuine GOE intention.
- Implement its own defined Environmental and Social Management Framework (ESMF)—after such is shaped with community consultations.[11]

Repeated assertions by the Ethiopian government, particularly EEPCO, that the 'benefits' of Gibe III generated electricity would serve the needs of Ethiopians so far without access to power contradicted by two realities. Firstly, a high proportion of Gibe III generated electricity is slated for export, as outlined in prior chapters and in the section below. Secondly, since electricity costs in Ethiopia are predicted to rise by a minimum of 200 % in the near term, according to the World Bank, the overwhelming proportion of Ethiopian citizens would have no means to purchase available electricity. This is true even with the unlikely scenario that the GOE subsidizes energy costs for large numbers of the poor. The major consumers of electricity—in Ethiopia, Kenya and elsewhere in eastern Africa—are projected to be industry, agroindustry, commerce and government entities, along with relatively better well-off domestic consumers.

Systematic actions taken by Ethiopia's institutional nexus of river basin development policy (see description in Chap. 2) over more than five decades for more than five decades reveal a clear pattern of ignoring the human and environmental consequences of major hydrodam development—the Gibe III being the worst case, to date. The close coordination and overlapping objectives among GOE Executive offices, international development banks (playing the key role among aid agencies in large capital projects) and global consulting industry members—both firms and individuals—form the functional core of bringing about such massive dam and infrastructural projects as the Gibe megadam.

The GOE's dismissive approach to the survival and well being of its own indigenous population within the lower Omo basin residents is evident from these two statements by Meles Zenawi, the late Prime Minister of Ethiopia, at Ethiopia's Annual Pastoralists' Day in 2011.

> *—On this occasion, I assure the people of South Omo, especially the pastoralists, that the time of losing your cattle or life because of the Omo flood is over. In the coming five years, there will be a very big irrigation project and related agricultural development in this zone. I promise you that, even though this area is known as backward in terms of civilization, it will become an example of rapid development. I also want to assure you that the work we have started*

---

[9]The former head of the EPA became an advisor to the new Environmental Protection and Forestry Ministry.
[10]See Sect. 4.1.3 of Proclamation No. 299/2000, for example.
[11]The Environmental and Social Management Framework, established in 2007 with World Bank funding of about USD 100 million, requires integrated consideration of social and environmental aspects and outcomes, including **with account of cumulative, additive and synergistic effects of projects in development**. The planning procedures actually enforced by the GOE were antithetical to such integral consideration.

in this area on infrastructure and social development will continue stronger than ever. I want to assure you again that all our development work will be in line with protecting the environment and the friends of backwardness and poverty, whatever they say or do, can't stop us from the path of development we are taking. [Emphasis added.][12]

—These people [dam critics] are concerned that butterflies will be disturbed by such projects and they will not allow the disturbance of butterflies even if this means millions of people have to be subjected to the deadliest killer diseases of all, poverty, in order to not disturb the butterflies.

Ethiopia's top officials frequently issue comparable statements regarding both the "backwardness" of the indigenous peoples in the Omo region and the 'colonial' perspectives or 'ignorance' on the part of organizations and individuals who have strongly objected to Gibe III dam and agricultural enterprises and the GOE's repressive policies in the Omo basin (see for example, the EEPCO website, http://www.gibe3.com/et).

GOE violations of ICESCR and African Charter recognized human rights in the lower Omo basin should be viewed within the broader Ethiopian context. GOE rights violations are abundantly documented in news accounts and at the websites of numerous Ethiopian diaspora groups and individuals. These violations detail the detention and imprisonment of political dissidents including journalists, scholars, artists and community activists; summary eviction of rural peoples for the privatization of their lands, and use of brutality in countering political resistance to these and other GOE actions—especially in minority pastoral and agropastoral areas. The western Ethiopia Gambella area is one of the most widely publicized areas of such actions within western and southwestern Ethiopia. Conditions there are well documented by Anuak and other local activists (see *Gambella Today* 2014, for example).

When viewed within the national context of GOE violations of the human right to water but also to health and sanitation, livelihood and freedom from political repression should be viewed within the broader national context. The reality of the GOE human rights violations among its citizens are abundantly documented—in news accounts, policy documents of international aid agencies and heir foreign governments, and numerous websites spearheaded by Ethiopian diaspora groups. These and other sources detail the detention and imprisonment of political dissidents, including journalists, including journalists, scholars, artists and community activists. Summary eviction of rural peoples by the GOE.

Non-governmental organizations (NGOs) and other groups have produced numerous detailed accounts of human rights violations by rights violations by the GOE in the lower Omo basin—reports that directly contradict government accounts. consistently contradict Ethiopian government Human Rights Watch, Survival International, the Oakland Institute, International Rivers, Mursi Online (a website) Oxford University's African Studies Center and the Africa Resources Working Group are among the most active in reporting on-the-ground conditions. Much of the reporting has focused on the Mursi and Bodi peoples who reside well upstream from the Dasanech and Nyangatom along the Omo River (Fig. 1.3), since travel from Addis Ababa and government restriction of independent investigators is less severe in the former. Field based investigation along the southernmost reaches of the Omo River in Ethiopia has become nearly impossible, due to strict government exclusion of independent restrictions. The GOE's allegedly 'voluntary' resettlement and 'community development projects' (CDPs) in the lowermost Omo basin, including those outlined in its Environmental and Social Management Plan for the Gibe III, are challenged by a number of these reports, which instead report citizens' accounts of forced relocation and riverine/delta land expropriation—often imposed with physical force and uniformly with suppression of any dissent.

GOE ministry and Ethiopian Electric Power Corporation (EEPCO) officials have consistently denied any such actions and generally dismiss even the most documented allegations of abuse as simply reflections of "ignorance" on the part of 'foreigners' (even researchers familiar with the area for many years) or a 'colonial' bias against Ethiopia's "national interest" and 'progress' through its Growth and Transformation Program.

Major donor countries, including the United States and the United Kingdom, have sent investigative teams for brief visits to region (especially the Mursi/Bodi region, where major sugar plantation development is advancing). These investigations, dubiously described as 'independent' of GOE influence, have issued only mild criticism of the GOE.

Meanwhile, billions of dollars of international aid continue to pour into the country each year. Much of this is 'direct aid' that is generally unmonitored and easily utilized in support of GOE policies in the lower Omo basin, as detailed in a section below.

---

[12] 13th Annual Pastoralists' Day celebrations, Jinka, South Omo, January 25, 2011.

> Hundreds of thousands of northern Turkana residents around the northern and central portions of Lake Turkana and in contiguous upland plains are facing a similar future to that of their northern neighbors, the Dasanech and Nyangatom. The commonality of full-scale livelihood destruction among transboundary ethnic groups, paradoxically, only intensifies cross-border armed conflict among them (Fig. 5.4).

Emerging Human Rights Violations in Kenya's Lake Turkana Region

➢ **The Kenyan state (GOK) is legally bound by the ICESCR treaty to take "deliberate and very specific steps" to ensure the realization of human rights. Yet the GOK's disregard for the Covenant is starkly evident in the Lake Turkana region**. Hundreds of thousands of indigenous Kenyan citizens face catastrophic level destruction of their means of survival from the Gibe III and associated developments. Chapter 9 and preceding sections of this book have detailed the mounting livelihood crisis among the northern Turkana. The Kenyan government's roles in these developments and their impacts on the transboundary region range from complicity and active collaboration, in the case of the Gibe III dam and the extensive irrigated farm enterprises, to full partnership as in the case of the Ethiopia-Kenya Eastern Electricity Project (EEHP)—the first phase of the planned Eastern Africa Power Pool (EAPP), funded by international aid (see below).

Despite public statements by the Kenyan government that it has investigated the possible impacts of the Gibe III dam on Lake Turkana and that it has 'negotiated' with the Ethiopian state to "guarantee" the safety of Kenya's indigenous population, the GOK has in fact failed to reach any such accord—which would be entirely unrealistic in any case, given the predictable lake retreat and agricultural enterprises without calling for a halt to the Gibe III dam and its linked development in order to conduct an investigation of such a high magnitude threat to its own citizens.. Instead, the GOK continued with full complicity throughout both planning and construction phases of the dam, without even minimal effort to study the lakeside population or the impacts on it by the predictable lake withdrawal and loss of fisheries.

Years after construction of the project began, the GOK began citing the deeply flawed African Development Bank reports (AFDB 2009, 2010; see Chap. 6) and sent "delegations" to meet with the Ethiopian government—with no reprieve for the Turkana and other ethnic groups residing around the lake.) Quite apart from the GOK's rhetoric about its concern, its insistence on an 'guaranteed artificial flood' to be managed by the GOE and appropriate 'compensation', the GOK failed in virtually all dimensions of such measures.[13] Moreover, GOK executive offices have clearly overruled opposition and even questioning of the Omo basin developments at all levels of administration; a policy explained to SONT investigators on numerous occasions.

The GOK's demonstrated indifference to the plight of its indigenous population around Lake Turkana is in violation of its 2010 Constitution—particularly Articles 43, 69 and 70; Kenyan national legislation; requirements set by the National Environmental Management Authority (NEMA), as well as the ICESCR. The previous chapters detail how the destruction of pastoral and fishing livelihood among Kenya's lakeside Turkana communities would result from from radical reduction of Omo River flow and lake retreat (see Figs. 5.2, 7.14 and 9.10)—even during what the GOE has described as a "brief" reservoir filling period and despite the 'promised' annual, or "regularized" artificial flood program—a program also repeatedly assured by the GOK, the development banks and their contracted consultants. The extreme improbability of this scenario is detailed in Chap. 6.

Kenyan citizens most affected by the dam and irrigated agricultural enterprises along the Omo River in Ethiopia are not only the hundreds of thousands of Turkana fishers and pastoralists residing around and dependent on the lake's resources, but also thousands of Dasanech settled in Kenya's portion of the Omo delta (see Fig. 1.2) and around the northeastern shoreline of the lake (Fig. 1.3), Gabbra pastoralists along its eastern shoreline, the small El Molo group along the southeastern shoreline, and other pastoral peoples seasonally dependent on the lake for livestock watering and 'last option' grazing (for example, the Rendille).

---

[13]It should be noted that at least two prominent NGOs changed their 'demands' from 'stop the Gibe' to a call for a guarantee of adequate artificial flood (despite clear evidence that an adequate artificial flood was a virtual impossibility and was without precedent in Africa)—in the process of accepting new large external funding. The new funds were in fact forthcoming and neither group lost GOK tolerance of its more limited demands.

Turkana villagers' urgent requests that the GOK stop the threat to the lake from the Gibe III have been sidelined or ignored altogether. Even widespread pleas by northern Turkana fishing communities for GOK support of their fishing rights in the face of Ethiopia-based commercial fishing fleets exploiting Kenya's northern Lake Turkana waters have gone unheeded. Members of the Beach Management Units (government controlled but locally based fisheries management bodies) at Lake Turkana continually complain about the government's refusal to respond to their extreme fear of their livelihood being destroyed by the Omo basin developments—and in immediate terms by the large, Ethiopia based commercial fishing boats taking major catch from the northern reaches of Kenya's lake.

- **The GOK failed entirely to inform its vast population in the lake region regarding the actual impacts of the Gibe III and its associated development on their communities, let alone earn for their "free and informed consent," as mandated by the United Nations, the African Union and the GOK itself**. Chap. 9 briefly described how "community consultations" with the AFDB's socioeconomic impact assessment consultants were orchestrated events with what were clearly predetermined outcomes. Villagers in two different communities where such 'consultations' were held reported to SONT (South Omo/North Turkana Research Project) researchers that the AFDB 'visitors' gave them assurances that the Gibe III dam would bring major amounts of food, clothing, health care and schooling to their communities. *No mention was made of lake retreat or its effects on their fishing, their livestock raising or their households*. To the contrary, the consultants repeatedly stressed that the lake *"would not be harmed"* by the dam. In their written report for the AFDB (in 2009), the socioeconomic consultants stated in their report (AFDB 2009) that their role was to describe the "benefits" of the dam for local communities.

By all accounts, Kenyan police, local government officials and 'translators' were highly intimidating in these meetings (one of which SONT researchers attended.) Despite the consultants' assurances to communities regarding the Gibe III and 'positive' conclusions regarding the socioeconomic impacts of the dam in their report back to the AFDB, the consultants did notate (although inconspicuously in the report's main text—see Chap. 6) extreme anger expressed by village elders over the dam project, which they had learned from other sources would bring destruction to their fishing and livestock raising at the lake.

> The exceptionally low quality of information gathering and analysis by the AFDB socioeconomic impact consultants for the Lake Turkana region, combined with their obvious bias (even self described) throughout the process, underscore the assessment's invalidity. There has been no adequate socioeconomic or environmental comprehensive impact assessment of the Lake Turkana region facing extreme crisis.

- **A major proportion of Turkana pastoralists and fishers settled in villages along Lake Turkana (see** Fig. 9.4 **and in nearby lands were either omitted—or in some cases. vastly underestimated—in the GOK's 2009 census**. These communities are therefore unlikely in the extreme to be accounted with the dismantlement of their fishing and pastoral from the effects of lake retreat. Those households and communities surviving destruction from lake retreat who do manage to relocate to a town with the hope of receiving assistance are likely to be regarded as *drought* victims rather than *development* victims. Given the highly aridic conditions of the region and the immense distances required, few would be able to reach Kenya's international borders to be counted as 'refugees'.

When discussing their situation and their fear of changes to come, comments such as, You will just come across our bones on the land" are commonplace among village elders at the lake. Ironically, these types of comments by the Turkana are also common among their northern neighbors, the Dasanech, with whom they are locked into increasing armed conflict in the face of their equally desperate efforts to survive (Fig. 5.3).

➢ Fear of Kenya's security forces prevails throughout the northernmost Turkana territory, as these forces are widely known to operate with impunity—including with brutality, in some areas. Although to date, the GOK has not exercised politically repressive measures equal to those of the GOE, the government is intolerant of opposition, whether to the Gibe III development or to the widespread incursion of oil corporations into Turkana lands (see Appendix A). Such threats have caused major curtailment of community voice throughout the area.

- **In 2012, Kenya and Ethiopia, along with the World Bank and the AFDB, announced their partnership to build a USD 1.2 billion energy transmission line from Ethiopia to Kenya: the Eastern Electricity Highway Project, or**

EEHP (**World Bank** 2012a). The Gibe III dam (along with planned additional dams in Ethiopia) was clearly envisaged from the beginning as a source of electricity for the broader Africa region, even though this partnership was announced was six years *after* the 2006 signing of a Memorandum of Understanding (MOU) between Ethiopia and Kenya for energy export from Ethiopia to its southern neighbor.

The transmission line is designed to extend from Sodo, Ethiopia—a short distance from the Gibe III dam—to a major electricity hub at Suswa, near Nairobi, Kenya (see map in the section below.) After intense negotiations between the two countries over pricing (ESI-Africa 2012), Kenya agreed purchase a continuous supply of 400 MW from Ethiopia—with an arrangement whereby Kenya would pay for all of the agreed upon amount of Ethiopia's hydro-generated electricity, *whether or not* it uses all of it.[14]

Like the Gibe III dam, the energy transmission system was actually planned years earlier. The World Bank, African Development Bank, key African state executives and international investors have long considered the 45,000 MW hydropower potential of Ethiopia as a 'tower of *hydropower*' for the broader African region, not just the 'tower of *water*' extolled by Ethiopia's Haile Selassie (see Chap. 2).

The EEHP constitutes the first phase of a planned power-sharing network throughout eastern Africa—the Eastern Africa Power Pool (EAPP)—a matter detailed further in the next section. The macro plan, in fact, is for an even more comprehensive network of energy systems—one linking entire regions of Africa (World Bank 2012b). This vision is reflected in many documents, among them the 100 % state-owned Kenya Electricity Transmission Company, Ltd. (KETRACO) report on the project, with this comment:

> *The proposed high voltage transmission line linking Ethiopia to Kenya will form a major part of the interconnection and will serve to close a major gap on the high voltage grid within the Eastern Africa Power Pool (EAPP) countries, because it will ultimately serve as one of the links between Northern and Southern Africa region.* (GOK, KETRACO 2012).

KETRACO was incorporated recently—in 2008, with limited liability, for the purposes of establishing a complex, high voltage electricity transmission infrastructure in line with the Kenya Vision 2030 program, building over 4000 km of transmission lines and establishing linkages for power export to the broader region. The 2012 document cited here focuses on 'mitigation' and resettlement—on the one hand, complying with the Operational and Safeguard policies of the World Bank and AFDB (the chief funders of the Ethiopia-Kenya energy 'highway' project) by conducting such a study, while on the other hand, concluding that mitigation measures, including resettlement and compensation, need be applied only "*when feasible*".

As in the Ethiopian context, a major gap exists between the Kenyan government's rhetoric concerning the EEHP as servicing Kenya's "national interest"—expressed in GNP growth terms—and "electricity for the poor", on the one hand, and the reality that hydroelectricity transmitted through the 'highway' would be largely purchased by commercial and industrial enterprises as well as middle to upper income groups—not the poorer elements without power, on the other. According to World Bank calculations, the price of electricity (in Kenya as well as Ethiopia) is expected to rise sharply. Certainly the electricity would not benefit the pastoralists and fishers in the Lake Turkana region and north central Kenya—those facing the direct impacts of the EEHP and its electricity supplier, Gibe III dam. Instead, all of the basic needs of the Turkana along the lake will be precipitously decreased or destroyed by the complex of developments.

International Development Bank Collaboration with Human Rights Violations

> While the Ethiopian and Kenyan governments are most immediately responsible for the human rights violations underway in transboundary region, the international development banks—particularly the World Bank and the African Development Bank—have been collaborators in these violations.

---

[14]The agreement was formally concluded between Kenya Power and the Ethiopian Electric Corporation (EEPCO).

➢ **Actions taken by the World Bank and African Development Bank with regard to the Gibe HI dam and its closely linked developments violate the International Covenant on Economic, Social and Cultural Rights (ICESCR) and other international standards, as well as their own internal operational policies.** While the development banks are not under formal obligation to observe United Nations declarations and covenants recognizing specific human rights, they have established safeguard policies and operational procedures *consistent* with U.N. derived principles (see Chap. 6). World Bank required procedures pertaining to the Omo River basin developments are delineated in the bank's Safeguard Policies and Performance Standards, its Operational Policies series (OP 4.0) and Bank Procedures (BP). These include requirements and procedures concerning environmental and social impact assessment, indigenous peoples, involuntary resettlement, participation and informed consent, and a host of related issues. AFDB safeguard policies largely parallel those of the World Bank.

The cumulative and synergistic effects between the Gibe III dam and dam-dependent irrigated agricultural enterprises along the lower Omo River, as well as between the Gibe III dam and the EEHP, are ignored in development bank assessments, despite the banks' own procedural requirements. *Large-scale irrigation systems are calibrated and inflexible systems that depend on effective regulation of the river by the megadam.* Previous chapters have documented how their combined radical reduction of Omo River flow volume and inflow to Lake Turkana would compromise already borderline potable water (due to lake salinity) for humans and livestock, eliminate annual flooding of recession agricultural lands as well last resort grazing for livestock, and eliminate a massive proportion of critical fish habitat. The mandate for *integral* consideration of these developments is explicit not only in the development bank's general operational policies, but also in the specific requirements for considering specific country requests for funding, and in the official interpretations of the ICESCR. The 2010 EIB report, as well as the 2009 and 2010 AFDB impact assessments, are fundamentally flawed in failing to draw these connections—compromising their validity as adequate assessments.

➢ **The World Bank and the AFDB have repeatedly asserted that they are not involved in the Gibe III dam and associated Omo basin development—a false assertion by any substantive examination of their roles in the complex of development activities.** It is true that no direct bank funding was finalized for the Gibe III dam's actual construction, since direct funding was precluded by the GOE's multiple violations of bank operational procedures. Both development banks were reportedly in the process of declining financing requests for the Gibe III by 2010 when the GOE secured alternative funding for dam construction—over USD 460 million from the state-owned Industrial and Commercial Bank of China (see Chaps. 2 and 6).[15] However, the AFDB and World Bank have supported the Gibe III dam project and its linked developments through a multiplicity of other means.

The two banks' collaboration with the Gibe III dam and its agricultural and electricity export and thus, with the human rights violations accompanying them, include the following activities.

(i) Technical and feasibility studies of hydrodam, irrigation agriculture and energy development potential within Ethiopia, including the lower Omo River basin.
The AFDB, for example, has funded multiple efforts regarding such development, including the more than USD 6 million direct funding for the 1996 Master Plan for development of the 'Omo-Gibe' river basin (Woodroofe & Associates 1996), as well as its recent impact assessments regarding the Gibe III dam and Kenya's Lake Turkana region (AFDB 2009, 2010). For its part, the World Bank has conducted numerous assessments of Ethiopia's hydropower potential since its first presence in Ethiopia, in the 1950s. It has been active in many aspects of the Gibe III dam's planning and legitimation, despite its preclusion from funding it directly. World Bank explicit support for the project has been explicit, as evidenced in these 2004 comments (notable for their major errors of fact):

—*The Omo River is particularly important, both for its large annual flow and its irrigation and hydroelectric potential, and its being one of the principal basins where there is un-likely to be any objection by downstream countries.*
—*There is no significant use of the Omo River by any other country and the river enters Lake Turkana within the boundaries of Ethiopia. It should, therefore, be relatively easy to negotiate a 'no objection' from Kenya(ibid.)d should that is required for multilateral/bilateral funding.* (World Bank 2004)

---

[15]The GOE issued public statements stating that the Ethiopian government 'withdrew' its requests for funding. For a brief time, non-governmental organizations in the U.S. and elsewhere, celebrated what they believed to be a victory when it was clear that development bank funding for the dam did not materialize, when in fact, the project was progressing on several fronts.

(ii) Environmental and socioeconomic baseline studies and impact assessments that exclude identification major human and environmental and social impacts (sometimes through specifications in the Terms of Reference, or scoping, such as in the AFDB 2009 and 2010 assessments) environmental and social—dimensions that are commonly 'defined out' of contract terms of reference (as, for example, in the AFDB 2009 and 2010 assessments (see Chap. 6).

(iii) Active reinforcement or promotion of false information 'justifying' the Gibe III development, including but not limited to the false characterization of (a) the Omo River as causing "disastrous" floods with major loss of human life as well as livestock and property destruction; (b) 'no seismic danger' in the Gibe III region; (c) 'no significant damage or loss of water in Kenya's Lake Turkana, (d) false characterization of the Omo River as only within Ethiopia, when it is clearly a transboundary river, including with part of its active delta within Kenya's national borders—among numerous other examples.

(iv) Planning and funding of the Eastern Electricity Highway Project (EEHP)—a roughly USD 1.3 billion project with major World Bank and AFDB loans, with false assertion of "no relation" to the Gibe III dam (see for example, AFDB 2013). Major coordination and organizational support for the East Africa Power Pool—a major distributional network for EHPP product.

(v) Major financial support for the Ethiopian regime, both the institutions directly engaged in building and managing the Gibe III dam and its linked developments, as well as political repression. This support involves multiple levels of funding, including 'direct aid' (see below).

➤ **The Ethiopia-Kenya Eastern Electricity Project (EEHP) is a major electricity export project planned as the first phase of an ambitious multi-nation, integrated power-trading network in eastern Africa—the East Africa Power Pool (EAPP).** This project, planned for completion by 2017 (briefly described in the previous section) is an electricity transmission line under construction for 1045 km, from Wolayta Sodo in Ethiopia (near the Gibe III dam) to Suswa outside of Nairobi, Kenya (Fig. 10.1).

Its power carrying capacity of 2000 MW is comparable to the 1600 MW generating capacity of the Gibe III dam. After years of planning the EAPP, the World Bank and AFDB announced their approval of slightly above USD 1.2 billion in finance for the Ethiopia-Kenya transmission and AFDB announced their approval of slightly above USD 1.2 billion in finance for the Ethiopia-Kenya transmission line.[16] This project cost includes minor funding by the government of France. The total loan package, not including ancillary developments and later measures, was set as follows.

World Bank—USD 684 million (USD 243 for Ethiopia, USD 441 for Kenya)
African Development Bank (AFDB)—USD 354 million
Agence Française de Developpement (AFD)—USD 118 million

The member countries of the EAPP are Ethiopia, Kenya, Tanzania, Uganda, Burundi, Sudan, Djibouti, Rwanda and the Democratic Republic of the Congo (Fig. 10.1).

• **The World Bank and the AFDB have repeatedly stated that their funding for the Ethiopia-Kenya transmission line (EEHP) is 'unrelated' to the Gibe III dam—a patently false statement.**

To support their claim of 'no connection' between the two projects, the banks point to the *approximately 50-km separation* between the Gibe III dam and the substation (point of origin) of the Ethiopia-Kenya transmission line at the town of Wolayta Sodo (Fig. 10.2) and contend that the Gibe III dam is not 'necessary' for the economic viability of the transmission line (World Bank 2012b).

In actual fact, electricity generated by the Gibe III dam would inevitably enter the transmission line for export of power to Kenya. The AFDB and the World Bank avoid two realities in making their disclaimers.

In the first place, a 51 km, 400 kV connecting line from the dam to the substation at Wolayta Sodo was planned *prior* to the EEHP announcement, so power generated at the Gibe III was assured for the export system (Fig. 10.2). Construction of the transmission line from the dam to Sodo substation was finalized with 85 % financing from the Chinese EXIM/Export-Import Bank. The banks *purport* to be free of any involvement in funding the Gibe III project, since GOE procedural violations of

---

[16]Chapter 1 lists the Megawatt (MW) capacity of hydrodams within Ethiopia.

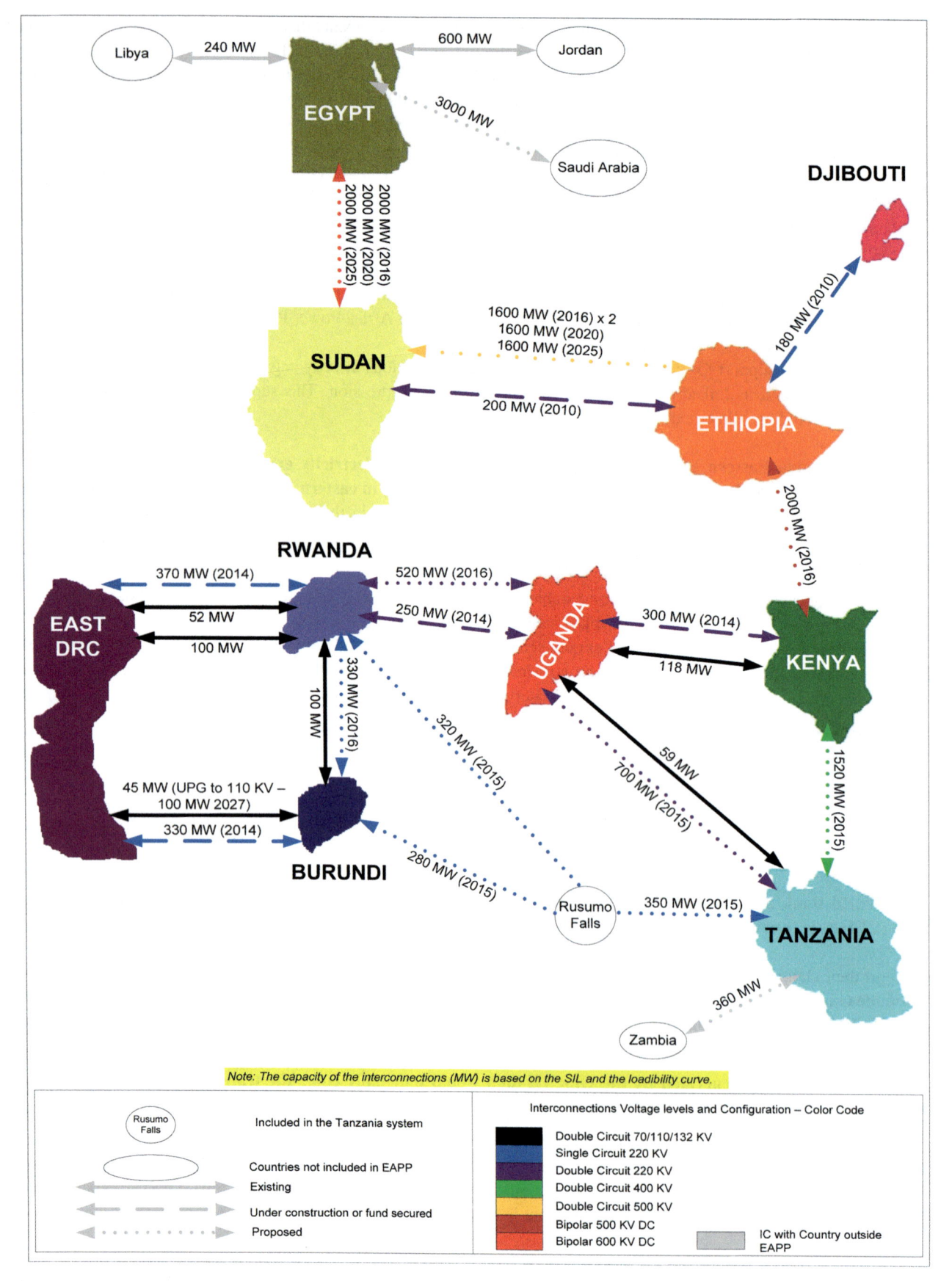

204   10 Human Rights Violations and the Policy Crossroads

Fig. 10.1 The planned eastern Africa electricity export and distribution system. *Source* Eastern Africa Power Pool (GOK/EAPP and EAC 2011)

Fig. 10.2 Chinese-financed 51-km transmission connection between the Gibe III dam and the Ethiopia-Kenya Energy Highway Project. *Source* Base map from World Bank map, reproduced in AFDB Project Appraisal Report (2012b). Labels for Gibe III to Sodo, Ethiopia, and Sodo to Kenya, along with Kenya transmission line labels, added by ARWG

their procedures prohibited them from such support. The fact that *Gibe III electricity would be integrated into both domestic and export transmission systems* was known to all agencies involved.

Moreover, the banks have long planned for a major proportion of Gibe III energy to be transported to export markets in Kenya, and beyond. Even prior to the launching of dam construction, the Ethiopian Power System Master Plan update by EEPCO (GOE, EEPCO 2006a) stated the following with regard to the significance of the dam in linking to the export system.

> *The link starts from Gilgel Gibe-III power plant and passes through Mega substation in Ethiopia and end [sic] at the towns of Nairobi or Eldorate in Kenya.* (Fig. 10.2) *The link from Gilgel Gibe-III to Mega might be double circuit 400 kV HVAC. HVDC link of at least 500 kV with transfer capacity of about 600 MW is anticipated from Mega substation to Kenya (Nairobi or Eldorate).*

- GOE officials have stated that the development banks, in financing the EEHP, insisted that the Gibe III dam be directed to major export markets; government spokespeople have routinely referenced the dam's generated electricity as for export. When specific proportions of Gibe III hydropower slated for export are specified at all, they have generally been 50-60 %. The following are representative of GOE statements on the matter.

  —"*The objectives of the project are the generation of 1870 MW electric power and enhancing economic integration in Africa through the export of surplus power by erecting a regional interconnection transmission system.*"[17] (GOE, EEPCO 2012)

  —"*The 1870 MW Gibe III hydropower plant is expected to nearly double Ethiopia's current power generating capacity. Ethiopia plans to export a portion of that electricity to Kenya with a power purchase agreement already signed between the two neighboring countries.*"[18] [Aiga Forum 2014]

  —"*We will definitely have surplus power for export when construction of Gibe III is finalized. It is a matter of only one year… and in order to secure loans for our power projects, we have to export power.* **When we ask for loans, financiers ask us if we will export power to neighboring countries**. *Our government subsidies power tariff rate as our people do not [sic] afford to pay.* **At the local market, the power purchasing power is minimal. So we should support our power development projects by exporting power**."[19][Mihret Debebe, CEO, EEPCO; Emphases added.]

  Of particular note is the 2013 statement by the CEO of EEPCO, in 2013:

  —"*Of the total power generated from Gilgel Gibe III, 900 Mw will be exported to foreign countries, such as Kenya, Sudan and Djibouti. Kenya will get 500 Mw, the largest amount of exported power and Gibe III will increase the generation capacity of Ethiopia by 234 % and makes the power export program of the country viable.*" (Mihret Debebe, CEO of EEPCO, Quoted in Addis Fortune, 2013).

- **Kenyan government ofcials have also released numerous statements identifying the Gibe III dam as the major source of their electricity import from Ethiopia**. Kenya's fully state-owned Kenya Electricity Transmission Co., Ltd. (KETRACO) was incorporated in 2008 to design, construct and maintain about 4000 km of transmission lines. The Sodo (Ethiopia) to Suswa (Kenya) line constituted a quarter of this total. KETRACO initiated an assessment of the GIBE III project's environmental impacts on Kenya's resources—a report that minimized the dam's environmental impacts. Like the GOE, EIB and AFDB impact assessments, this GOK assessment was anything but 'independent' of Gibe III development interests. The assessment (ESIA) contract was awarded to the global consulting industry rm, Panafcon—an Africa-focused consulting company that was formed in the early 1990s as a subsidiary of the Netherlands-based global firm DHV International Group of Companies and later bought out by a group of Kenyan investors.[20] Panafcon is partnered with Italian-based *ELC-Electroconsult*—itself a major contractor in Ethiopian river basin development. ELC, for example won contracts for Gibe dam projects where it won contracts (along with France based Coyne et Bellier) to represent the GOE in a supervisory role for quality control of the design and construction of the Gibe III.

The United States has recently moved into a key role among multilateral and bilateral promoters of hydroelectricity generation and distribution with in Africa. In its July 2012 'U.S. Position' statement concerning the EEHP, the U.S.

---

[17]News at the government website specifically for the Gibe III project.
[18]Aiga Forum is an Ethiopian government focused website.
[19]Stated in GOE (EEPCO) (Reported in: https://www.thereporterethiopia.com).
[20]http://panafcon.net/index.php/about-panafcon.

Department of Treasury uncritically supported both of the World Bank claims. With the launching of 'Power Africa' by President Obama in 2013, the U.S. took a key role in actively promoting donor groups for launching an unprecedented major system of interconnecting electricity grids, with the EAPP (and therefore, the IHPP) occupying a prominent role. The U.S. role, detailed in its 2015 'Power Africa' report, explicitly demonstrates the operating assumption within the U.S. Executive that developing such electricity networks is *intrinsically* positive, with no substantive qualification regarding segments of society—and natural resources—that bear the brunt of this development.

➢ **The World Bank and AFDB have major loans to the Ethiopian government in recent years—loans that both reinforce and legitimize GOE policies in the lower Omo River basin and their devastating impacts on transboundary peoples and environments**. The two banks have long played a key role in promoting Ethiopia as a 'showcase' country for the 'positive' effects of aid: assistance totaled USD 4 billion in 2014, for example. World Bank loans alone increased markedly to Ethiopia: from USD 988 million in 2010 to USD 1.64 billion in 2014, for example. The AFDB is also a dominant funder—ranking fourth amongst donors. Ethiopia has remained the second highest recipient of Ofcial Development Assistance (ODA) in recent years, garnering about 7 % of the total (OECD 2014). It has also ranked second as a recipient of debt relief in Africa.

Coordinating multi-donor financing—both as general support and as loans for specific projects within Ethiopia—has been a World Bank function since its initial presence there, following WW II. Bilateral agencies including USAID and DFID, along with the EU are also lead funders of the Ethiopian government. Their assistance basically conforms to that of the development banks in 'turning a blind eye' to the GOE's undemocratic, highly repressive policies, while pursuing their own economic, political and military interests.[21]

In view of the geopolitical and economic interests of the major donor nations to the eastern Africa/Horn region and the importance they attach to Ethiopia as a 'partner' in those interests, such studies are likely generated to contribute evidence for the positive nature of such partnership. Consideration of these interests, moreover, also provides at least partial explanation for the 'blind eye' approach to GOE policies both governmental and bank failure to conduct adequate social and environmental impact assessments that address the cumulative and synergistic effects of the Omo River basin and hydroelectric transmission developments.

There are abundant efforts in World Bank and AFDB documents to rationalize their exceptionally high level of nancial and other support for the Ethiopian government. Some of these 'empirical' studies freely incorporate 'rights' language and references to quantitative 'data', implying 'scientific' investigation. The African Development Bank's recently updated study—*2004–2013 Country Policy and Institutional Assessment (CPIA)* is one of many such efforts (AFDB 2014). The classicatory system designed for this assessment assigns values to five different 'clusters' of policy measures for forty African nations. These 'measures' include social equity, property rights, environmental and social protection and corruption.[22] Using this system, the AFDB ranks Ethiopia among the highest in the continent— almost across the board. For example, the AFDB concludes that Ethiopia is the top-rated nation for 'Environmental Policy and Regulation', among the top three nations for 'Equity of Public Resource and 'Policies for Social Inclusion/Equity,' and among the top four nations for 'Transparency or Accountability.'

The consistency of gaps between such 'measures' described and conclusions drawn in the 2014 AFDB report, on the one hand, and widely available evidence of social inequity, seizure of property, lack of implementation of regulations and 'protections, lack of transparency and repression of political dissent, on the other, is fundamental and suggests both the arbitrariness of such measures—however quantitatively expressed—and the bank's predetermined design for generating 'evidence' in support of its policies within Ethiopia.

➢ **Much of the international aid to the GOE is allocated as direct budget support, or direct aid—a form of aid largely 'decentralized' and 'unspecied' by donors**. Prior to the GOE's no-bid tendering of contracts for the Gibe III dam construction, direct budget support to Ethiopia from major agencies totaled about a third of all aid. Since direct aid is *unmonitored* by international funders, relying instead on Ethiopia's "self-reporting," this type of aid is highly fungible—and therefore easily used by the GOE to support its preferred policies.

---

[21]U.S. national interests, for instance, involve strong geopolitical, military and investment objectives in the Horn region and the broader East Africa region—in no small part, involving oil and gas and related energy development.
[22]The policy clusters measured (to hundredth values on a scale of 1 to 5) are: Economic Management, Structural (both primarily macro economic measures), Social Inclusion/Equity, Governance and Infrastructure and Regional Integration.

The origin of the direct aid program to Ethiopia rests with the GOE's repressive policies and domestic popular resistance to them. Following social uprisings in 2005—sparked by election corruption and GOE arrests of more than 30,000 people, the 27-member Development Advisory Group (DAG) for Ethiopia suspended funding for the government. The World Bank, AFDB and their partners quickly installed the new 'direct aid' program, a centerpiece of which was the **Promoting Basic Services Program (PBS)**—for the first years of its existence, termed the Protecting Basic Services Program.

The PBS was launched only months after the DAG suspension was imposed—in 2006, as a multi-donor form of direct aid. In the words of a senior World Bank staff member, as a *"creative solution"* whereby aid to Ethiopia could be continued despite criticisms of the GOE's repressive policies. In its carefully nuanced statement, the Bank stated, in an interim report (a draft provided to this writer by a World Bank staff member for the Bank's Implementation Status & Results Report in 2014): *In the uncertain environments following a contested general election in Ethiopia in 2006, Development Partners recognized the imperative to continue supporting the Government's significant progress towards achieving the MDGs [Millennium Development Goals] through strengthened decentralized basic services.*

Often cited by the development banks and major bilateral agencies as justification for their extraordinary level of nancial support for the GOE, the PBS program—in reality—exemplifies the lack of accountability for billions of dollars received in loans to the Ethiopian government. A little known World Bank report issued in 2012, entitled *Diagnosing Corruption in Ethiopia* (World Bank 2012c), raised this problem in at least muted terms, but there is no evidence of any impact on the exceptionally strong support for the regime, which continues to be regarded in aid circles as simply "needing improvement." The 'new' PBS program constituted a symbolic change rather than a substantive one and it has been described as such in numerous aid accounts—even within major aid agencies (Knoll 2008).[23]

Dispersal of PBS aid within Ethiopia involves block grants to 'decentralized' political units for "expanding and protecting basic services, or "pro-poor policies," as they are dubbed. These "services" include road-building, 'modernization' of agriculture, education, water supply, and payment of local salaries—all under the rubric of "strengthening local financial management systems," supporting local civil society organizations of the GOE's choice, and even—"continuing to develop transparency and accountability" (World Bank 2013). Block grants are issued to ten rural dominated regions within Ethiopia—including the Southern Nations, Nationalities, and Peoples' Region, or SNNPR. which is one of the nine ethnic divisions (kililoch) within Ethiopia and where the lower Omo River basin is located, as noted in earlier chapters.

Phase I of the PBS Program was replaced by Phase II in 2009—the latter version receiving highly mixed reviews even by limited World Bank and AFDB criteria, and finally by Phase III in September of 2012 (activated in January of 2013). when it was both refunded at a higher level and given its present name —*Promoting* Basic Services.[24]

According to World Bank figures, the most recent approved financing for these programs is as follows.

PBS III—USD 4.887 billion
PBS II—USD 3.364 billion
PBS I—USD 2.562 billion

- The overall PBS regional block grants increased after 2006 and this increase reflected development bank and major bilateral agency support for the GOE. Of the USD 4.88 billion for direct aid to Ethiopia, the World Bank portion is USD 600 million, with terms common to the its 'soft loan' window, the International Development Association (IDA)—that is, 40 year loan terms, with negligible interest a 10 year grace period. AFDB finance terms, issued by its own soft loan window, basically parallel those of the World Bank. While the World Bank and the AFDB are overwhelmingly dominant in aid PBS funding (as in Ethiopia's aid portfolio overall), the U.K.'s DFID has been a major contributor among the bilateral agencies—providing more than £ 740 million in support.[25] Other co-financers of Phases II and/or III, although

---

[23] Knoll (2008) details the PBS program within the context of 'budget support' in African aid more generally, as it unfolded from the mid 1990s.
[24] Unless otherwise specified, information regarding the development bank expenditures for the PBS programs are drawn from reports by the AFDB (2012a) and the World Bank (2013, 2014), as well as annual reports.
[25] Major political pressure has been exerted on DFID by non-governmental organizations (e.g., Survival International and Human Rights Watch) to terminate its funding for the PBS III program. This pressure has been largely fueled by HRW (and other) reports concerning the GOE's eviction and political repression policies in the Gambella region (see below)

much smaller contributors, are the European Union and the bilateral agencies of Austria, Italy, Ireland, Canada, Spain, Germany and the Netherlands. While not a direct contributor to the PBS III, the United States remains the largest bilateral donor to Ethiopia.[26]

> All aspects of the PBS are implemented at the discretion of the GOE, without effective oversight—in spite of periodic development bank public assurances of 'independent' evaluation. A 2013 analysis of PBS program funding and GOE human rights violations by Inclusive Development International (IDI 2013) quotes one World Bank officer's disclaimer regarding the GOE's usage of funds:
>
> *"The PBS itself has no direct mechanism to influence choices made at the local government level.*

- **The SNNPR receives a substantial amount of the PBS III funds and program support, which is readily used by local officials for a multiplicity of activities that further the major impacts on not only Ethiopia's lowermost Omo River Basin population, but also on the major population residing around Kenya's Lake Turkana**. These GOE activities include: road building and other infrastructural construction, for political structures and security apparatus, expropriating resources of indigenous communities, and serving the needs of the large commercial plantations that replacing indigenous communities. In the name of "decentralized development"—effectively unmonitored, PBS funds can be used for the eviction process itself and the political repression that accompanies it.

AFDB and World Bank appraisal documents make frequent assertions of the PBS Program's "success." Such statements are explicit in the AFDB's (2012a) PBS appraisal, for example:

"The 84.3 million citizens of Ethiopia will be the main beneficiaries of the PBS III program. The direct beneficiaries will be the citizens in woredas from 10 regions that receive federal block grants."

The report continues with a summary dismissal of PBS: *"The programme has been classified under category III and is not expected to have any negative environmental impact."*

Such blatant exclusion of even the possibility of environmental and social negative outcomes of the program, let alone highly probable and disastrous ones, has avoided triggering of development bank Safeguard and Operational Policy procedures. The development banks (and large bilateral agencies) steadfastly maintain the rhetoric of an "in place accountability mechanism" for the PBS program and insist that the programs are for the "promotion of civil society organizations" and the "engagement of civil society."

The reality, however, is altogether different. The 'Social Accountability' sub-component of PBS III is under the control of the GOE's Ministry of Finance and Economic Development (MOFED)—one of the key ministries directly promoting the Omo River basin developments. 'Participating' civil society organizations (CSOs) are selected by the GOE, so only those conforming to government policies are selected. This is even true for 'independent' auditing of PBS policies, the results of which are submitted to the banks for their appraisal. Predictably, bank appraisers of this direct aid are generally complicit with their client institutions' perspectives. So-called '*external*' peer reviewers selected by the AFDB for its PBS III Project appraisal, for example, were selected from three other PBS funding agencies—the World Bank, DFID and Irish Aid.[27]

- Human Rights Watch produced a report on the PBS program (HRW 2012a) focusing on the relatively more accessible Gambella region of western Ethiopia, where strong local resistance to major expropriation and political repression has transpired and an active Anuak diaspora-based organization (the Anywaa Survival Organisation, or ASO) has been effective. The HRW report underscores the World Bank's failure to consider its responsibility for the debacle underway there, even by the Bank's own internal procedural requirements. The Gambella accounts underscore the seriousness of

---

[26]For details regarding U.S. assistance, see the U.S. State Department, USAID and CIA websites; also see Oakland Institute (2013).
[27]The measures used for assessing the PSB are comparable to those used for Ethiopia's general economic advancement of GTP and MDG goals—namely, *aggregate* and per capita indicators—but applied to the *regional* level. Data for regional level 'services delivered,' 'infrastructure constructed', water resource developments undertaken, local personnel newly salaried, and so forth exclude on-the-ground realities and certainly the matter of who 'benefits' and who bears the 'costs' of the GOE's development actions—in the SNNPR, the Dasanech, Nyangatom, Kara, Bodi, Mursi and neighboring indigenous peoples in the lower Omo River basin.

this problem as it applies to the less documented crisis unfolding in the lowermost Omo basin within the SNNPR—certainly, in the for the Dasanech and Nyangatom region.[28]

Specific reports on the SNNPR PBS III and related projects, including the 'One Water, Sanitation and Hygiene National Program (OWNP) in the region reflect such omission and misrepresentation. The OWNP (the 'ONE WASH' program) is the world's largest sector-wide approach (SWAP) to 'Water, Sanitation, and Hygiene (WASH)—sponsored by UNICEF and extolled as a major "success" already in 'achieving' Ethiopia's Growth and Transformation Program and MDG goals. The recent OWNP social assessment report, written on behalf of the Ministry of Water Resources (GOE, MOWR 2013), is illustrative of the bias at play.[29] While identifying the 'Dassench' [sic] Woreda district as having a human population of more than 56,000, with the 'Dassench' constituting more than 97 % of that number, the report barely mentions "traditional water resource management" and "recession agriculture" and makes mention of the Gibe III dam or irrigated agricultural schemes and their full-scale impact on the region's water resources.

> While the OWNP assessment outlines what it terms "progress" underway from the program, the reality of tens of thousands of indigenous residents being evicted and expropriated with collapse of their livelihoods already underway, is omitted altogether. Such reporting, if not directly unethical, is irresponsible in the extreme and reinforces the massive scale human destruction already beginning to manifest in the region due to the Gibe III dam and its linked developments.[30]

➤ **The World Bank commonly objects to complaints regarding their human rights record—in general as well as within Ethiopia—as well as their 'obligation' to observe U.N. recognized human rights, in two ways.**

Firstly, the Bank refers to its own Articles of Agreement (Sect. 10) that define it as a *"non-state actor,"* meaning that it should not *"interfere in the political affairs of any member state."*

Secondly, the Bank contends that while demands for it to observe human rights are issued in the abstract, the specifics of exactly how it might implement such concerns remain vague.

With regard to the first assertion, it is true that international law applies only to individual states. A considerable literature and discourse has emerged, however, which leads to a different interpretation of non-state actors and internationally recognized human rights—namely, one whereby the development banks *should* be held accountable.[31]

> The past thirty years have witnessed repeated clashes between the World Bank and civil society organizations from both 'developed' and 'developing,' (or Global South) countries with respect to the human rights dimensions of bank policies. The World Bank's elaborate set of Standards and Operational Policies—largely replicated in the regional development banks, including the AFDB—are now challenged at a far more detailed and informed level.

Regarding the Bank's second contention—that human-rights concerns are generally formulated only in abstract terms, *the human rights issues in the developments impacting the peoples of the lower Omo River Basin and Lake Turkana region are certainly not abstract*. They center upon the denial of ICRCS recognized rights for hundreds of thousands of indigenous peoples—one articulating with the violation of rights to health, livelihood and freedom from political repression.

> Because the World Bank and AFDB have actively collaborated in (and have even encouraged, with funding and promotional activities) Ethiopian and Kenyan governmental human rights violations in the transboundary region, informed challenge to the banks' breach of their own Safeguard and Operational Policies is entirely appropriate.

---

[28] The World Bank has responded to non-governmental and other protest of its PBS support for the GOE's policies in Gambella, with an Inspection Panel investigation. Released in November, 2014, the Bank's report was equivocal and fragmentary (World Bank 2014b), and was countered by NGO critics (including Human Rights Watch and IDI).

[29] This social assessment was carried out with the support of U.K.'s DFID.

[30] An April, 2014 AFDB assessment of the 'One Water' program notates (in its summary tables and briefly in text) the issue of 'dam' development (without mention of the Gibe III), and even "adverse impacts of pastoral water supply systems" in general. The points remain abstract ones, however, so its omissions and misrepresentations largely parallel those of the GOE's ministry report.

[31] See for example the writings of Skogly (2001), Ghazi (2003), Alston (2005) and Clapham (2005). Moreover, the U.N. Committee on Economic, Social and Cultural Rights (CESCR), described earlier in this chapter, in its General Comment No. 14, asserted that non-State actors were obligated to observe the international human right to health (Coomans 2007).

The Stark Policy Choice: Catastrophic Level Destruction *or* Sustainable Development Within a Human Rights Framework

> If the Ethiopian and Kenyan governments continue their present course of violating internationally defined human rights, particularly those recognized in the International Covenant on Economic, Social and Cultural Rights Treaty (ICESCR), their actions may well be regarded as criminal.

➢ The crisis unfolding in the Ethiopia-Kenya-South Sudan transboundary region is fundamentally one of policy *objectives* and *accountability*. Ethiopian government and international development bank officials have consistently described the purpose of the Gibe III dam and its associated developments as "essential" to Ethiopia's "national interest" and its 'growth and transformation' program and Ethiopia's electrification needs, including electrification for the "rural poor." Yet the on-the-ground effects of this development in the lower Omo basin and Lake Turkana region—already in evidence during the early phases of this development—suggest the makings of a humanitarian crisis with massive scale hunger and vulnerability to disease, cross-border inter-ethnic armed conflict, major government expropriation and political repression of citizens, and irreversible environmental destruction—outcomes that can hardly be regarded as 'progress.' Despite these predictable effects already emerging in the transboundary region, however, both government and bank statements continue to label the development a sure 'success,' based on GNP and *per capita* income, energy and other projections for the nation as a whole.[32] Sorting out the differences between such macro level projections and ground level realities for the half million citizens struggling to survive in three-nation transboundary area is key to a rational discourse concerning the economic and political future of the region.

> At the most general level, the policy choice is an unambiguous one. Either continue along the present road to social and natural resource devastation—a road rife with human rights violations—or bring a halt to the destruction underway in order to initiate a new pathway for action—one consistent with internationally recognized human rights and geared to creating a sustainable future for the peoples of the transboundary region as well the three nations involved.

Continuing along the present road is an unacceptable option by any measure of project outcomes that includes the transboundary region's indigenous peoples. Even if the alternative pathway defined above were to be pursued, major questions emerge as to how to proceed. For example, how is it possible to take account of the increasingly desperate survival needs of the pastoral, agropastoral and fishing peoples in the transboundary region that is most impacted by the development, while simultaneously addressing the aspirations of citizens in the region as a whole?

Although insufficient as full answers to this question, two matters require immediate attention if a solution to this challenge is to be constructed.

First of all, there is legal imperative for action. Based on violations of human rights in the transboundary region—rights recognized by the ICESCR that must be restored. This requires restoring indigenous communities' access to food and the necessary means to produce it as part of their right to an adequate standard of living. *Compliance with the ICESCR is not a choice—it is a legally binding treaty.*

This action could be accomplished with at least a *temporary suspension* of the plans for Gibe III hydrodam electricity generation, an end to the withholding of downstream river flow and the reinstatement of Dasanech and Nyangatom communities to their lands and resources along the Omo River. Reinstatement of the Omo River's flow would allow inflow to Lake Turkana, bringing at least temporary reprieve to hundreds of thousands of pastoralists and fishers depending on the lake's resources. Such a suspension should trigger at least a pause in temporary suspension of large-scale irrigated plantation construction as well, while an acceptable future scenario is being constructed.

---

[32]Because of the high capital outlay for major dams, the international development banks are by far the key funders and coordinators of these developments. See Chap. 2 for a history of river basin development within Ethiopia—dubbed by Emperor Haile Selassie himself as the water tower of Africa.

> Restoration of the specific human rights involved should also be a triggering force for major evaluation and assessment of the Gibe III dam and its linked developments which were never responsibly dealt with by the Ethiopian or Kenyan governments, nor by the international development banks that have substantively backed the development from early planning phases to the present.

—Given the actions already taken by the Ethiopian Government, it is most unlikely that the GOE would cease its human rights violations of its own accord or even pause its headlong pursuit of the Gibe III dam and its linked developments. Their reversal of these actions would be even less likely following the inception of actual electricity generation at the dam. Interruption of the human rights abuses and humanitarian disaster unfolding in the transboundary region, coupled with the GOE's prohibition of independent 'eyes-on-the-ground' in the lowermost Omo basin and its policies of political repression would almost certainly require *leverage* by external institutions

—The reality is that the World Bank and the AFDB, along with other major aid agencies and their donor states, *have* such leverage and could exert it by taking any of a broad range of action, including include withholding of funds allocated, imposition of 'conditions' for new project lending or continued institutional support, and a host of other political and economic measures. The fact that major donor countries have strong economic interests (in oil and gas development, for example) and geostrategic objectives of in the region, however, and their political judgment that supporting the present Ethiopian regime adequately serves those interests suggests that any leverage applied by them would require substantial pressure from their own public constituencies.

Secondly, in-depth and accurate information in the form of environmental and socioeconomic baseline studies or impact assessments (EIAs, or ESIAs) for the Gibe III dam and its linked developments are *essential* to sound decision making regarding the future of the region's peoples and environments. Both national governments and development banks, however, have failed to make any such effort—in breach of international standards, the two countries' Constitutions and domestic laws and the development banks' own operational principles. In addition to the pressing human rights violations requiring attention, two *required* ESIA failures—even considered individually—merit leveraging a suspension of Gibe III dam, irrigated agriculture and energy transmission line completion and operationalization, at least until such time that they are resolved. (See Chaps. 2 and 6 for detailed discussion).

- **The Gibe III dam and its closely linked developments have clear transboundary impacts, so any adequate environmental and socioeconomic impact assessment (ESIA) must be conducted within *transboundary* framework**. None of the assessments produced—by either African government, the EIB or the AFDB, have produced such an ESIA. Instead, all assessments have addressed only fragments of the impacts requiring attention and even these with multiple failings (see Chap. 6) and only for Ethiopia or Kenya. It is striking that the predictable negative impacts of the Gibe III project are strongest in the tri-nation border region—the area most neglected in all reports. All assessments done by the governments and development banks have addressed only fragments of the region's environmental and social dimensions concerns in one or another country, despite the fully verifiable transboundary character of the project, not to mention the reality that the strongest impacts are in the border region of Ethiopia, Kenya and the Ilemi Triangle/South Sudan.[33]
- *The cumulative and synergistic impacts* **of all three components of development must be dealt with as one unitary system**. The Gibe III dam has been documented in preceding chapters (as well as by Oxford University scholars and several NGOs cited earlier) as having cumulative effects with the irrigated agricultural enterprises along much of the Omo River downstream from the Gibe III dam. The Gibe III's linkage with the energy highway system from Ethiopia to Kenya (the EEHP) is documented in the previous section, World Bank and AFDB statements to the contrary notwithstanding.

➢ **Considering two internationally established commitments by States—namely, to protect ICESCR recognized human rights (including the right to food, an adequate standard of living, and water) and to promote environmentally and socially sustainable development—questions inevitably arise as to who actually *controls* environmental and socioeconomic impact assessments and related studies, and who *participates* in them.**

---

[33]Ironically, all the ethnic groups in the region—although excluded from substantive participation in the assessments produced (GOE 2009a, 2009b; EIB 2010; AFDB 2009, 2010) conduct their lives with full cognizance of transboundary social and environmental dy live with full cognizance of transboundary social and environmental dynamics.

- It has become commonplace among development specialists and researchers with extensive field-based experience in some bilateral aid agencies (particularly in Scandinavia) and non-governmental organizations to regard *cooperative investigation with communities* as essential to the production of accurate, thorough and useful analysis of a region's socioeconomy and environmental status, as well as identification of the local residents' most pressing needs. Multiple forms of community/investigator cooperative approaches to considering livelihood systems, natural resource use patterns in relation to possible development (or conservation) policies have transpired in ways that actually assist communities in improving their life circumstances, rather than undermine them.

Local communities likely to be most affected by the development in question need to be included as *active* participants in all phases of impact assessments and related studies, including:

—identifying the socioeconomic and environmental parameters, or 'bounding' of the task at hand,
—gathering information and data (field derived components, at least), with ongoing interpretation, and
—drawing conclusions and summarizing the key issues derived from the effort, including their significance for the future of local communities and their environment.

Much has been learned about cooperative processes between external investigators and local communities that increase accurateness of results. In pastoral and agropastoral contexts, for example, it is clear that local participants need to be selected through dialogue with traditional leaders, rather than government appointed ones. Broad-based and community-held discussions are a critical component and these need to be held in all phases of the work and throughout the geographic area being considered. At community gatherings, local participants, including community leaders, male and female elders and both genders of youth need to actively engage in the above dimensions of the collective effort. Those community members involved in core roles of the investigation have a major roles in planning items for open community discussion and community feedback, while working closely on a daily basis with external investigators. The results from open meetings (as well as from 'feedback surveys', etc.) need to be incorporated into the final assessment, or report—*including* its summary and conclusions. Results and conclusions need to be made presented to communities *prior* to releasing them to policymakers and client institutions. Finally, all such documents must be fully transparent and available to the public.

Constructive baseline studies and environmental/socioeconomic impact assessments (ESIAs) and the like need to be unrestricted in their ability to question the project or development approach in question, based upon its findings. This contrasts sharply with the existing arrangement for the Omo basin projects detailed in Chap. 6, for example, where implementation of the project is *presumed to be a result of the ESIA*—a situation permitting opportunity only for 'suggested' after-the-fact mitigation or monitoring actions, or additional studies, while the project moved to completion. The Gibe III assessment scenario, unfortunately, is indication of the predominant pattern in assessment related study contracting—one involving major 'reciprocity' and 'revolving door' relationships.

- **Changes in both perspective and method of this order obviously necessitate a significant shift in *accountability* relations**. Clearly, investigators have some basic accountability to those who contract them. Yet to assure a cooperative outcome, with accuracy of the work produced, they must also have genuine accountability to local communities which have sufficient authority to ensure that the content and outcomes of investigation represent the realities of their lives and their environments, and authority to guarantee that their voice will be heard well beyond their immediate localities—in fact in the distant offices of policymakers and politicians. It should be clear that *global consulting industry* corporations and individuals—parties well-equipped and experienced in accepting accountability only to their client institutions in a system that amounts to a 'contract treadmill' and too often, a 'complicity treadmill' involving very large sums of money, are ill-equipped for such an orientation.
- A participatory, community-inclusive approach to studies and impact assessments for consideration of development programs is best implemented by non-governmental organizations, academics (research units or groups, individuals), relatively small consulting organizations, rather than by global consulting industry representatives. Lead investigators should be able to demonstrate qualifications including:

—In-depth familiarity with the geographic region concerned, including with substantive field-based experience in the region or one closely paralleling it.
—Appropriate professional training credentials in socioeconomic or human ecological (broadly defined) fields.
—Experience and commitment to studies weighted to field-based investigation rather than 'desk studies' with utilization of available printed information, etc., and brief field 'visits'.

—A clear and demonstrated record of genuinely participatory work with local communities.

—Absence of financial contracts or agreements with governments, development agencies or private firms engaged in any substantive way with the project or program under consideration—for example, for a five-year period. Additional participants in studies, such as technical experts (e.g., for seismic, geological, hydrologic, agricultural or cultural studies) should to be held to the same standard.

—Full public transparency of funds received for study and the major parameters of the study underway.

➤ **The approaches outlined above for impact assessment and related investigations a may well be objectionable to in mainstream development circles, as "impractical" and "unachievable" in remote areas and among culturally traditional peoples.** To the contrary, there are plentiful accounts by seasoned investigators and aid practitioners that support the conclusion that communities in highly remote and marginalized contexts are not only receptive to genuine inclusion in such processes that help determine their own future—their participation is in fact *essential* to the success of development or conservation planning with equitable and sustainable objectives.

Such understanding is long overdue in major agencies and government policymaking offices.

> **Changing the course of existing development and broader social policies directed to the Ethiopia-Kenya-South Sudan transboundary region offers the possibility of sparing the lives and livelihoods of hundreds of thousands of citizens and rethinking the region's social and environmental future within a human rights framework. Embracing humane and sustainable development prospects for upcoming generations in eastern Africa remains a possibility—even at this 'eleventh hour.'**

Literature Cited

African Commission for Human and Peoples' Rights (ACHPR). 2004. http://www.achpr.org/about/achpr/.
African Commission for Human and Peoples' Rights (ACHPR). 2015. Resolution on the Right to Water Obligations - ACHPR/Res.300. http://www.achpr.org/sessions/17th-eo/resolutions/300/.
African Development Bank (AFDB). 2009. A. S. Kaijage, N. M. Nyagah, Report, socio-economic analysis and public consultation of Lake Turkana Communities in Northern Kenya. Tunis.
African Development Bank (AFDB) 2010 (Nov.), S. Avery, Assessment of Hydrological Impacts of Ethiopia's Omo Basin on Kenya's Lake Turkana Water Levels, Final Report, 146 pp.
African Development Bank (AFDB), African Development Fund. 2012a (July). Ethiopia. Promoting Basic Services Programme (PBS III). Appraisal Report/.
African Development Bank (AFDB). 2012b (Sept). Ethiopia-Kenya Electricity Highway. Project Appraisal Report. 41 pages.
African Development Bank (AFDB). 2013 (July 5). AfDB launches $ US 1.26 billon Kenya-Ethiopia Electricity Highway. http://www.afdb.org/en/news-and-events/article/afdb-launches-us-1-26b-kenya-ethiopia-electricity-highway-11733.
African Development Bank (AFDB). 2014 (Apr). 2004–2013 Country Policy and Institutional Assessment (CPIA). http://www.afdb.org/en/documents/project-operations/country-performance-assessment-cpa/country-policy-and-institutional-assessment-cpia/.
Africa Resources Working Group (ARWG). 2009 (Jan). A Commentary on the Environmental, Socioeconomic and Human Rights Impacts of the Proposed Gibe III Dam in the Lower Omo River Basin of Southwest Ethiopia. http://www.arwg-gibe, 29 pages.
Aiga Forum. 2014 (June 12). Ethiopia to fill Gibe III dam, rejects renewed calls for halt. http://aigaforum.com/news/gibe3-news-061214.php.
Alston, P. (Ed.). 2005. *Non-State Actors and Human Rights*. Oxford: Oxford University Press.
Carr, C. J. 2012 (Dec). Humanitarian Catastrophe and Regional Armed Conflict Brewing in the Border Region of Ethiopia, Kenya and South Sudan: The Proposed Gibe III Dam in Ethiopia, Africa Resources Working Group (ARWG). https://www.academia.edu/8385749/Carr_ARWG_Gibe_III_Dam_Report, 250 pages.
Clapham, A. 2005. *Human Rights Obligation of Non-State Actors*. Oxford: Oxford University Press.
Coomans, F. 2007. Application of the International Covenant on Economic, Social and Cultural Rights in the Framework of International Organisations. In: von Bogdandy and Wolfrum, R. (Eds.), Max Planck Yearbook of United Nations Law, Vol. 11, 359–390. Koninklijke Brill N. V.
ESI-Africa. 2012 (Jan 23). Kenya and Ethiopia sign power purchase deal. ESI-Africa.com.
Ethiopia, Government of (GOE), Ethiopian Electric Power Corporation (EEPCO). 2006. Ethiopian Power System Master Plan.
Ethiopia, Government of (GOE), (Ethiopian Electric Power Corporation). 2009a. CESI and Mid-Day International Consulting Engineers 2009, Gibe III Hydroelectric Project, Environmental and Social Impact Assessment, Report No. 300 ENV RC 002C.
Ethiopia, Government of (GOE), Ethiopian Electric Power Corporation (EEPCo). 2009b Agriconsulting S.p.A. and Mid-Day International Consulting 2009 , Level 1 Design, Environmental and Social Impact Assessment, Additional Study of Downstream Impacts. Report No. 300 ENV RAG 003B.

Ethiopia, Government of (GOE), Ethiopian Electric Power Corporation (EECPO). 2009c. Salini MDI Consulting, Environmental and Social Management Plan.

Ethiopia, Government of (GOE), Ethiopian Electric Power Corporation (EEPCO). 2012 (April), Gibe III News http://www.gibe3.com.et/EEPCo.html

Ethiopia, Government of (GOE), Ministry of Water Resources (MOWR). 2013. Social Assessment of the Water Supply, Sanitation and Hygiene Program—OWNP. Ethiopia. http://www.mowr.gov.et.

Ethiopia, Government of (GOE), Ministry of Foreign Affairs (MoFA). 2014 (April). The Fifth and Sixth Periodic Country Report (2009–2013) On The Implementation Of The African Charter On Human And Peoples' Rights In Ethiopia. Addis Ababa.

European Investment Bank (EIB). 2010 (Mar). Sogreah Consultants, Independent Review and Studies Regarding the Environmental & Social Impact Assessments for the Gibe III Hydropower Project. Final Report, 183 pages.

Gambella Today. 2014. Ethiopia: Lives for land in Gambella.

Ghazi, B. 2003. *The IMF, The World Bank Group and the Question of Human Rights*. Transnational Publishers.

Hall, R.P., B. Van Koppen and E. Van Houweling. 2014. The Human Right to Water: The Importance of Domestic and Productive Water Rights. Sci Eng Ethics (2014) 20:849–868.

Human Rights Watch (HRW). 2012a (Jan). Waiting here for death: Forced Displacement and "Villagization" in Ethiopia's Gambella Region. www.hrw.org, 13 pages.

Human Rights Watch (HRW). 2012b. What will happen if hunger comes? Abuses Against the Indigenous Peoples of Ethiopia's Lower Omo Valley. www.hrw.org, 91pages.

Human Rights Watch (HRW). 2013. Abuse-Free Development. How the World Bank Should Safeguard Against Human Rights Violations. http://www.hrw.org. 60 pages.

Human Rights Watch (HRW). 2014a (Feb. 18). Ethiopia: Land, Water Grabs Devastate Ethiopia. https://www.hrw.org/news/2014/02/18/ethiopia-land-water-grabs-devastate-communities.

Human Rights Watch (HRW). 2014b. World Report: Ethiopia. https://www.hrw.org/world-report/2014/country-chapters/ethiopia.

Inclusive Development International (IDI). 2013 (Oct). Human Rights and the World Bank Safeguards Review: Lessons from Ethiopia. Forced Villagization and the Protection of Basic Services Project. www.inclusivedevelopment.net.

Kenya, Government of (GOK), Eastern Africa Power Pool (EAPP) and East African Community (EAC). 2011. SNC Lavalin, Final Master Plan Report Vol. IV. Regional Power System Master Plan and Grid Code Study.

Kenya, KETRACO (Kenya Electricity Transmission Company Limited). 2012 (Jan). Proposed Ethiopia-Kenya Power Interconnection Project. Final Environmental and Social Impact Assessment (ESIA) Report.

Knoll, M. 2008. Budget Support: A Reformed Approach or Old Wine in new Skins? UNCTAD/OSG/DP/2008.5.

Knox, J. 2014. United Nations Mandate on Human Rights and the Environment. http://srenvironment.org/tag/un-mandate/.

Oakland Institute. 2013. Overlooking Violence, Marginalization, and Political Repression. http://www.oaklandinstitute.org.

Organisation for Economic Co-operation and Development (OECD). 2014. http://www.oecd.org/countries/ethiopia/aid-at-a-glance.htm.

Skogly, S. 2001. *The Human Rights Obligations of the World Bank and the International Monetary Fund*. Cavendish Publications.

Survival International. 2013a (Mar 24). Concern mounts over humanitarian crisis in Lower Omo, Ethiopia. http://www.survivalinternational.org/news/10094.

Survival International. 2013b (Apr 15). Aid agencies turn blind eye to 'catastrophe' in Ethiopia. http://www.survivalinternational.org/news/9125.

Survival International. 2013c (Nov 27). Top human rights watchdog investigates Ethiopia and Botswana. http://www.survivalinternational.org/news/9777.

Survival International. 2014. The Omo Tribes. http://www.survivalinternational.org/tribes/omovalley/gibedam

United Nations, General Assembly. 2007. The U.N. Declaration on the Rights of Indigenous Peoples (UNDRIP).

United Nations, General Assembly. 2010 (July 28). Resolution GA Res 64/292. The human right to water and sanitation. http://www.un.org/en/ga/64/resolutions.html.

United Nations, OHCHR and UNEP. 2012. Human Rights and the Environment. Joint Report. http://www.unep.org/environmentalgovernance/Portals/8/publications/JointReport_OHCHR_HRE.pdf.

United Nations, Committee on Economic, Social and Cultural Rights (CESCR) 2002 (Nov). General Comment No. 15, The Right to Adequate Food. http://www.refworld.org/docid/4538838c11.html

United Nations, Committee on Economic, Social and Cultural Rights (CESCR) 2002 (Nov). General Comment No. 15, http://www.unhchr.ch/tbs/doc.nsf/0/a5458d1d1bbd713fc1256cc400389e94.

United Nations, Office of the United Nations High Commissioner for Human Rights (OHCHR). 2012. Human Rights Treaty System: An introduction to the core of human rights treaties and the treaty bodies. Fact Sheet No. 30. http://www.ohchr.org/Documents/Publications/FactSheet30Rev1.pdf.

United Nations, General Assembly. 2013 (Dec). The human right to safe water and sanitation. Resolution A 68/157. http://www.un.org/en/ga/search/view_doc.asp?symbol=A/RES/68/157.

United States, USAID. 2015. Power Africa. https://www.usaid.gov/powerafrica.

Winkler, Inga. 2012. The Human Right to Water: Significance, Legal Status and Implications for Water Allocation. Hart Publishing, Oxford and Portland, Oregon.

Woodroofe, R. & Associates, with Mascott Ltd. 1996. Omo-Gibe River Basin Integrated Development Master Plan Study, Final Report. Vols. I–XV.

World Bank. 2004. Ethiopia's Path to Survival and Development: Investing in Water Infrastructure. Background Note for FY04CEM. World_Bank_CP-WaterMila01.pdf.

World Bank. 2012a (Jul). AFCC2/RI-The Eastern Electricity Highway Project under the First Phase of the Eastern Africa Power Integration Program. http://www.worldbank.org/projects/P126579/regional-eastern-africa-power-pool-project-apl1?lang=en.

World Bank. 2012b (Jul). Fact Sheet: Ethiopia-Kenya Eastern Electricity Highway Project. First Phase of Regional Eastern Africa Power Integration Program. http://www.worldbank.org, http://search.worldbank.org/all?qterm=EEHP.

World Bank. 2012c (June). Diagnosing Corruption in Ethiopia: Perceptions, Realities and the Way Forward for Key Sectors. (Janelle Plummer). http://dx.doi.org/10.1596/978-0-8213-9531-8.

World Bank. 2013 (Dec). Implementation Completion and Results Report on IDA Grants to the Federal Democratic Republic of Ethiopia for a Ethiopia Protection of Basic Services Program Phase II Project. Report No. ICR2574.

World Bank. 2014a (Mar.). Ethiopia - Promoting Basic Services Program Phase III Project: P128891 - Implementation Status Results Report.

World Bank. 2014b (Nov 21). Ethiopia, Promoting Bassic Services Phase II Projet (P128891). Inspection Panel. Report No. 91854-ET.

Open Access This chapter is distributed under the terms of the Creative Commons Attribution-NonCommercial 2.5 International License (http://creativecommons.org/licenses/by-nc/2.5/), which permits any noncommercial use, duplication, adaptation, distribution and reproduction in any medium or format, as long as you give appropriate credit to the original author(s) and the source, provide a link to the Creative Commons license and indicate if changes were made.

The images or other third party material in this chapter are included in the work's Creative Commons license, unless indicated otherwise in the credit line; if such material is not included in the work's Creative Commons license and the respective action is not permitted by statutory regulation, users will need to obtain permission from the license holder to duplicate, adapt or reproduce the material.

Appendix A: Activation of Oil Exploration and Development in the Ethiopia–Kenya–South Sudan Transboundary Region[1]

Joshua S. Dimon[2] with Claudia Carr

While the Gibe III component of the Omo River basin development constitutes the most urgent, intensive and extensive threat to livelihoods along the lower Omo River and Lake Turkana, the peoples in this region are also under threat from other large-scale development programs. These too promise further restrictions of access to livelihood resources, including land grabs for large agri-business plantations along the lower Omo and, on an even larger scale, oil development. There has been a major expansion of oil exploration and development activities along Lake Turkana and the lower Omo basin in the last five years. The impacts of these developments will only compound Gibe III's negative impacts on livelihoods in the region by further restricting pastoral territories, further abstracting water from the Omo and Lake Turkana for drilling, further reducing the water quality of the remaining water in the Omo River and Lake Turkana, and increasing the militarization of the region and potential for armed conflict.

Current concessions for oil exploration in the region cover the entirety of Lake Turkana and the south Omo River, extending along the so-called tertiary rift valley zone of Ethiopia and the eastern branch of the East African Rift Valley. Exploration and, more recently, development of oil has been accelerating for the last five years in the region, although it is part of a much longer history of oil exploration.

Contrary to news of brand-new interest/discoveries, exploration in this region has been going on for many years, without the prior knowledge of those living in the zones of exploration. The first exploration in the region was in the 1950s, with the first hints of oil resulting from shallow boring for water in Northwest Kenya in 1952 (Hedburg 1953). Exploration picked up again in the 1980s with help from the World Bank in both Kenya and Ethiopia (Rachwal and Destefano 1980; McGrew 1982; Hartman and Walker 1988). This zone is part of a larger regional exploration history starting in the 1940s. The Africa Resources Working Group (ARWG) has been investigating this history across thirteen countries covering seventy years and over three hundred companies, and has documented this in several forthcoming papers.

With decades of exploration history in the zones, oil companies currently investing in this region generally already know what exists in the area, and where, and accelerate quickly to exploration drilling, which is a far more damaging phase than seismic exploration, and requires a greater amount of auxiliary support infrastructure, including security, where the company deems the area a potential risk to their property or employees. Given the growing tensions due to the Gibe III impacts, and the increasing conflict in South Sudan, private security forces are likely in use here, further heightening the potential for violent conflict in this region.

---

[1]This Appendix was written in July of 2014. Since that time, major expansions of Tullow Oil, Africa Oil and other petroleum and associated interests have been underway in the region. This discussion offers an outline and perspective regarding the genesis and basic character of the fast-moving development now underway.

[2]Joshua S. Dimon is completing a doctoral degree in Environmental Science, Policy and Management at the University of California, Berkeley—specializing in extractive industry development within Africa.

Currently, Africa Oil and Tullow are the two dominant companies involved in the region, although over 15 companies are involved in the tertiary rift valley all together. Tullow has been very active in the broader region, with oil discoveries in the Ugandan portion of the Western branch of the rift valley. Africa Oil is a more recent addition to the region, but includes individuals with many years of geologic experience in the region. Perhaps more problematic, Africa Oil gained its concessions along the Omo River in Ethiopia through an intermediary that initially gained the concessions as agricultural plantation project land. The inter-changing roles of the investors engaged both across industries, and across countries in the region demonstrates both the unified front facing the pastoral peoples of the Southern Omo, northern Lake Turkana region, and the strategic advantage these companies have in negotiations with regional governments.

> Given the evidence from the past seven decades of exploration, and current upsurges in interest in the broader northern Rift Valley Area and Horn of Africa, the oil companies are arguing this region is likely to become next West Africa. The convergence of this massive expansion of the oil industry in the same impact zone as that of the Gibe III dam will decimate livelihoods in this region. The impacts from all of these developments together far exceed the sum of the individual impacts, as livelihoods will already be stressed beyond their coping points by the Gibe III dam.

Seismic exploration, while progressing for decades, has rapidly expanded in recent years. Additionally, two wells have already been drilled by Tullow Oil in the lower Omo area near the Ethiopia–Kenya border by Tullow Oil, and many more South of the Lake in Kenya where development drilling is already in progress. These developments will directly and severely impact pastoral livelihoods in the region. Seismic exploration requires the clearing of vast swaths of land from flora, fauna and people including for the placement of explosives or large "thumper trucks" every 100 m. This is carried out for distances of hundreds of kilometers in order to generate seismic waves. These operations involve security measures to prevent people from approaching the seismic lines during clearing, placing of trucks and explosives, and seismic testing itself which can extend for months. For pastoral peoples, these actions mean major disruption of settlements, livestock movements and access to grazing and water sources.

Drilling involves the same securitization of the location of the drilling rigs. It also involves the production of a very large amount of drilling waste, including toxic drilling muds, drill cuttings (rock) contaminated with drilling muds and possibly hydrocarbons, and produced water (from injection during drilling) also contaminated with drilling muds and hydrocarbons. If the drilling is on land, in areas such as this region, the wastes will be left near the drill site in waste ponds, likely unlined, risking contamination of any groundwater that may be accessed in the region, as well as nearby land. If the drilling is on water, the wastes are generally dumped directly into the water system next to the drilling platform. This would be concern enough without the impacts on Lake Turkana water levels from the dam, as it can seriously impact benthic organisms, chemical oxygen demand, and fish life (Patin 1999; Satterly 2003). However, with any reduction in Lake level from the Dam itself, this would be far worse.

Each of these phases of oil and gas development involve security, generally from either private security forces, or the military forces of the hosting country. In a region already experiencing the livelihood tensions noted in throughout this book, as well as the impending tensions from the impacts of the Dam itself, this increased militarization of the region will only augment armed conflict and humanitarian concerns. The exploration programs that have been progressing in the South Omo, North Turkana regions have already sparked conflict when residents of the region first encountered oil company trucks that said they had authorization from the government to commence work in the region, despite community members knowing nothing about it.

Literature Cited

Hartman, J.B., and T.L. Walker. 1988. Petroleum developments in Central and Southern Africa in 1987 (Part b). *Bulletin of the American Association of Petroleum Geographers* 72(10b):196–227.

Hedberg, H. 1953. Petroleum developments in Africa in 1952.

McGrew, H.J. 1982. Petroleum developments in central and Southern Africa in 1981. *Bulletin of the American Association of Petroleum Geographers* 66(11):2251–2320.

Appendix A: Activation of Oil Exploration

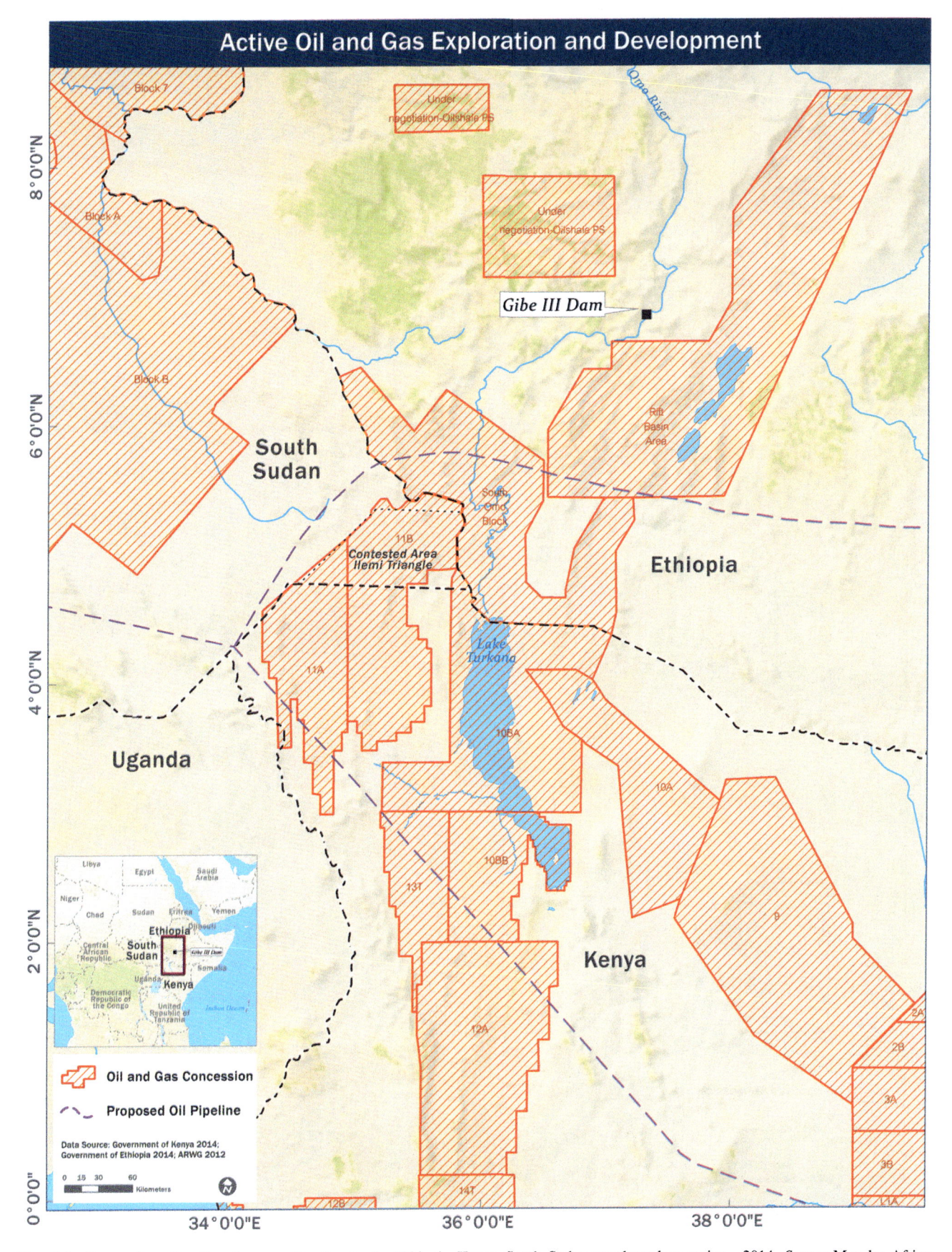

Fig. A.1 Concessions for oil and gas exploration in the Ethiopia–Kenya–South Sudan transboundary region—2014. *Source* Map by Africa Resources Working Group (ARWG), compiled from relevant oil industry documents/websites

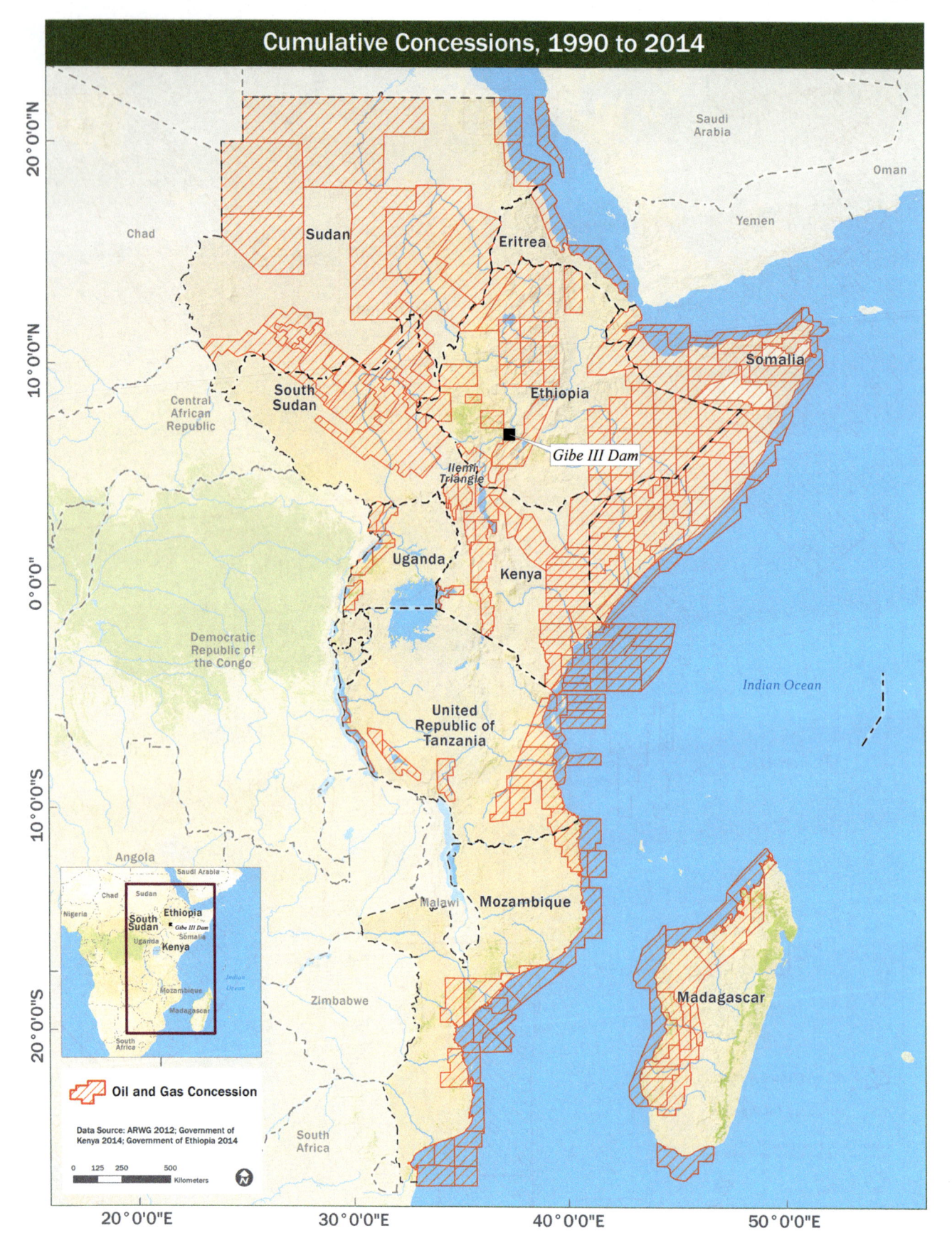

Fig. A.2 Cumulative concessions for oil and gas exploration in Eastern Africa. *Source* Mapping by J. Dimon, A. Gray and C. Carr of Africa Resources Working Group (ARWG) with data from relevant petroleum literature and oil industry websites

Patin, S.A. 1999. *Environmental impact of the offshore oil and gas industry*. New York: Eco Monitor Publishing.

Rachwal, C.A., and E.R. Destefano. 1980. Petroleum developments in central and Southern Africa in 1979. *Bulletin of the American Association of Petroleum Geographers* 64(11):1785–1835.

Satterly, N. 2003. *PCR inhibition and toxic effects by sediment samples exposed to drilling muds*. Masters Thesis, submitted to Louisiana State University and Agricultural and Mechanical College.

Appendix B: Species Collected in the Lower Omo River Basin and Transborder Region

Collection by C. Carr (Taxonomic Update by F.H. Brown)

Acanthaceae
Barleria acanthoides Vahl
Barleria eranthemoides R. Br.
Barleria linearifolia Rendle
Blepharis persica (Burm. f.) Kuntze (syn of B. ciliaris L.) B.L. Burtt
Crossandra nilotica Oliv.
Ecbolium anisacanthus (Schweinf.) C.B.Cl.
Ecbolium revolutum (L.) C.B.Cl.
Hypoestes verticillaris R. Br.
Justicia anselliana T. Anders.
Justicia caerulea Forssk.
Justicia flava Forssk.
Justicia odora (Forssk.) Vahl (syn. *J. fischeri* Lindau)
Justicia sp.
Justicia striata (Flotsch.) Bullock
Peristrophe bicalyculata (Retz.) Nees
Ruellia patula Jacq.

Actiniopteridaceae
Actiniopteris radiata (Sw.) Link

Aizoaceae
Corbichonia decumbens (Forssk.) Exell
Trianthema triquetra Willd.
Zaleya pentandra (L.) Jeffrey

Amaranthaceae
Achyranthes aspera L.
Aerva javanica (Burm.f.) Schult.
Celosia argentea L.
Celosia populifolia (Forssk.) Moq.
Celosia schweinfurthiana Schinz
Cyathula orthacantha (Hochst.) Schinz
Dasysphaera prostrata (Gilg) Cavaco
Digera muricata (L.) Mart.
Psilotrichum elliottii Bak.
Psilotrichum gnaphalobryum (Hochst.) Schinz
Pupalia lappacea (L.) A. Juss.
Sericocomopsis pallida (S.Moore) Schinz
Heeria reticulata (Bak. f.) Engl.
Lannea floccosa Jacl.
Rhus natalensis Krauss

(continued)

Annonaceae

Uvaria leptocladon Oliv.

Apocynaceae

Adenium obesum (Forssk.) Roem. & Schult.
Saba florida (Benth.) Bullock

Araceae

Pistia stratiotes L.

Arecaceae

Hyphaene compressa H. Wendl. (H. thebaica is syn.)

Aristolochiaceae

Aristolochia bracteolata Lam.

Asclepiadaceae

Calotropis procera (Aiton) W.T. Aiton
Caralluma acutangulata (Decne.) N.E. Br.
Caralluma somaliea N.E. Br.
Curroria volubilis (Schtr.) Bullock
Gomphocarpus fruticosus (L.) Ait
Leptadenia hastata (Pers.) Decne
Pergularia daemia (Forsk.) Chiov.
Sarcostemma viminale (L.) R. Br.
Tacazzea apiculata Oliv.

Asparagaceae

Asparagus sp.

Balanitaceae

Balanites aegyptiaca (L.) Del.
Balanites orbicularis Sprague
Balanites sp. (= Carr 859)
Balanites zeylanicum (Burm.) R. Br.

Boraginaceae

Cordia crenata Delile
Cordia sinensis Lam.
Heliotropium indicum L.
Heliotropium ovalifolium Forsk.
Heliotropium somalense Vatke
Heliotropium steudneri Vatke
Heliotropium supinum L.
Trichodesma zeylanicum (Burm. f.) R. Br.

Burseraceae

Boswellia hildebrandtii Engl.
Commiphora africana (A. Rich.) Engl.
Commiphora edulis Engl. (C. boiviniana is syn.)
Commiphora kua J.F. Royle (Vollesen) var. *kua* (C. madagascariensis is syn.)
Commiphora sp.

Capparaceae

Boscia angustifolia A Rich. var. *angustifolia* *Boscia coriacea* Pax
Cadaba farinosa Forssk. subsp. *farinosa*
Cadaba gillettii R.A. Graham
Cadaba glandulosa Forssk.
Cadaba rotundifolia Forssk.
Capparis fascicularis DC. var. *fascicularis*
Capparis tomentosa Lam.
Cleome brachycarpa DC.
Cleome parvipetala R.A. Graham
Crateva adansonii DC.
Maerua crassifolia Forssk.
Maerua decumbens (Brongn.) De Wolf (*Maerua subcordata* (Gilg.) De Wolf is syn.)
Maerua oblongifolia (Forssk.) A. Rich.

(continued)

Celastraceae

Hippocratea africana (Willd.) Loes.
Maytenus senegalensis (Lam.) Exell

Ceratophyllaceae

Ceratophyllum demersum L.

Chenopodiaceae

Suaeda monoica J.F. Gmel

Combretaceae

Combretum aculeatum Vent.
Terminalia brevipes Pamp.

Commelinaceae

Commelina benghalensis L.
Commelina forsskaolii Vahl

Compositae (Asteraceae)

Delamerea procumbens S. Moore
Helichrysum glumaceum DC.
Kleinia longiflora DC.
Kleinia squarrosa Cufod. (Syn is *Kleinia kleinioides* (Sch. Bip.) M.R.F. Taylor)
Pluchea dioscoridis DC.
Pluchea ovalis DC.
Sphaeranthus ukambensis Vatke & O. Hoffm.
Vernonia cinerascens Sch.-Bip
Vernonia sp. (= Carr 333)

Convolvulaceae

Hildebrandtia obcordata S. Moore
Ipomoea aquatica Forsk.
Ipomoea sinensis (Desr.) Choisy subsp. *blepharosepala* (A. Rich) Meeuse
Seddera hirsuta Hallier f. var. *hirsuta*

Cucurbitaceae

Coccinia grandis (L.) Voigt
Cucumis dipsaceus Spach.
Cucumis figarei Naud.
Kedrostis foetidissima (Jacq.) Cogn.
Kedrostis gijef (J.F. Gmel.) C. Jeffrey
Luffa ?echinata Roxb.
Momordica rostrata A. Zimm.

Cyperaceae

Cyperus alopecuroides Rottb.
Cyperus articulatus (Cav.) Steud.
Cyperus laevigatus L.
Cyperus longus L.
Cyperus maritimus L.
Cyperus rotundus L.
Cyperus teneriffae Poir.
Scirpus maritimus L.

Dichapetalaceae

Tapura fischeri Engl.

Dracaenaceae

Sansevieria ehrenbergii Schweinf. ex Baker

Ebenaceae

Diospyros scabra (Chiov.) Cufod.
Diospyros sp.

Elatinaceae

Bergia suffruticosa (Del.) Fanzl

Euphorbiaceae

Acalypha fruticosa Forssk.
Acalypha indica L.
Euphorbia grandicornis Goebel

(continued)

Euphorbia heterochroma Pax
Euphorbia hypericifolia L.
Euphorbia tirucalli L.
Euphorbia triaculeata Forsk.
Jatropha ellenbeckii Pax (J. fissispina is syn.)
Phyllanthus amarus Schumach. and Thonn.
Phyllanthus maderaspatensis L.
Phyllanthus reticulatus Poir.
Phyllanthus sp. (= Carr 411)
Ricinus communis L.
Securinega virosa (Willd.) Pax and K. Hoffm.
Tragia hildebrandtii Muell. Arg.

Fabaceae (Caesalpinioideae)

Cassia didymobotrya Fres.
Cassia italica (Mill.) F.W. Andr. var. *micrantha* Brenan
Cassia nigricans Vahl.
Cassia occidentalis L.
Delonix elata (L.) Gamble
Tamarindus indica L.

Fabaceae (Faboideae)

Canavalia cathartica Thou.
Canavalia virosa Wight and Arn.
Crotalaria polysperma Kotschy
Crotalaria pychnostachya Benth.
Indigofera arrecta Hochst. ex A. Rich.
Indigofera ciferrii Chiov.
Indigofera coerulea Roxb.
Indigofera hochstetteri Baker
Indigofera oblongifolia Forsk.
Indigofera schimperi Jaub. & Spach
Indigofera spicata Forsk.
Indigofera spinosa Forssk.
Indigofera tinctoria L.
Indigofera volkensii Taub. forma
Ormocarpum trichocarpum (Taub.) Engl.
Rhynchosia minima (L.) DC.
Rhynchosia pulverulenta Stocks
Sesbania sericea (Willd.) Link
Sesbania sesban L. Merr. var. *nubica* Chiov.
Sesbania somalensis Gillett
Tephrosia purpurea (L.) Pers. var. *pubescens* Bak
Tephrosia uniflora Pers.
Vigna luteola (Jacq.) Benth.
Vigna radiata (L.) Wilczck. var. *sublobata* (Roxb.) Verdc.
Vigna unguiculata (L.) Walp. var. unguiculata (*Syn. of Vigna unguiculata* (L.)) Walp. subsp. *cylindrica* (L.) Van Eselt
Acacia drepanolobium
Dichrostachys cinerea (L.) W. and A.
Mimosa pigra L.
Senegalia (Acacia) mellifera (Vahl) Benth.
Senegalia (Acacia) senegal (L.) Willd.
Senegalia brevispica Harms
Vachellia (Acacia) drepanolobium Harms ex Y. Sjöstedt
Vachellia (Acacia) horrida (L.) Willd.
Vachellia (Acacia) nubica Benth.
Vachellia (Acacia) paolii Chiov.
Vachellia (Acacia) reficiens Wawra
Vachellia (Acacia) sieberiana DC.
Vachellia (Acacia) tortilis (Forssk.) Hayne subsp. *spirocarpa* Chiov. (Hochst. ex A. Rich.) Brenan
Vachelliia (Acacia) seyal Del.

Gentianaceae

Enicostema axillare (Lam.) A. Raynal subsp.*axillare* (*Syn of Enicostema hyssopifolium* (Willd.) Verdoor

(continued)

Hyacinthaceae
Urginea indica (Roxb.) Kunth
Lamiaceae
Basilicum polystachon (L.) Moench
Hyptis pectinata (L.) Poit.
Leonotis nepetifolia (L.) R. Br.
Leucas ?glabrata R. Br.
Leucas nubica Benth.
Ocimum americanum L.
Ocimum forsskaolii Benth (*syn. = O. hadiense* Forssk.)
Ocimum kilimandscharicum Baker ex Gürke
Orthosiphon somalensis Vatke
Plectranthus hadiensis (Forssk.) Schweinf. ex Spreng. (*Syn. of Ocimum hadiense*) Forssk.
Plectranthus sp. (= Carr 739)
Loranthaceae
Loranthus sp. (= Carr 880)
Plicosepalus sagittifolius (Sprague) Danser
Tapinanthus aurantiacus (Engl.) Danser
Malvaceae
Abelmoschus esculentus (L.)
Abutilon figarianum Webb
Abutilon fruticosum Guill. and Perr.
Abutilon graveolens W. and A.
Abutilon hirtum (Lam.) Sweet
Fabaceae (Mimosoideae)
Abutilon pannosum (Forsk. f.) Schlecht.
Hibiscus micranthus L.f.
Hibiscus sp.
Pavonia patens (Andr.) Chiov.
Pavonia zeylanica Cav.
Senra incana Cav.
Sida rhombifolia L.
Trichilia roka (Forsk.) Chiov.
Menispermaceae
Cissampelos mucronata A. Rich.
Cocculus hirsutus (L.) Diels
Moraceae
Ficus sycomorus L.
Nyctaginaceae
Boerhavia erecta L.
Commicarpus helenae (Romer & Schultes) Meikle syn of *C. stellatus* (Wight) Berhaut
Commicarpus plumbagineus Standl.
Nymphaceae
Nymphaea lotus L.
Olacaceae
Ximenia americana var. *caffra* (Sond.) Engl.
Jasminum abyssinicum Hochst. ex DC.
Onagraceae
Ludwigia adscendens subsp. *diffusa* (Forrsk.) P.H. Raven (*Syn. of Ludwigiastolonifera*) (Guill. & Perr.) P.H. Raven
Ludwigia leptocarpa (Nutt.) H. Hara
Passifloraceae
Adenia venenata Forssk.
Pedaliaceae
Pterodiscus ruspolii Engl.
Sesamothamnus busseanus Engl.
Sesamum latifolium J.B. Gillett

(continued)

Poaceae

Aristida adscensionis L.
Aristida kenyensis Henrard
Aristida mutabilis Trin. & Rupr.
Cenchrus ciliaris L.
Cenchrus setigerus Vahl
Chloris roxburghiana Schult.
Chloris virgata Sw.
Chrysopogon aucheri (Boiss.) Stapf var. *aucheri*
Cymbopogon schoenanthus (L.) Spreng.
Cynodon dactylon (L.) Pers.
Dactyloctenium giganteum Fischer and Schweickt.
Dactyloctenium sp. nov.
Digitaria macroblephara (Hack.) Stapf
Dinebra retroflexa (Vahl) Panz.
Echinochloa haploclada (Stapf) Stapf
Enneapogon brachystachyus (Jaub. and Spach) Stapf
Enneapogon cenchroides (Roem. and Schult.) C.E. Hubb
Enteropogon macrostachyus (A. Rich.) Benth.
Eragrostis cilianensis (All.) Vignolo ex Janch.
Eragrostis namaquensis Nees
Eriochloa fatmensis (Hochst. & Steud.) Clayton (Syn is E. nubica)
Heteropogon contortus (L.) Roem. & Schult.
Lintonia nutans Stapf
Loudetia phragmitoides (Peter) CE Hubbard
Panicum coloratum L.
Panicum maximum Jacq.
Panicum monticola Hook. F (*Syn* = *P. meyeranum* Nees)
Panicum poaeoides Stapf
Perotis patens Gand. var. *parvispicula* Robyns
Phragmites ?karka (Retz) Steud.
Phragmites australis (Cav.) Steud.
Schoenefeldia transiens (Pilger) Chiov.
Sehima nervosum (Rottl.) Stapf
Setaria acromelaena (Hochst.) Dur. and Schinz
Sorghum verticilliflorum (Steud.) Stapf
Sorghum virgatum Stapf
Sporobolus consimilis Fresen.
Sporobolus fimbriatus Nees var. *latifolius*
Sporobolus helvolus (Trin.) T. Durand & Schinz
Sporobolus ioclados (Trin.) Nees (Syn. is *S.marginatus*)
Sporobolus pellucidus Hochst.
Sporobolus pyramidalis Beauv.
Sporobolus spicatus (Vahl) Kunth
Stipagrostis hirtigluma (Trin. & Rupr.) de Winter
Tetrapogon cenchriformis (A. Rich.) Clayton
Tetrapogon tenellus (Roxb.) Chiov.
Tragus berteronianus Schult.
Urochloa setigera (Retz.) Stapf
Vossia cuspidata (Roxb.) Griff.

Polygalaceae

Polygala erioptera DC.
Persecaria senegalensis f. *albotomentosa* (R.A. Graham) K.L. Wilson (*Syn. is Polygonum senegalense* Meisn. f. *albotomentosum* R.A. Graham)
Portulaca foliosa Ker-Gawl.
Portulaca oleracea L.
Portulaca quadrifida L.
Talinum portulacifolium (Forssk.) Asch. ex Schweinf.

Rhamnaceae

Ziziphus mauritiana Lam.
Ziziphus mucronata Willd.
Ziziphus pubescens Oliv.

(continued)

Rubiaceae
Kohautia caespitosa Schnizl.
Tarenna graveolens (S. Moore) Brem.
Rubiaceae
Fagara chalybea (Engl.) Engl. (Syn of *Zanthoxylum chalybeum*)
Salvadoraceae
Dobera glabra (Forssk.) Poir.
Salvadora persica L.
Sapindaceae
Allophylus ferrugineus Taub. var. *ferrugineus* (Syn = *A. macrobotrys* Gilg.)
Cardiospermum halicababum L. var. *halicababum*
Glenniea africana (Radlk.) Leenh. (Syn =*Melanodiscus oblongus* Radlk. Ex Taub., in Engl.
Haplocoelum foliolosum (Hiern) Bullock
Lepisanthes senegalensis (Juss. ex Poir.) Leenh. (is Syn. of *Aphania senegalensis*)
Scrophulariaceae
Stemodia ?serrata Benth.
Striga hermonthica (Delile) Benth.
Simaroubaceae
Harrisonia abyssinica Oliv.
Solanaceae
Lycium sp.
Nicotiana tabacum L.
Solanum hastifolium Dunal
Solanum incanum L.
Solanum nigrum L.
Solanum sepicula Dunal
Withania somnifera (L.) Dunal
Sterculiaceae
Melochia corchorifolia L.
Sterculia sp.
Tiliaceae
Corchorus olitorius L.
Corchorus trilocularis L.
Grewia bicolor Juss.
Grewia fallax K. Schum.
Grewia tenax (Forssk.) Fiori
Grewia villosa Willd.
Typhaceae
Typha sp.
Ulmaceae
Celtis integrifolia Lam.
Vahliaceae
Vahlia goddingii E.A. Bruce
Verbenaceae
Chascanum laetum (Walp.) (Syn = *Svensonia laeta* (Walp.) Moldenke)
Phyla nodiflora (L.) Greene
Premna resinosa (Hochst.) Schauer
Priva adhaerens (Forsk.) Chiov.
Vitaceae
Cayratia ibuensis (Hook f.) Suess.
Cissus cactiformis Gilg.
Cissus quadrangularis L.
Cissus rotundifolia (Forssk.) Vahl
Cyphostemma sp.
Zygophyllaceae
Tribulus cistoides L.
Tribulus terrestris L.
Zygophyllum simplex L.

Appendix C: Reference set of Selected Major Figures

Content by C. Carr represented in graphic form by Laura Daly

Fig. 1.3

Appendix C: Reference set of Selected Major Figures

Fig. 4.6

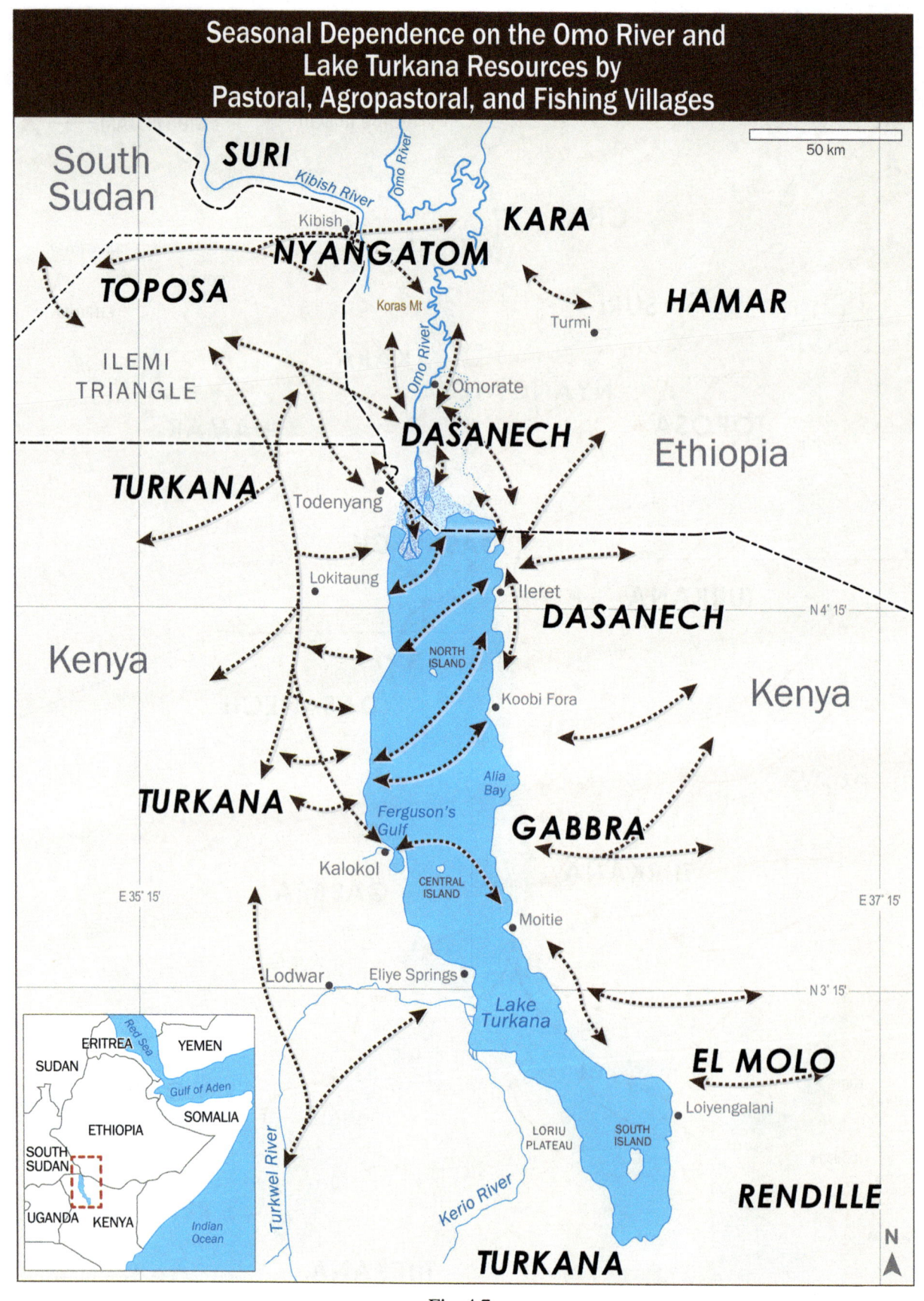

Fig. 4.7

Appendix C: Reference set of Selected Major Figures

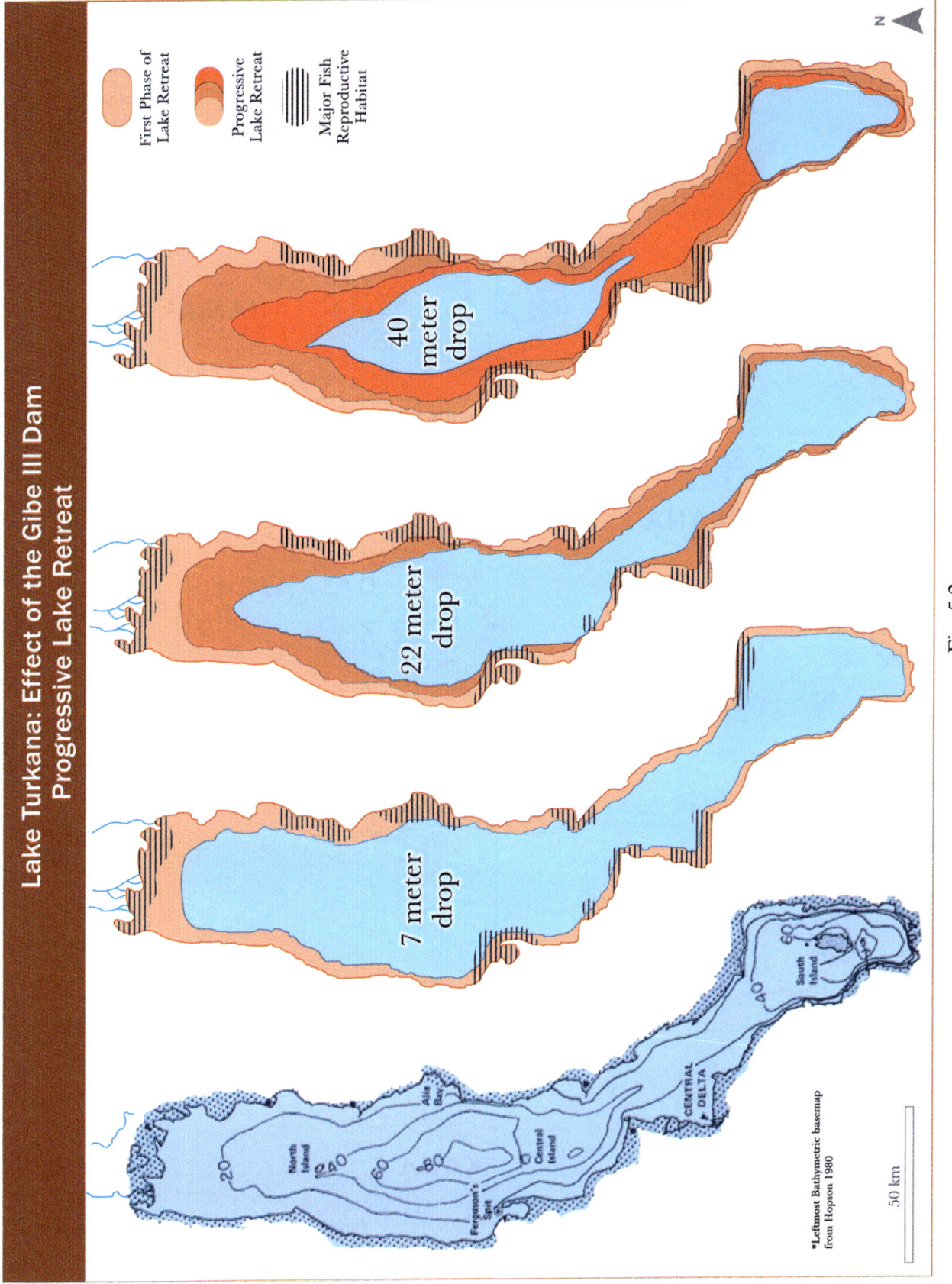

Fig. 5.2

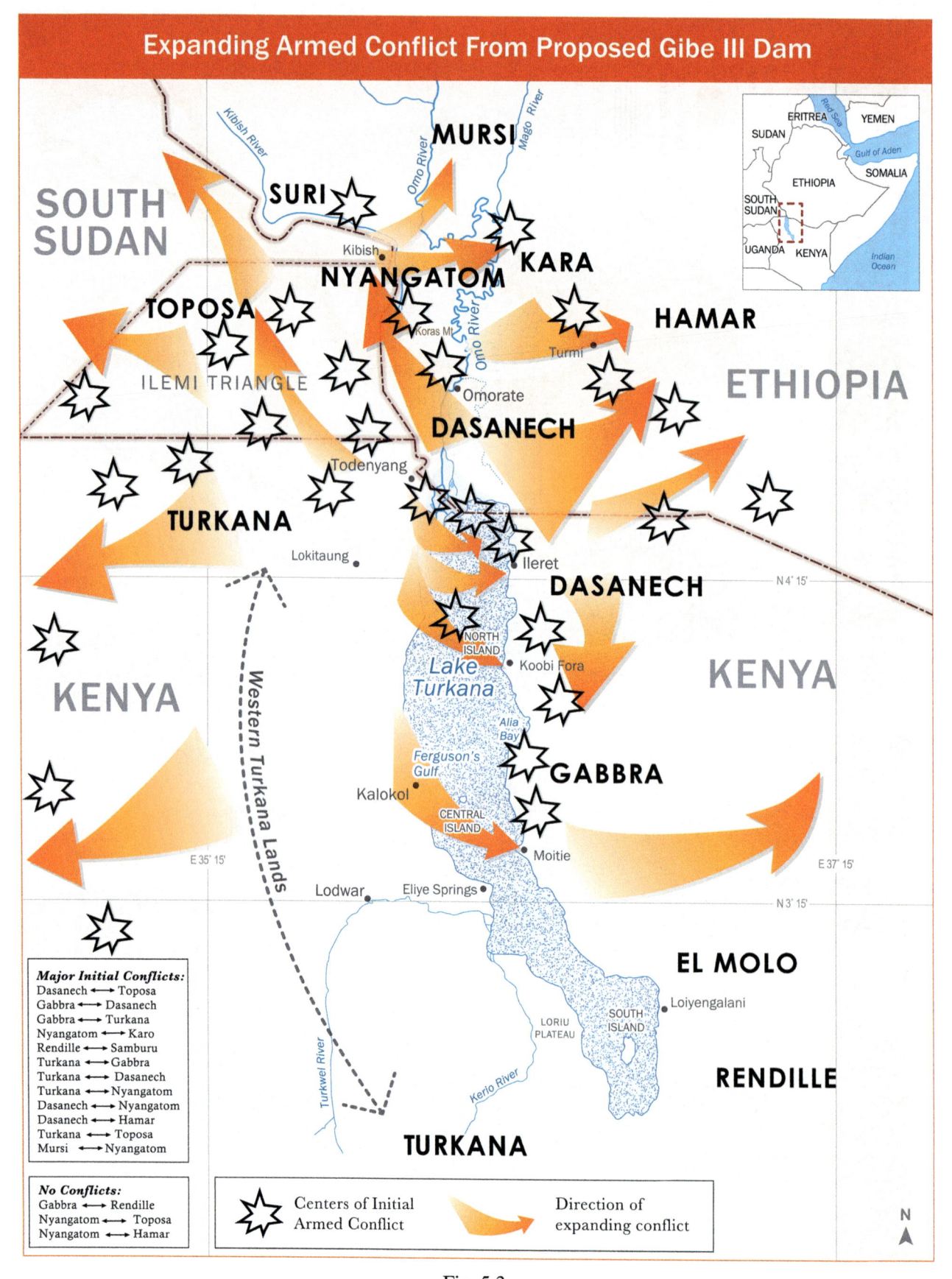

Fig. 5.3

Appendix C: Reference set of Selected Major Figures

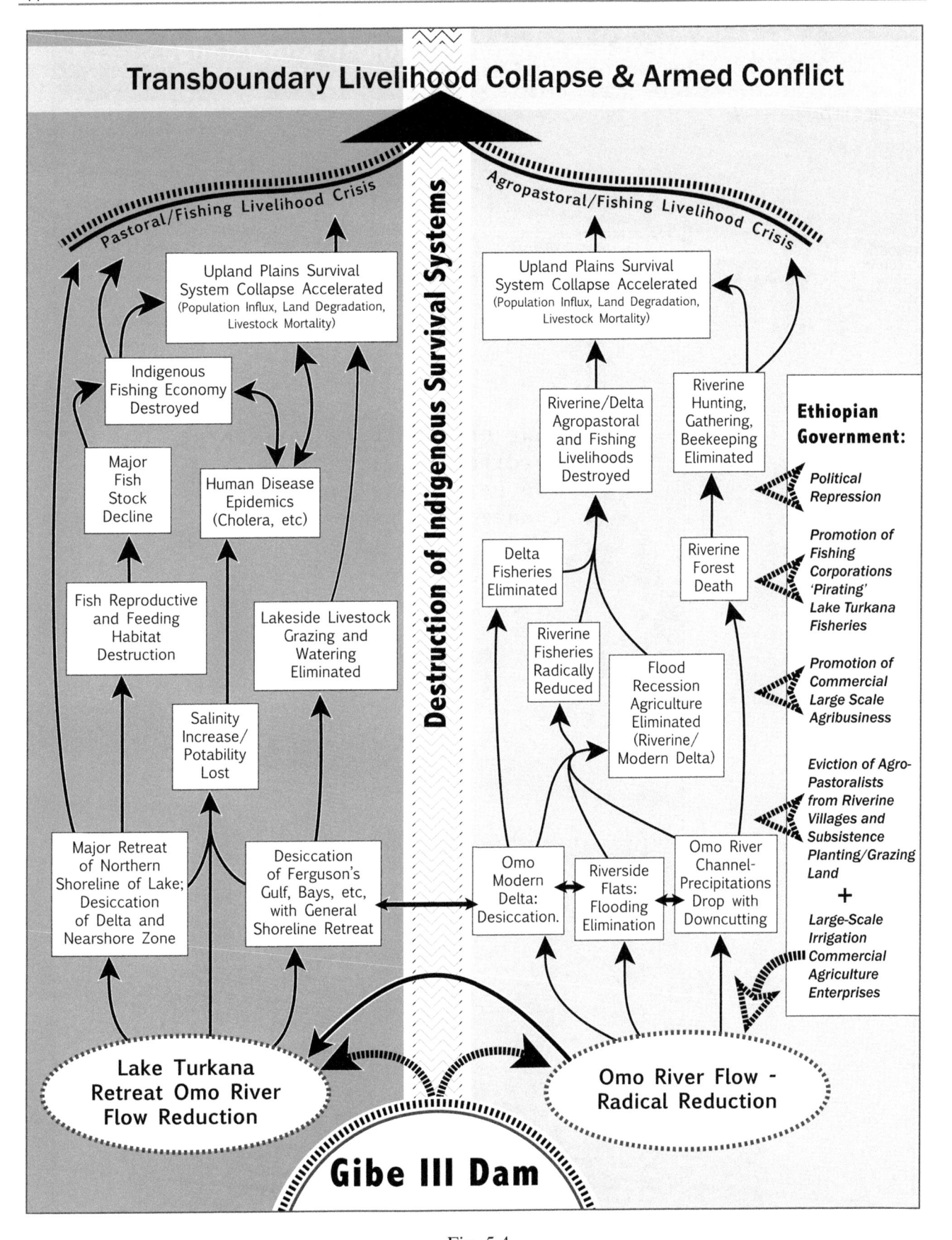

Fig. 5.4

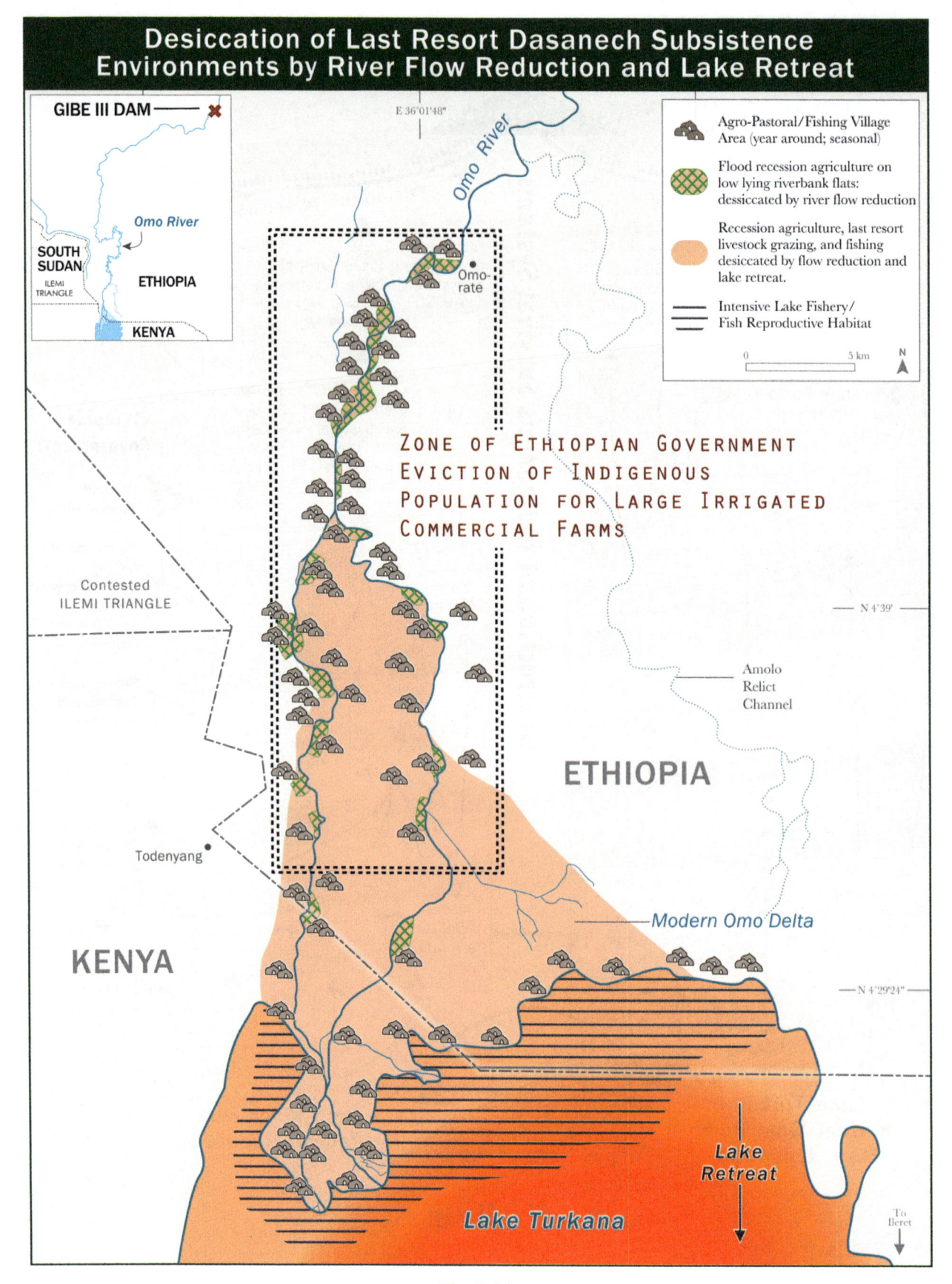

Fig. 7.14

Appendix C: Reference set of Selected Major Figures

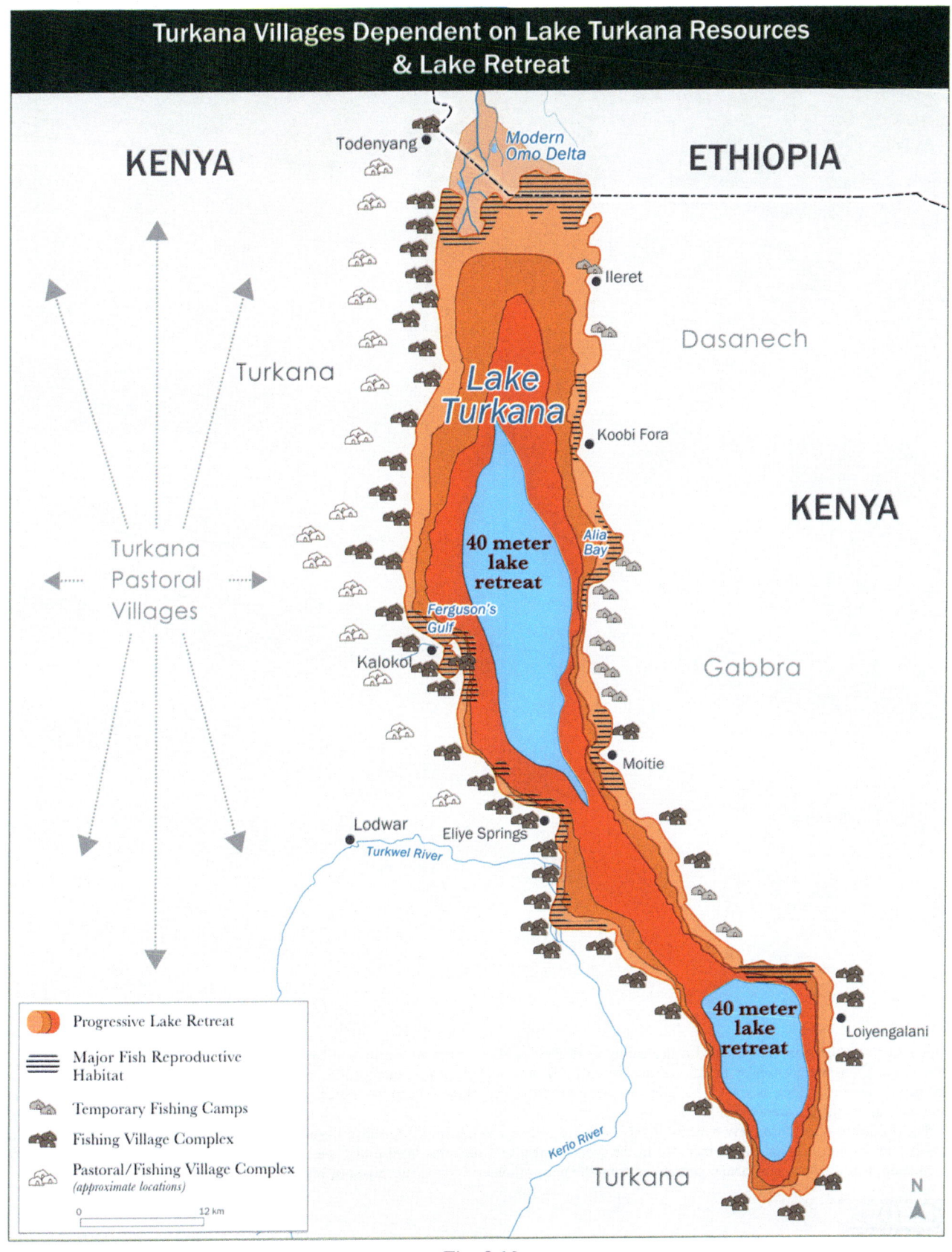

Fig. 9.10

Open Access This book is distributed under the terms of the Creative Commons Attribution-NonCommercial 2.5 International License (http://creativecommons.org/licenses/by-nc/2.5/), which permits any noncommercial use, duplication, adaptation, distribution and reproduction in any medium or format, as long as you give appropriate credit to the original author(s) and the source, provide a link to the Creative Commons license and indicate if changes were made.

The images or other third party material in this book are included in the work's Creative Commons license, unless indicated otherwise in the credit line; if such material is not included in the work's Creative Commons license and the respective action is not permitted by statutory regulation, users will need to obtain permission from the license holder to duplicate, adapt or reproduce the material.

Printed by Printforce, the Netherlands